TENSOR CALCULUS

TENSOR CALCULUS

by

STANISŁAW GOŁĄB

Translated from the Polish
by
Eugene Lepa

ELSEVIER SCIENTIFIC PUBLISHING COMPANY
AMSTERDAM · LONDON · NEW YORK
PWN—POLISH SCIENTIFIC PUBLISHERS
WARSZAWA
1974

Graphic design: Jacek Neugebauer

Distribution of this book is being handled by the following publishers:

for the U.S.A. and Canada
AMERICAN ELSEVIER PUBLISHING COMPANY, INC.
52 Vanderbilt Avenue, New York, N. Y. 10017

for Albania, Bulgaria, Chinese People's Republic, Czechoslovakia, Cuba, German Democratic Republic, Hungary, Korean People's Democratic Republic, Mongolia, Poland, Rumania, Democratic Republic of Vietnam, the U.S.S.R., Yugoslavia
PWN—POLISH SCIENTIFIC PUBLISHERS
Warszawa

for all remaining areas
ELSEVIER SCIENTIFIC PUBLISHING COMPANY
335 Jan van Galenstraat, P.O. Box 211
Amsterdam, The Netherlands

ISBN 0-444-41124-0
Library of Congress Card Number 73-78250

PRINTED IN POLAND

In fond memory of my teacher, Professor Antoni Hoborski, the author of the first Polish exposition of tensor calculus, on the twenty-fifth anniversary of his martyr's death in the Sachsenhausen concentration camp.

CONTENTS

FROM THE PREFACE TO THE FIRST EDITION

The development of mathematical theory is influenced by many factors. In addition to the fundamental factors, i.e. stimuli from other sciences, other factors such as the choice of appropriate methods or merely the choice of appropriate notations also exert an influence. A history of the exact sciences would reveal many interesting instances when an unfortunate choice of symbols resulted in periods when a theory was held back as well as instances when, on the other hand, a happy choice of symbols initiated a period of vigorous development of a theory.

Within the exact sciences, various mathematical disciplines have in recent years penetrated ever new areas of other sciences. In particular, the engineering sciences, which have always been an inexhaustible source of mathematical problems, have been reaching out for more and more new tools of mathematical analysis. However, this has not only been a matter of solving new problems: the vast quantity of scientific material accumulated calls for continual improvement in didactic techniques so that a more lucid presentation of the material is possible. Hence the need of new, more effective methods. Tensor calculus, the far-reaching generalization of vector calculus which long ago won a place for itself in physics as well as in engineering, has proved to be one such effective method. With not much more than half a century behind it, tensor calculus has recently begun to enter all areas of technology, following its initial success in Einstein's revolutionary theory of relativity. One of the most successful applications of tensor calculus has been to geometry, and more especially to non-Euclidean geometries. It has not only enabled well-known theorems of differential geometry to be substantially simplified, but has also made possible the generalization of many theorems to spaces of higher dimensions and has, finally, led to the creation of many new geometries and new spaces.

It is no exaggeration to say that the great era of multidimensional differential geometry in the present century owes its development to the methods of the absolute calculus invented by C. Ricci.

An indication of how tensor calculus and its multiple applications have developed is provided by the extensive foreign literature in this field. The fact that an international Congress devoted to tensor calculus and its geometric and physical applications was held in Moscow in 1934 is a measure of the growth of tensor calculus into an autonomous, major mathematical theory. The journal "Proceedings of the Seminar On Vector and Tensor Analysis and Applications to Geometry, Mechanics, and Physics", edited by V. F. Kagan, also appeared for the first time in Moscow in 1933; this journal resumed publication after the war and its several volumes contain the best papers in this field. Similarly, since 1938 the journal "Tensor" has been published in Japan by A. Kawaguchi in Sapporo as an organ of The Tensor Society which is itself of an international character. The self-declared aim of this journal is to publish papers not only on tensor calculus itself but also on its applications to physics, mechanics, statistics, astronomy, technology, geology, physiology, and other sciences.

The role of tensor calculus in the exact sciences is so powerful today that even in areas, where it has not entered directly as a method of investigation, it has made itself felt at least through the introduction of the convenient and clear tensor symbols, as in the theory of groups or in theoretical chemistry.

On the other hand, it is also possible to discern a tendency towards the "abuse" of tensor calculus. Some authors of otherwise good textbooks are inclined to lump sets of quantities together quite arbitrarily under the collective title of "tensor", without studying the rules governing the transformation of these quantities under a change of the system of reference. Indeed these authors sometimes make the mistake of using the name "tensor" for things which in fact are not tensors.

The choice of material and of symbols presented great difficulties. The author was faced with an alternative: either to expound the theoretical foundations superficially and less rigorously (as is done in most textbooks), putting the main emphasis on applications, or else to devote much space to a thorough exposition of the fundamental concepts, and thus limiting the discussion of applications to the main ones only. Having long held the view that the most important thing in any theory is to lay firm foundations (for it is easier to provide the superstructure than it is to correct or alter the foundations), I have opted for the second course. In fact as far as applications to physical theories are concerned it is true to say that a single work could no longer accommodate them all. Individual works on particular applications, making use of the modern vector-tensor method, should appear in this field.

Notwithstanding this decision to give a rigorous presentation of the theory, the present author did not want to exceed the prescribed scope of this book and the proofs of the theorems are thus omitted while appropriate references for the student are given instead.

The choice of method, and especially of the symbols used, was made after much consideration and hesitation. The purpose above all was to make the student sufficiently familiar with tensor calculus for it to become a highly efficient and reliable tool in his hand. I have chosen the Schouten method (Kernindexmethode) as the most suggestive one which appeals to the definition of geometric object and which, above all, is excellent from the mnemotechnical point of view.

A special monograph by Ślebodziński, *Formes exterieures et leurs applications* [76], is devoted to the exposition of another method, namely one developed first and foremost by the great recently deceased geometer, E. Cartan, and known as the theory of exterior forms.

In the dozen-odd years I have taught tensor calculus in various courses (either in universities or for technical universities), I have sometimes managed to make an original contribution to the theory of tensors. The relevant chapters, containing original material not previously published or known material presented or derived in an original manner, have been indicated by asterisks.

I did not think it advisable to begin this textbook with a long introduction expounding vector calculus as this would have added considerably to the size of the book.

The student is assumed to have mastered the theory of determinants and matrices, and the fundamentals of differential geometry, in addition to having a good knowledge of mathematical analysis. Furthermore, the student familiar with vector calculus in three-dimensional Euclidean space will find it much easier to study this book, while a grasp of the basics of group theory is also desirable.

Readers who are not mathematicians (i.e. physicists or engineers) will in general want to make their way through the thickets of theory as quickly as possible in order to reach the applications of interest to them. Their tendency to skip the proofs or calculations is thus understandable. Nevertheless, they should in my opinion summon the patience to digest the basic concepts and definitions thoroughly, and especially to master the symbols and technique of tensor calculus so that they are able to handle techniques of calculation with ease instead of finding them a hindrance. It is with these readers in mind that the principal formulae and symbols have been listed.

Readers generally prefer the theory to be interspersed with applications or problems. In this particular case, however, it has not been possible to conform to this style of writing.

Tensor calculus is a geometrical theory through and through: in fact, as far as two-index tensors are concerned, it is distinguished from matrix theory primarily by the fact that the latter has a purely algebraic and analytic basis.

It is natural, therefore, for geometric applications to be scattered throughout the exposition of the theory, even though certain of the geometric aspects regarded as applications of tensor calculus have been collected together in Part Three.

This textbook has had to be published quickly, and consequently contains no problems for the student to solve. The reason for this is that much care and time is needed to set up problems and this has therefore been postponed to a future date.

One of the purposes this textbook should serve is that of promoting the early publication of detailed monographs dealing with important areas of physics and written in lucid tensor symbols applicable to all theories employing concepts which are invariant under changes of systems of reference: physical theories are the foremost examples of such theories.

The primary purpose of this book, mathematical rigour aside, is to make the reader aware of the fact that underlying the theory of tensors — or more generally, the theory of affinors and other geometric objects — is the concept of the coordinate system and the possibility of employing an infinite number of coordinate systems.

Mathematicians are perhaps not fully aware of how much they owe Descartes in this respect. There is little exaggeration in the words of Lamé [44] who wrote that “Sans l’invention des coordonnées rectilignes, l’algèbre en serait peut-être encore au point où Diophante et ses commentateurs l’ont laissaient, et nous n’aurions, ni le Calcul infinitésimal, ni la Mécanique analytique. Sans l’introduction des coordonnées sphériques, la Mécanique céleste était absolument impossible... Alors viendra nécessairement le règne des coordonnées curviligne quelconques, qui pourront seules aborder les nouvelles questions dans toutes leur généralité.

Oui, cette époque définitive arrivera, mais bien tard: ceux qui, les premiers, ont signalé ces nouveaux instruments, n'existeront plus et seront complètement oubliés, à moins que quelque géomètre archéologue ne ressuscite leurs noms.

Eh! qu'importe, d'ailleurs, si la science a marché!"[1].

This clearly shows that Lamé foresaw the profound impact that the method of the most general coordinate systems would have on mathematics and its applications although he did not live to see it take effect.

In conclusion, I should like to express my heartfelt thanks to Dr. T. Trajdos–Wróbel who has undertaken to be editor of this publication. The presentation has gained much in matters of content and form as a result of his comments and discussions.

Thank are also due to Professor S. Drobot for a number of valuable critical comments.

Cracow, Dec. 1953 S. GOŁĄB

---

[1] "Without the invention of rectilinear coordinates, algebra might perhaps still be at the point where Diophantes and his commentators left it, and we would not have had either infinitesimal calculus or analytical mechanics. Without the introduction of spherical coordinates, celestial mechanics would have been absolutely impossible.

And so, there will necessarily come the reign of general curvilinear coordinates which alone will be capable of grappling with the new problems in their full generality.

Yes, this definitive era shall come, but rather late: those who were the first to draw attention to these new instruments, will no longer be in existence and will have been utterly forgotten, unless some archeologist-geometer revives their names.

But this is of no importance, so long as science has forged ahead!" G. Lamé, *Leçons sur les coordonnées curvilignes et leurs diverses applications*, Paris 1859, p. 367–368.

FROM THE PREFACE TO THE SECOND EDITION

Tensor calculus today is an area of mathematics which is enjoying vigorous growth and increasing popularity as its methods and applications conquer ever new fields of mathematics and other sciences. Moreover, these circumstances also give rise to new treatments of the subject. Such treatments may emerge either from premisses of a logical nature, or for reasons concerning the purposes of the exposition: whether it is to be addressed to the mathematician and geometer, to the physicist, or to the engineer. Hence the difficulty encountered in writing a textbook of tensor calculus which is to attain all possible goals simultaneously and satisfy the needs of all possible readers. A survey of the contemporary literature on the subject reveals that widely differing treatments and presentations are possible, beginning with the very definitions of the fundamental concepts. Simplicity of the early introductory chapters does not always go hand in hand with the rest of the text being accessible. A similar comment could also be made with reference to the choice of notations which play such a major role here.

The exposition of tensor theory (in the present edition as in the first) has been based on the fundamental concept of geometric object which had undergone far-reaching evolution and generalization since the first edition (i.e., in the past ten years). I regard this approach well-founded even though many recent textbooks of tensor calculus (e.g. that of Lichnerowicz [48]) adopt a completely different approach in which the fundamental concept is that of the vector (contravariant) — the primary concept in axiomatic theory, while the concepts of covariant vector and further concepts of tensor are defined. It seems to me that the tensor theory based on the geometric object is more compact, more uniform. Accordingly, the arrangement of this textbook has not been altered to any great extent. This is so despite criticisms that this textbook is too difficult, especially for those interested first and foremost in the applications of tensor calculus. This book could, of course, be made more didactic along these lines at the cost of a considerable increase in size, as other authors have done (e.g. Rashevskii [60]). Tensor calculus in affine spaces could be presented first and only then followed by the concept of the tensor in spaces in which a more general (in general, infinite) group of transformations is assumed. However, I have always proceeded on the understanding that every reader (hence, physicist and engineer as well) today has a sufficient grasp of curvilinear coordinate systems to be able to accept a very general group of transformations from the outset without experiencing any fundamental conceptual difficulties.

I should like to point out that the whole of the book (with the exception of the chapter on integral theorems and Section 16) deals with local aspects of tensor calculus. The transition to global properties is today associated with the theory of fibre spaces which has been treated in only a few monographs as yet.

As the reader will notice, the changes introduced since the first edition have affected the initial chapters and sections. Major additions have been made mainly to Chapter VIII, dealing with special spaces. A fundamental change concerns terminology. I have dropped the term affinor (whether correctly so I do not know; I merely wished to conform with the majority) since in the meantime Schouten, who had used this term for so many years, also discarded it [67].

The many suggestions of colleagues and pupils who drew my attention to errors in the first edition have enabled me to avoid these errors in the second edition. It is not possible to mention all those critics by name. In particular, however, I am indebted to my colleague, T. Czarliński, who read the text very painstakingly and spotted most of the errors.

Cracow, April 1965 S. Gołąb

PREFACE TO THE THIRD EDITION

The third edition differs from the second only in that it contains some minor additions and corrects errors noticed in the second edition. I wish to thank Dr. Z. Żekanowski who carefully read the text and suggested some improvements.

Cracow, September 1970 S. Gołąb

Part I

The Algebra of Tensors

CHAPTER 1

INTRODUCTORY CONCEPTS

0. Introductory Remarks. *Space*, *coordinate system*, and *change of coordinate system* are the fundamental concepts in tensor calculus. Today these concepts are defined very precisely, but in the course of historical development they underwent various stages of evolution. In these intermediate stages, they were defined with greater or lesser exactness, but always in a manner which appealed more to intuition and which was thus perhaps easier for beginners to grasp.

Accordingly, physicists and engineers who find the first chapter too difficult are urged before tackling it to read up on Riemannian geometry and tensor analysis in three-dimensional Euclidean space, employing only Cartesian, or rectangular, coordinate systems with which the reader is already familiar from the elements of analytic geometry. After such a preliminary study the reader will be in a better position to cope with the present chapter with its concepts which although less intuitive and more abstract, are at the same time formulated in a more rigorous and modern fashion.

1. Analytic Spaces. The concept of space underwent a succession of long processes of abstraction before forming the basis for the construction of various theories. The first stage, very early and yet extremely mature, was the work of Euclid, familiar to all of us. The setting for Euclid's work is a three-dimensional point space. Points, i.e. idealized versions of the smallest possible objects, are the elements of this space. We use these points to construct objects called figures and we study their properties. By intention, Euclidean geometry was meant to be a theory suited to the world around us and its theorems were intended to give us a reasonable indication of some relationships which occur in the reality surrounding us.

By virtue of the perfection it achieved, Euclidean geometry weighed heavily upon the further development of the concept of space, in regard both to the development of geometry itself and the applications in the natural sciences, above all in physics.

Although Euclidean geometry underwent great, fruitful generalizations in the 19th and 20th centuries, the fundamental property of its structure — continuity — has left its mark on most generalizations up to the present time, and it would seem that we are only now on the threshold of possible applications of spaces which do not have a continuous structure. This appears to be mainly due to the fact that although atomistic and continuity theories of the structure of the universe have competed with each other since ancient times, the mathematical instrument for description and analysis, in the form of real numbers and mathematical analysis, has unequivocally assigned continuous space a privileged position as the theoretical picture.

In Euclidean geometry, the point (i.e. the smallest element of space) is actually a primary concept (even though Euclid does try to explain it). The discovery and development of analytic geometry have enabled us to replace the point by a real number, or a set of two or three numbers, and it has thus been possible to arithmeticize the whole of geometry. Accordingly, it has been possible to effect a generalization of Euclidean geometry in an analytic setting, as well as in the setting established by Lobatchevski and Bolyai de Bolya (1823). It seemed that the definition of geometry and the classification of geometries formulated in 1872 by Klein in his famous Erlangen Program [132] would for a long time be the fixed program following which various geometries would develop. It has turned out, however, that science and its needs have gone far beyond the limits of this elegantly defined program.

On the one hand, the needs of theoretical physics as it developed resulted in further striking generalizations of geometry. On the other hand, abstract considerations associated with the concept of space itself have also led to new generalizations.

Thus the concept of *point spaces* came into being. Their elements or points, are not analytically describable, and yet many concepts analogous to those found in analytic geometries can be defined in these spaces. Point spaces, which were introduced by Fréchet[1] (1906) and studied further by

(1) Following Fréchet (1906), we say a point space is a *metric space* if to each pair of points p, q of that space there corresponds a unique non-negative number $\varrho(p, q)$ which satisfies the conditions:

Menger and his school, by Appert, Blumenthal, Pauc, and many present-day investigators, are beginning to play a role of increasing significance.

These general spaces are not yet of major importance in tensor calculus, in which — as before — apart from analytic spaces only Banach vector spaces are used.

Finite-dimensional analytic spaces, which some authors refer to as *Cartesian spaces,* will form the basis for our discussion. We shall call the elements of these spaces *points.* The concept of point will be a fundamental one for us. Space will be assumed to be made *topological* in the sense first defined by Hausdorff [30](2). This means that every point p of space possesses *neighbourhoods* that is, sets of points containing the point p itself and, furthermore, satisfying several simple postulates (Kuratowski [41], Vol. I p. 33).

A further important assumption we make about our space is the postulate that local coordinate systems exist (the term reference frame will be used in a somewhat different sense; cf. p. 136).

DEFINITION. The statement that the *point space* Σ_n *admits a local coordinate system* (*n-dimensional*) *at the point* p means that there exists a neigh-

---

I. $\varrho(p,q) \geqslant 0$; $\varrho(p,q) = 0$ if and only if $p = q$.

II. $\varrho(p,q) = \varrho(q,p)$ for every p, q.

III. $\varrho(p,q)+\varrho(q,r) \geqslant \varrho(p,r)$ for every p, q, r (the triangle inequality).

To be precise, one should say that space is equipped with a distance metric. Metric spaces in the Fréchet sense are Hausdorff spaces.

(2) The concept of *topological space* was first introduced in 1914 by Hausdorff [30], and in axiomatic form at that. The primary concept is that of the point. Certain subsets of the entire space are called "neighbourhoods". This primary concept of neighbourhood must satisfy the following postulates:

I. To each point p there corresponds at least one neighbourhood containing that point, called the *neighbourhood* of the point p.

II. If U_1 and U_2 are two neighbourhoods of the point p, there exists a third neighbourhood U_3 of the point p which is contained in the common part $U_1 \cap U_2$ of the neighbourhoods U_1, U_2.

III. If U is a neighbourhood of the point p and $q \in U$, then there exists a neighbourhood V, of the point q, which is contained in the neighbourhood U.

IV. If $p \neq q$, there exist neighbourhoods U of p and V of q such that $U \cap V = 0$.

V. For every point p there exists a countable sequence of neighbourhoods U_n with the property that each neighbourhood of the point p contains at least one neighbourhood of the sequence.

An Euclidean space becomes a Hausdorff space if, for example, the interior of a ball is taken to be a typical neighbourhood. An affine space (p. 14) is also a Hausdorff space when an appropriate definition of neighbourhoods is given.

bourhood $U(p)$ of the point p with the property that a continuous one-to-one correspondence can be set up between the set of points q of this neighbourhood $U(p)$ and the set of all n-tuples $x = (x_1, \ldots, x_n)$ of real numbers satisfying the conditions

$$|x - \mathring{x}| < \delta$$

or more explicitly

$$|x_i - \mathring{x}_i| < \delta \qquad (i = 1, 2, \ldots, n),$$

where δ is a positive number; the element $\mathring{x}$ corresponds to the point p.

REMARK 1. Since the analytic space X_n of elements $x = (x_1, \ldots, x_n)$ can also be made topological the concept of continuous correspondence is meaningful.

REMARK 2. It has become the custom to use the term points when referring to elements which are sequences of numbers. In doing so, one should distinguish between *analytic points* (denoted above by x) and *geometric points* (denoted above by q). The adjectives analytic and geometric will be omitted wherever there is no danger of ambiguity.

REMARK 3. The number n appearing in the definition above is called the *dimension* of the space (in the neighbourhood of p). By a theorem due to Brouwer [96], this dimension is a uniquely determined natural number. It should be noted that the dimension of a topological space can also be defined in a purely topological fashion (Kuratowski [41], Vol. I, p. 162 et seq.).

REMARK 4. Not all authors confine themselves to sequences of real numbers $(x_1, \ldots, x_n)$ and allow complex values as well. In this case, however, the space can scarcely be said to be n-dimensional since its dimension (in the Menger–Urysohn sense) would be $2n$.

The numbers $x_1, \ldots, x_n$, appearing in the definition above, which are assigned to a point q are called the *coordinates* of q (in the coordinate system $B(q)$ under consideration).

Thus, the *coordinate system* $B(q)$ is, in short, the correspondence between points q and elements x, which can be described concisely by the equation

$$x = f(q). \tag{1.1}$$

There is, of course, an infinite number of coordinate systems $B(q)$. Suppose we take a particular continuous mapping of the neighbourhood $|x - \mathring{x}| < \delta$ onto the neighbourhood $|y - \mathring{y}| < \eta$, where $y = (y_1, \ldots, y_n)$

and η is a given positive number; this mapping can be written analytically as

$$y_i = \varphi_i(x_1, \ldots, x_n) \qquad (i = 1, 2, \ldots, n) \tag{1.2}$$

or, briefly,

$$y = \varphi(x). \tag{1.3}$$

If we then set $y = \varphi[f(q)]$, we obtain a new coordinate system $B(q)$ for the points q of the neighbourhood $U(p)$.

In addition to local coordinate systems, we shall in certain spaces also consider integral coordinate systems, i.e. systems in which the correspondence (1.1) is given between the whole of the space of all points q and the whole, or part, of the space of analytic elements x.

Note, however, that the postulate that integral coordinate systems exist is also a condition on the topological structure of the space under consideration. This is a rather restrictive assumption and, accordingly, we deal mainly with local coordinate systems.

2. Groups and Subgroups. As stated above, if we have a local coordinate system $B(q)$ for the neighbourhood $U(p)$, we can obtain another local coordinate system $\bar{B}(q)$ by means of a given continuous transformation of the type (1.2) (a homeomorphism, in the terminology of Poincaré). In both geometric and physical considerations it is necessary to use all sorts of coordinate systems and to go over from one coordinate system to another. It is also in general necessary to have an infinite number of coordinate systems.

It turns out that we obtain different geometries in any particular space, depending on what coordinate systems we allow in our studies of that space. Accordingly, we must establish what coordinate systems are to be regarded as allowable (preferred, adopted, admissible) coordinate systems. In the presence of certain a priori assumptions concerning our space, certain coordinate systems can be singled out (Gołąb [108]). Without such a priori assumptions, however, it is not possible a priori to select a particular set of allowable coordinate systems in an invariant manner.

Such a set can be distinguished only if, while presupposing the existence of one (local) group of coordinates $B_0(q)$, we choose set G (or group or pseudogroup, as we shall also say) of transformations (1.2) taking the original coordinate system (which we shall call the *protosystem*) over into other coordinate systems. The resultant set of coordinate systems will be referred to as the *set of allowable coordinate systems.*

It is both important and necessary, however, that it be possible to go from each allowable coordinate system $B_1(q)$ to any other allowable coordinate system $B_2(q)$ by means of some transformation (1.2) belonging to the set G. For this reason, the choice of the set G cannot be arbitrary; the set G must have certain properties, and in fact must be what we today call a *pseudogroup*.

It turns out (this was first noted by Veblen and Whitehead [84]) that the classical concept of a group of transformations is too restrictive and for the purposes of differential geometry must be replaced by a more general concept, which we now describe in detail [110].

DEFINITION. Suppose that the set $\mathfrak{G}$ of transformations T has the following properties:

I. Each transformation T has an open non-empty set for its domain D.

II. If a transformation T_1 of domain D_1 belongs to the set $\mathfrak{G}$ and if T_1 is restricted to an arbitrary non-empty domain D_2 which is part of D_1, the restricted transformation T_2 also belongs to $\mathfrak{G}$.

III. If a transformation T_1 of domain D_1 and the corresponding counter-domain D_2 belongs to $\mathfrak{G}$ and if the transformation T_2 of domain D_2 also belongs to $\mathfrak{G}$, then the superposition of transformations T_2T_1 (of domain D_1) also belongs to $\mathfrak{G}$.

IV. If a transformation T_1 of domain D_1 belongs to $\mathfrak{G}$ and if x is an arbitrary point of the domain D_1, there exists a domain D_0 containing x and contained in D_1 and a transformation T_2 whose domain is an image of the domain D_0 under the transformation T_1 such that $T_2 \in \mathfrak{G}$ and $T_2T_1 \equiv I$ in D_0 (I is the *identity transformation*).

A set $\mathfrak{G}$ with these four properties is called a *pseudogroup*. The implication of these properties is that the superposition of any two transformations of the pseudogroup $\mathfrak{G}$ again belongs to $\mathfrak{G}$, provided that the common part of the image of the domains of the transformations T_1 and the domain of the transformation T_2 is not empty.

Property IV postulates that the transformations of the pseudogroup $\mathfrak{G}$ possess local invertibility and is thus weaker than the Veblen–Whitehead postulate requiring global invertibility.

It is interesting that it is not necessary to stipulate that the transformations of the pseudogroup $\mathfrak{G}$ be continuous since continuity follows from the four properties assumed.

The concept of a pseudogroup of transformations took shape gradually. The original definition (O. Veblen and J. H. C. Whitehead [84]) was made somewhat more rigorous by J. A. Schouten and J. Haantjes [149], but the first precise definition did not make its appearance until 1939 (Gołąb [110]). In the first edition of this textbook, I adopted a somewhat narrower definition, excluding empty transformations from the pseudogroup, i.e. transformations with an empty set as their domain. This definition was subsequently the subject of generalization (Dubikajtis [99]). It seems that insofar as the requirements of differential geometry are concerned, the following generalization due to Dubikajtis will prove to be most useful.

A set $\mathfrak{G}$ of transformations T is called a *pseudogroup* if it possesses the following properties:

I′. Every transformation $T \in \mathfrak{G}$ has an open set (or, possibly, an empty set) for its domain.

II′. If a transformation T_1 of domain D_1 belongs to $\mathfrak{G}$ and $D_2 \subset D_1$, where D_2 is an open set, the image $T_1(D_2)$ of the set D_2 under the transformation T_1 is an open set.

III′. If the transformations T_1, T_2 of domains D_1, D_2 belong to $\mathfrak{G}$, and if the point p belongs to D_1, and the point $T_1(p)$ belongs to D_2, there exists an open set $D_3 \subset D_1$ and a transformation T_3 of domain D_3 such that $T_1(D_3) \subset D_2$ and $T_3 = T_2 T_1$ for all points belonging to D_3.

IV′. If $T_1 \in \mathfrak{G}$ and $p \in D_1$, there exists an open set D_2 containing p and contained in D_1 and a transformation $T_2 \in \mathfrak{G}$ such that $T_1(D_2) \subset D(T_2)$ and such that we have $T_2 T_1 = I$ for all points of D_2.

An important generalization of the definition due to Dubikajtis in relation to the original definition [110] consists in making the transformation subject to local instead of global composition. A crucial reason for introducing the concept of pseudogroup and not using the concept of group alone, is that the transformations we shall make use of will be local in character whereas the classical concept of the group of transformations is, in fact, global in character.

The first generalization of the concept of abstract group, it should be noted, was due to H. Brandt [95] and at first went unnoticed by geometers. This concept, called a *groupoid*, was first used for the purposes of geometry by A. Nijenhuis [55]. Ch. Ehresmann [103] arrived independently at the concept of grupoid and even its generalization and applications to geometry. A more profound analysis of the concepts of grupoid, pseudogroup of transformations, and analytic structures can be found in the work of W. Waliszewski [87].

Let a pseudogroup $\mathfrak{G}$ of transformations (1.2) be given. It follows from properties III and IV that this pseudogroup always contains identity transformations (of various domains).

If we proceed from some proto-system of coordinates $B_0(q)$ defined by the relation $x = f_0(q)$ and if $Z(B_0, \mathfrak{G})$ we denote the set of all coordinate systems obtained from B_0 by the transformation $y = \varphi(x)$, where $\varphi \in \mathfrak{G}$, the set Z has the following fundamental property:

If $B_1 \in Z$ and $B_2 \in Z$, where the systems B_1 and B_2 are defined by the relations $y_1 = f_1(q)$ and $y_2 = f_2(q)$, respectively, there exists (one and only one) transformation $y_2 = \psi(y_1)$, which belongs to the set $\mathfrak{G}$ and which is such that $f_2(q) \equiv \psi[f_1(q)]$ for all q lying in the part common to the domains of the functions f_1 and f_2.

In brief, *it is possible to go over from every coordinate system of the set Z to any other coordinate system by means of a transformation belonging to the pseudogroup* $\mathfrak{G}$.

The set $Z(B_0, \mathfrak{G})$ will be called the *set* or *group of allowable coordinate systems*. The set Z depends not only on the pseudogroup $\mathfrak{G}$ but also on the choice of the protosystem B_0. However, inasmuch as the choice of the protosystem B_0 is completely arbitrary, the set Z in fact depends on the pseudogroup $\mathfrak{G}$ adopted a priori.

Subsequently, we shall see that it is possible to construct for the concept of pseudogroup many concepts analogous to those of group theory, particularly the extremely important one of the pseudo-subgroup.

Henceforth, unless there is danger of misunderstanding, we shall use the word *group* instead of *pseudogroup* and we denote pseudogroups by the more convenient letter G instead of $\mathfrak{G}$.

3. Two Points of View Concerning Transformations in Analytic Space. The transformation T of the analytic space X_n, abbreviated to equation

$$y = \varphi(x) \tag{1.3}$$

or defined in more detail by the set of equations (1.2)

$$y_i = \varphi_i(x_i, \ldots, x_n) \qquad (i = 1, 2, \ldots, n),$$

may be interpreted (owing to the one-to-one correspondence between the points of space and the elements of the analytic space) in two ways.

I. Consider the coordinate system defined by means of relation (1.1)

$$x = f(q), \qquad \text{where } q \in U(p).$$

Suppose that in the domain which is the image of the neighbourhood $U(p)$ equation (1.2) is an invertible transformation of that domain into itself. A unique point r such that $f(r) = \varphi(x)$ then exists in the neighbourhood $U(p)$ and hence the transformation φ uniquely determines the mapping $r = F(q)$ of points of the neighbourhood $U(p)$ to the points of that same neighbourhood. Relationship (1.2) may be called a *point-to-point transformation* of the space under consideration, or rather of a certain neighbourhood of that space.

II. Putting by definition

$$g(q) = \varphi[f(q)], \tag{1.4}$$

we obtain the relation

$$y = g(q), \tag{1.5}$$

which represents a new coordinate system in the neighbourhood $U(p)$, namely one obtained from the system defined by equation (1.1) by means of transformation (1.2).

To put it more succinctly, transformation (1.2) may be regarded either as a transformation from one part of the point space to another, or as a transition from one coordinate system to another. Under the first interpretation, the transformation determines the transition from one point to another; from the second point of view, the change of coordinates does not change the point but merely alters its coordinates in the transition to the new coordinate system.

In our geometrical discussion we adopt the second approach.

Consider, for example, a two-dimensional Euclidean plane. The set of equations

$$y_1 = x_1 \cos\alpha + x_2 \sin\alpha, \qquad y_2 = -x_1 \sin\alpha + x_2 \cos\alpha \tag{1.6}$$

may be interpreted (for α fixed) in two ways.

I. Let (x_1, x_2) be the coordinates in the rectangular Cartesian system of any given point q of our plane. Then the transformed point with coordinates (y_1, y_2) determined by relations (1.6) is arrived at by rotating the line Oq (O denotes the origin of the coordinate system) through an angle α about O in the positive direction. Equations (1.6) may thus be said to represent a rotation of the entire plane through an angle α about O.

II. Now consider a point q which in the original Cartesian system has coordinates (x_1, x_2). Going over to a new coordinate system in which the coordinates of the point q are (y_1, y_2) defined by relations (1.6), we find that the new system is also a rectangular Cartesian system which is the result of a rotation through the angle α about the origin.

Note that under the first interpretation the coordinate system remained at rest, so to speak, while the entire plane rotated, whereas in the second case the plane stayed at rest and the coordinate system was rotated.

DEFINITION. *Transformation* (1.2) *is said to be of class C^n if the functions φ_i have continuous derivatives up to and including the n-th order throughout their domain.*

Suppose that transformation (1.2) is of class C^1. By J let us denote the Jacobian of that transformation, that is $J = |\partial\varphi_i/\partial x_k|$ $(i, k = 1, 2, \dots, n)$.

THEOREM. *If $J(x) \neq 0$, transformation* (1.2) *is locally invertible.*

The proof of this theorem can be found in any textbook of mathematical analysis.

The converse theorem is not true: we may take as a counter-example the transformation

$$y_1 = x_1^3 - x_2^3, \quad y_2 = x_1^3 + x_2^3,$$

which is of class C^1 (in fact it is of class C^n for n arbitrary) and is even globally (and not only locally) invertible

$$x_1 = \sqrt[3]{\frac{y_2+y_1}{2}}, \quad x_2 = \sqrt[3]{\frac{y_2-y_1}{2}}.$$

Nevertheless, the Jacobian $J = 18x_1^2x_2^2$ is zero for, say, $x_1 = x_2 = 0$.

REMARK. The fact that a transformation has a non-zero Jacobian everywhere does not guarantee, as is well known from mathematical analysis, that the transformation is globally invertible.

THEOREM. *Suppose that we are given a set $\mathfrak{G}'$ of transformations $y = \varphi(x)$ of class C^r, where r is an arbitrary natural number, with the property that*

$$J = \det \frac{\partial\varphi_i}{\partial x_k} \neq 0 \tag{1.7}$$

and with domains which are open sets. We furthermore regard empty trans-

formations as belonging to $\mathfrak{G}'$. The set $\mathfrak{G}'$ then constitutes a pseudogroup in the sense of Dubikajtis [(1)].

The proof is left to the reader.

DEFINITION. If in a point space Σ_n the allowable coordinate systems we adopt for the neighbourhood $U(p)$ of a point p are all those which can be obtained from a certain protosystem by means of transformations of class C^r belonging to the pseudogroup defined in the theorems above, then we shall say that the space Σ_n is locally a space G_r (in the neighbourhood of the point p) [(2)].

4. The Geometry of the Space G_r. We shall now give a definition of geometries originally formulated in 1872 by Klein [132]. This definition takes as its starting-point a group G (or pseudogroup) of transformations (1.3)

$$y = \varphi(x),$$

which are interpreted in the first way, i.e. as point transformations of the space Σ_n into itself.

Let the letter F denote a *figure*, i.e. an arbitrary set of points of the space Σ_n; and let F_φ denote the image of F under transformation (1.3).

The set of all those properties of such figures F which are invariant under the transformations of the group G is called the *geometry based on the (pseudo-) group G*. This means that if these properties are enjoyed by certain figures F, they are also enjoyed by all figures of the form F_φ for any transformation $\varphi \in G$ [(3)].

As we stated earlier, the Klein formulation of geometry as the theory of the invariants of a particular group of transformations has turned out to be insufficient and not general enough. Further on we shall learn a more general formulation of geometry, based on the concept of geometric object and employing the second interpretation of transformations (1.3).

Note that if G' is a subgroup of the group G, the geometry based on the group G' will be richer in theorems than will that based on the group G since every property which is invariant under transformations of the

(1) This theorem could be generalized by weakening the assumptions (as to the transformations being of class C^1 and as to the Jacobian of the transformation not being zero), but we shall not go into the details here.

(2) This notation has nothing in common with that used by some authors, wherein G_r denotes an r-part group, that is, one dependent on r essential parameters.

(3) This definition is not sufficiently general, but we do not go into the details here.

group G will also be invariant under transformations of the group G', although the converse is not necessarily true. However, if we have some theorem in a geometry based on the group G' and if it turns out to be valid also in the geometry based on the group G, the theorem thus acquires a more profound and more general meaning. This accounts for the tendency to generalize geometries by extending group G, that is by seeking a group $\bar{G}$ such that the group G is a proper subgroup of $\bar{G}$ and then checking to see what theorems of the geometry based on the group G remain valid in the geometry based on the group $\bar{G}$.

5. Subgroups of the Group G_1. An extremely important subgroup of the group G_1 is the *affine group* which in fact should more properly be called the *linear group*. This is a group (in the proper sense of the word, and therefore not only a subgroup) of linear transformations of the form

$$y_i = \sum_{k=1}^{n} \alpha_{ik} x_k + \beta_i \qquad (i = 1, 2, \ldots, n) \tag{1.8}$$

which are non-singular, i.e. transformations whose determinant

$$J = |\alpha_{ik}| \neq 0$$

(the determinant of coefficients here coincides with the Jacobian of the transformation).

This group will be denoted by the symbol G_a (1). The geometry based on this group will be denoted by the symbol E_n and will be called *affine geometry*.

Note that the group G_r ($r = 1, 2, \ldots$) is an infinite group (in the terminology of Lie) since the transformation depends on arbitrary (in a sense at least) functions, in contrast to the group G_a which belongs to the category of finite groups, where the transformation depends on the choice of a finite number of values of certain constants, namely the parameters of the group. The parameters in the group G_a are the coefficients α_{ik}, β_i which may assume values independent of one another (the constraining condition $J \neq 0$ does not place a limit on the number of independent parameters). The group G_a is called an $n(n+1)$-parameter group.

(1) The symbol G is used here together with a letter subscript, which does not denote any natural number, to mean a finite subgroup. On the other hand, $G_1, G_2, G_3, \ldots, G_\infty$ always denote infinite groups in keeping with the Lie terminology.

The group G_a in turn has many subgroups, the most important of which we now describe.

The centred affine group G_c. This is a subgroup of the group G_a given by taking, in transformation (1.8),

$$\beta_i = 0 \quad (i = 1, 2, \ldots, n).$$

The name of the group is due to the fact that the *centre* or origin $(0, \ldots, 0)$ is invariant under the transformations of the group. This group is n^2-parametric.

The unimodular (equivoluminar) group G_u. This is obtained from the group G_a by imposing on the parameters the additional condition

$$J^2 = 1.$$

The group G_u thus has one parameter less than the group G_a. A subgroup of G_u and also of G_c is the *centred affine and unimodular group* G_{cu} which is defined by the two conditions

$$\beta_i = 0 \quad \text{and} \quad J^2 = 1.$$

In a geometry based on the unimodular group, one can introduce the concept of the *volume measure* of a bounded domain.

Apart from the unimodular group G_u, an important role is played by the *special unimodular group* G_{su} which consists of those affine transformations for which not only the condition $J^2 = 1$, but also the more special condition $J = 1$, is satisfied. This means that in addition to the unit of volume being established, so also is the orientation of the space (coordinate systems of the same orientation are allowable). The concept of orientation of space will be made more precise further on (Section 31).

REMARK. A set of affine transformations for which $J = -1$ does not constitute a group since the Jacobian of a compound transformation is equal to the product of the Jacobians of the component transformations, and $(-1)(-1) = 1 \neq -1$.

The orthogonal, or metric, group G_m. If the conditions

$$\sum_{k=1}^{n} \alpha_{ik}\alpha_{jk} = \delta_{ij} \quad (i, j = 1, 2, \ldots, n), \tag{1.9}$$

where the δ_{ij} are *Kronecker symbols* [1], are imposed upon the coefficients α_{ik} of transformation (1.8), we obtain a subgroup of transformations. For if we take any two transformations of the type (1.9)

$$y_i = \sum_k \alpha_{ik} x_k + \alpha_i, \qquad z_i = \sum_k \beta_{ik} y_k + \beta_i$$

and form the compound transformation

$$\begin{aligned} z_i &= \sum_k \beta_{ik} \Big(\sum_j \alpha_{kj} x_j + \alpha_k \Big) + \beta_i \\ &= \sum_{k,j} \beta_{ik} \alpha_{kj} x_j + \sum_k \beta_{ik} \alpha_k + \beta_i = \sum_k \gamma_{ik} x_k + \gamma_i, \end{aligned}$$

where, for brevity, we have put

$$\gamma_{ik} = \sum_j \beta_{ij} \alpha_{jk}, \qquad \gamma_i = \sum_j \beta_{ij} \alpha_j + \beta_i,$$

using the conditions

$$\sum_k \alpha_{ik} \alpha_{jk} = \delta_{ij}, \qquad \sum_k \beta_{ik} \beta_{jk} = \delta_{ij},$$

we find that

$$\begin{aligned} \sum_k \gamma_{ik} \gamma_{jk} &= \sum_k \sum_l \beta_{il} \alpha_{lk} \sum_m \beta_{jm} \alpha_{mk} = \sum_{k,l,m} \beta_{il} \alpha_{lk} \beta_{jm} \alpha_{mk} \\ &= \sum_{k,l,m} \beta_{il} \beta_{jm} \alpha_{lk} \alpha_{mk} = \sum_{l,m} \beta_{il} \beta_{jm} \sum_k \alpha_{lk} \alpha_{mk} \\ &= \sum_{l,m} \beta_{il} \beta_{jm} \delta_{lm} = \sum_m \beta_{im} \beta_{jm} = \delta_{ij}, \end{aligned}$$

which shows that the coefficients γ_{ik} of the compound transformation also satisfy condition (1.9) and, hence the set of transformations satisfying (1.9) constitute a group.

---

[1] Kronecker introduced a concise notation, viz.

$$\delta^i_j = \delta^{ij} = \delta_{ij} = \begin{cases} 0, & \text{when} \quad i \neq j, \\ 1, & \text{when} \quad i = j. \end{cases}$$

This symbol, which is particularly convenient in tensor calculus, has come into general use. We also introduce the so-called generalized Kronecker symbols defined by the equations

$$\delta^i_{j'} = \delta^{i'}_j = \begin{cases} 0, & \text{when} \quad i \neq j, \\ 1, & \text{when} \quad i = j. \end{cases}$$

REMARK. It may be readily demonstrated that relations (1.9) imply the relations

$$\sum_{i=1}^{n} \alpha_{ik}\alpha_{ij} = \delta_{kj} \quad (k, j = 1, 2, \dots, n) \tag{1.10}$$

and the relation

$$J^2 = 1, \tag{1.11}$$

so that summing up we have the following theorem:

THEOREM. *The group G_m is a subgroup of the unimodular group G_u.*

The use of the name metric for the group under discussion is justified by the following facts. Consider two points p_1 and p_2 having the coordinates $x_i^{(1)}$ and $x_i^{(2)}$, respectively, in some allowable coordinate system.

Next, form the expression

$$\bar{\varrho}(p_1, p_2) = \sum_i [x_i^{(2)} - x_i^{(1)}]^2 \tag{1.12}$$

and consider the values it takes for a pair of transformed points.

We have

$$\sum_i [y_i^{(2)} - y_i^{(1)}]^2 = \sum_i \Big[\sum_k \alpha_{ik} x_k^{(2)} + \alpha_i - \sum_k \alpha_{ik} x_k^{(1)} - \alpha_i\Big]^2$$

$$= \sum_i \Big\{\sum_k \alpha_{ik}[x_k^{(2)} - x_k^{(1)}]\Big\}^2 .$$

Now calculate the component of the outer sum:

$$\Big\{\sum_k \alpha_{ik}[x_k^{(2)} - x_k^{(1)}]\Big\}^2 = \sum_k \alpha_{ik}(x_k^{(2)} - x_k^{(1)}) \sum_j \alpha_{ij}(x_j^{(2)} - x_j^{(1)})$$

$$= \sum_{k,j} \alpha_{ik}\alpha_{ij}(x_k^{(2)} - x_k^{(1)})(x_j^{(2)} - x_j^{(1)}).$$

Hence,

$$\sum_i [y_i^{(2)} - y_i^{(1)}]^2 = \sum_{i,k,j} \alpha_{ik}\alpha_{ij}[x_k^{(2)} - x_k^{(1)}][x_j^{(2)} - x_j^{(1)}]$$

$$= \sum_{k,j} (x_k^{(2)} - x_k^{(1)})(x_j^{(2)} - x_j^{(1)}) \sum_i \alpha_{ik}\alpha_{ij}$$

$$= \sum_{k,j} (x_k^{(2)} - x_k^{(1)})(x_j^{(2)} - x_j^{(1)})\,\delta_{kj}$$

$$= \sum_k (x_k^{(2)} - x_k^{(1)})^2 = \bar{\varrho}(p_1, p_2).$$

It is thus seen that the expression on the right-hand side of equation (1.12) is invariant under the transformations of the orthogonal group.

If the value

$$\varrho(p_1, p_2) = [\bar{\varrho}(p_1, p_2)]^{1/2} \tag{1.13}$$

is interpreted as the *distance* between the points p_1, p_2 we now have a (*distance*) *metric* in our space G_m.

Using the distance metric ϱ, one can introduce an *angular metric* as well and the geometry based on this group is therefore called *metric geometry.* The space based on the group G_m will henceforth be denoted by R_n.

The name for the orthogonal group stems from the fact that with the angular metric introduced, any two lines

$$\begin{aligned} x_i &= \tau, \quad x_j = 0 \quad \text{for} \quad j \neq i, \\ x_k &= \sigma, \quad x_l = 0 \quad \text{for} \quad l \neq k, \end{aligned} \qquad \text{where } \tau, \sigma \text{ are independent parameters,}$$

are perpendicular to each other (orthogonal), provided that $k \neq i$.

The *orthogonal centred affine group* G_{mc} is a subgroup of the orthogonal group G_m when the additional condition $\beta_i = 0$ is imposed in equation (1.8). If we assume further that $J = 1$, we obtain the *group of rotations* G_0.

The group of rotations in R_n space is $\binom{n}{2}$-parametric since the number of independent relations (1.9) to which the choice of the n^2 coefficients α_{ik} is subject is $n+\binom{n}{2} = \binom{n+1}{2}$. In three-dimensional Euclidean space R_3 the group of rotations is threedimensional. For the three independent parameters we can choose, for instance, the three Eulerian angles as is frequently the practice in mechanics, or the directional parameters λ_1, λ_2, λ_3 of the axis of rotation (determined up to a positive factor) together with the angle ω about that axis. The coefficients α_{ik} are then given by the formulae

$$\begin{aligned}
\alpha_{11} &= \cos\omega + \lambda_1^2(1-\cos\omega), \\
\alpha_{12} &= \lambda_1\lambda_2(1-\cos\omega) - \lambda_3\sin\omega, \\
\alpha_{13} &= \lambda_1\lambda_3(1-\cos\omega) + \lambda_2\sin\omega, \\
\alpha_{21} &= \lambda_1\lambda_2(1-\cos\omega) + \lambda_3\sin\omega, \\
\alpha_{22} &= \cos\omega + \lambda_2^2(1-\cos\omega), \\
\alpha_{23} &= \lambda_2\lambda_3(1-\cos\omega) - \lambda_1\sin\omega, \\
\alpha_{31} &= \lambda_1\lambda_3(1-\cos\omega) - \lambda_2\sin\omega,
\end{aligned}$$

$$\alpha_{32} = \lambda_2\lambda_3(1-\cos\omega)+\lambda_1\sin\omega,$$

$$\alpha_{33} = \cos\omega+\lambda_3^2(1-\cos\omega)$$

$(\lambda_1^2+\lambda_2^2+\lambda_3^2 = 1)$, which can be rewritten more simply as

$$\alpha_{kk} = \cos\omega+\lambda_k^2(1-\cos\omega),$$

$$\alpha_{ij} = \lambda_i\lambda_j(1-\cos\omega)+\varepsilon\lambda_k\sin\omega,$$

where $i \neq j$, $k \neq j$, $\varepsilon = \pm 1$, depending on whether the permutation (i, j, k) is odd or even (Hoborski [34], p. 12).

6. Linear Subspaces of the Space E_n. The concept of linear subspace can be introduced in the space E_n, even though the latter is ametric (i.e. lacking a metric). Consider in the space X_n a subset of elements defined by means of the (parametric) equations

$$x_i = \sum_{k=1}^{p} \lambda_{ik}\tau_k+\mu_i \qquad (i = 1, 2, \ldots, n), \tag{1.14}$$

where the parameters τ_k are treated as independent quantities (assuming all values from $-\infty$ to $+\infty$), and the coefficients λ_{ik}, μ_i are regarded as constants. The natural number p is assumed to obey the inequality $1 \leqslant p \leqslant n-1$. Assume that the matrix $[\lambda_{ik}]$ $(i = 1, 2, \ldots, n;\ k = 1, 2, \ldots, p)$ is of order p. The set of points of the space X_n deflned by equations (1.14) will then be of dimension p.

Now perform the affine transformation

$$y_i = \sum_{k} \alpha_{ik}x_k+\beta_i \qquad (i = 1, 2, \ldots, n),$$

to obtain

$$y_i = \sum_{k=1}^{n} \alpha_{ik} \sum_{j=1}^{p} (\lambda_{kj}\tau_j+\mu_k)+\beta_i = \sum_{j=1}^{p} \nu_{ij}\tau_j+\varrho_i.$$

We thus see that in the new coordinates the set is again defined by equations of the same type (linear). Moreover, the rank of the matrix $[\nu_{ij}]$ is p, since this matrix is the product of the matrices $[\alpha_{ik}]$ and $[\lambda_{ik}]$ of rank n and p, respectively. We therefore conclude that the definition of the set of points of the space E_n by means of equations of the type (1.14) is invariant in character. Such sets will be called *linear subspaces of dimension p*. For $p = 1$ we shall speak of a *straight line*; for $p = 2$, of a *plane*; and for $p = n-1$, of a *hyperplane*. In the last case, the equation of the hyperplane can be written as

$$\sum_{i=1}^{n} \omega_i x_i + \omega_0 = 0, \quad \text{where} \quad \sum_{i=1}^{n} \omega_i^2 > 0.$$

In the subspace (1.14) the parameters τ_k may be regarded as the coordinates of a point in p-dimensional space and, by subjecting the parameters τ_k to linear and non-singular (i.e. affine) transformations, we can realize it as the affine space E_p. The space E_p will be said to be *embedded in space* E_n.

7. Choice of Notation. The vigorous development enjoyed by tensor calculus has in great measure been due to a fortunate, appropriate system of notations. If a notation is to be legible and conducive to the development of a given theory, it must be reasonably concise. A symbol denoting a concept cannot be too abbreviated since its form must reflect the entire essence of the concept; on the other hand, it cannot be too extended since a very lengthly notation leads to confusion. It is extremely difficult, and at the same time very important, to find the happy medium in this respect. The notation of tensor calculus has undergone far-reaching changes and modifications and has not been unified to this day. A pioneer in this field has been the contemporary Dutch geometrician and physicist, J. A. Schouten ([67], [69]), author of several textbooks on the subject. A number of ingenuous "tricks" employed by Schouten to modernize the notation have made this calculus much more lucid and, furthermore, automated and formalized to such an extent that the possibility of computational errors is very remote indeed. We shall therefore in principle base ourselves on the Schouten notation which he himself called the *method of kernel letters*. For didactic reasons we shall introduce the definitive notation gradually, in several stages, in order to make the great advantages and successes of this method clear to the reader.

Both points and other geometric objects (a rigorous definition of a geometric object follows below) will be denoted briefly by means of a single letter (kernel letter), and in more detail by sequence of so-called coordinate numbers, where the terms of the sequence are indicated by the addition of an index to the kernel letter. However, inasmuch as the coordinates of objects will depend on the coordinate system, it will be necessary to introduce some distinguishing symbols, which will depend on the allowable coordinate systems.

8. Points of Space. Coordinate Systems. Coordinates of a Point. Types of Indices. In two-dimensional analytic geometry (Euclidean and Cartesian) we usually denote the coordinates of a point by x, y (in three-dimensional geometry by x, y, z). In n-dimensional space, of course, it is not possible to use different letters to denote the different coordinates. We then use a single letter and provide it with a numerical index which runs consecutively through all values from 1 to n. Tradition behoves us usually to write the index as an inferior suffix. Instead of writing out a whole sequence

$$x_1, x_2, \ldots, x_n, \tag{1.15}$$

we abbreviate it to

$$x_k \quad (k = 1, 2, \ldots, n). \tag{1.16}$$

Of course, the notation (1.16) is much shorter than (1.15) only if the symbol $(k = 1, 2, \ldots, n)$ can be omitted as being tacitly understood.

The notation for the coordinates of a point in the form (1.15) or (1.16) is inconvenient for the beginner in that in classical analytic geometry or mathematical analysis we use the last few letters of the alphabet to denote variables, and providing a variable with an index usually means that we have assigned a specific numerical value to that variable. When we pass to n-dimensional space, it is necessary to drop this convention. The symbol x_k will denote the k-th variable, and the particular numerical values assigned to this variable should be indicated (by adding an index above or below the letter) by the symbols

$$\overset{0}{x}_k, \quad \overset{1}{x}_k, \quad \underset{0}{x_k}, \quad \text{etc.}$$

In doing this, we introduce the convention once and for all that an index denoting that a certain variable is being singled out is to be added above, or below, the letter, and not to the right of it. This convention is meaningful and eminently practical.

For certain very profound reasons which will become clear in a little while, we shall in future denote the n coordinates of a point not by means of the symbol (1.16) but by the symbol

$$x^k, \tag{1.17}$$

that is, we shall write the running index as a superior suffix. As we shall see the reason for this modification as compared to traditional notations) lies in the fact that in the symbol for a partial derivative $\partial y/\partial x$, which is in common use and therefore difficult to modernize, the independent variable appears at the bottom and not at the top, and this is due to the fact that in the quotient symbol the dividend is on top and the divisor is on the bottom. We therefore see that the "reason" for the notation which we are introducing is rooted deeply in the traditional notation of algebra and analysis.

Finally, for a much less important reason, viz. that Roman letters are reserved for a certain type of geometric objects, we replace the letter x by ξ.

Accordingly, the coordinates of points in a particular system of coordinates will, finally, be denoted by the symbol

$$\xi^k \qquad (k = 1, 2, \ldots, n). \tag{1.18}$$

However, as stated above, we allow in our discussion a whole (pseudo-) group of coordinate systems and in many cases we shall at one and the same time deal with several (two or three) coordinate systems; accordingly an agreement must be reached as to how the coordinates of one and the same point in the various coordinate systems are to be denoted. There are several possibilities open to us: we may either (i) change the letter and denote the coordinate in other coordinate systems by η^k, ζ^k, etc., or (ii). add one or two bars, or an asterisk, above the letter, writing $\bar{\xi}^k$, $\bar{\bar{\xi}}^k$, $\overset{*}{\xi}{}^k$; or, finally, (iii) we may add primes $'\xi^k$, $''\xi^k$, $'''\xi^k$ (this would be in keeping with the traditional procedures of classical analytic geometry or mechanics).

The most suitable method turned out to be one which does not come to mind immediately, a method which was proposed by Schouten. Namely, the coordinates of a point in another coordinate system are denoted by a change of index, or rather kind of index, for example, by ξ^λ, if letters belonging to one and the same alphabet are regarded as indices of the same kind. The Greek index cannot then, of course, run through the sequence $\lambda = 1, 2, \ldots, n$ since this would lead to ambiguity. The indices of another alphabet must then necessarily run through a different sequence of n elements. If n were a concrete natural number, the Greek indices could run through, say, the sequence of natural numbers in the Roman alphabet, but this method cannot be used in theory, where the precise value of n is not given. A way out of this difficulty is to make the indices of the Roman alphabet, for instance, run through a sequence written in italics $k = \mathit{1}, \ldots, n$ while the Greek indices run through a corresponding sequence in Roman type $\lambda = 1, \ldots, \mathrm{n}$.

However, another notation, described below, will be more convenient. Let one coordinate system be assigned, say, indices of the Greek alphabet, running through the sequence $\lambda, \mu, \nu = 1, \ldots, n$, while another coordinate system is assigned primed indices, running through the sequences

$$\lambda', \mu', \nu' = 1', 2', \ldots, n'.$$

Thus, if the coordinates of a point in the original coordinate system are

$$\xi^\lambda \qquad (\lambda = 1, 2, \ldots, n), \tag{1.19}$$

then in a new coordinate system, obtained from the original one by means of the transformation

$$\xi^{\lambda'} = \varphi^{\lambda'}(\xi^1, \xi^2, \ldots, \xi^n) \qquad (\lambda' = 1', 2', \ldots, n') \tag{1.20}$$

the same point will have coordinates

$$\xi^{\lambda'} \qquad (\lambda' = 1', 2', \ldots, n'). \tag{1.21}$$

With this method of notation, we can denote the point itself by the concise symbol ξ, omitting the index and hence indicating that this is a geometric object which is independent of the coordinate systems. At the same time we are able to introduce a succinct, but eloquent, symbol for coordinate systems. A letter enclosed in parentheses, (λ) or (λ'), or (λ'') for instance, denotes the corresponding coordinate system.

Adopting all the above conventions, we see that the sentences

"the point ξ has coordinates ξ^λ in the coordinate system (λ)",

"the point η has coordinates $\eta^{\lambda'}$ in the coordinate system (λ')"

have a well-defined meaning expressed in a simple symbolic notation. The transformation (1.20) which leads from the system (λ) to the system (λ') can be indicated by means of the abridged symbol

$$\varphi[\lambda \to \lambda']. \tag{1.22}$$

We now introduce a very convenient symbol for the partial derivatives of transformation (1.20). Consider the notation

$$A_\lambda^{\lambda'} \stackrel{\text{def}}{=} \partial\xi^{\lambda'}/\partial\xi^\lambda \qquad (\lambda' = 1', 2', \ldots, n';\ \lambda = 1, 2, \ldots, n). \tag{1.23}$$

Since the allowable set of transformations (1.20) will always constitute a group, or pseudogroup, there will always exist (locally at least) a transformation

$$\xi^\lambda = \psi^\lambda(\xi^{1'}, \ldots, \xi^{n'}) \qquad (\lambda = 1, 2, \ldots, n) \tag{1.24}$$

that is inverse to transformation (1.24). Similarly, for the inverse transformation we have

$$A^{\lambda}_{\lambda'} \stackrel{\text{def}}{=} \partial\xi^{\lambda}/\partial\xi^{\lambda'}. \tag{1.25}$$

If we now denote the Jacobian of the transformation (1.20) briefly by

$$J = |A^{\lambda'}_{\lambda}|, \tag{1.26}$$

it follows from the well-known theorem of analysis that

$$|A^{\lambda}_{\lambda'}| = J^{-1}. \tag{1.27}$$

Since in tensor calculus we are constantly dealing with sums of products, and since it is extremely tedious to have to keep writing summation signs with summation indices, we shall follow Einstein's example and adopt the convention by which we omit the summation sign (which must, however, be tacitly-understood) wherever a summation index occurs simultaneously at two levels, i.e. once at the top and once at the bottom. Thus, in the sums

$$\sum_{k=1}^{n} \alpha_k, \qquad \sum_{k=1}^{n} \beta_k^2, \qquad \sum_{k=1}^{n} \alpha^i \beta_k, \qquad \sum_{n=1}^{n} \alpha_k \beta_k$$

we cannot omit the summation sign. However, the sums

$$\sum_{k=1}^{n} \alpha_k \beta^k, \qquad \sum_{k=1}^{n} \alpha^{ik} \beta_i, \qquad \sum_{i,k=1}^{n} \alpha^{ik} \beta_{ik}$$

can be abbreviated (omitting the summation sign) as

$$\alpha_k \beta^k, \qquad \alpha^{ik} \beta_i, \qquad \alpha^{ik} \beta_{ik}.$$

Omission of the summation sign is allowable clearly only when the limits between which the summation indices run are known.

Since the essence of the tensor method may be said to be a matter of indices, we now define several relevant terms.

1. We distinguish between fixed and running indices. In the symbol ξ^k the index k is a running index (this is simply a variable with the sequence of natural numbers from 1 to n as its range). Each of the numbers from 1 to n, on the other hand, will be a fixed index.

2. We distinguish between living and dead indices. If we are given, say, n points $\underset{1}{\xi}, \ldots, \underset{n}{\xi}$ which in the coordinate system (λ) have coordinates

$$\underset{1}{\xi}^{\lambda}, \ldots, \underset{n}{\xi}^{\lambda} \qquad (\lambda = 1, 2, \ldots, n),$$

then this set of points is written briefly as

$$\underset{\mu}{\xi^{\lambda}} \qquad (\lambda, \mu = 1, 2, \dots, n). \tag{1.28}$$

Here λ is a living index which is subject to transformations under the transition to the system (λ'), whereas the index μ is here a dead index. It is tied, as it were, to the kernel letter ξ and does not transform under transition to another coordinate system, system (1.28) is thus written in coordinates (λ') as follows:

$$\underset{\mu}{\xi^{\lambda'}} \qquad (\lambda' = 1', 2', \dots, n'; \ \mu = 1, 2, \dots, n). \tag{1.29}$$

3. We distinguish between dummy (apparent, saturated) and free indices. In the symbol $\alpha_\lambda \beta^\lambda$ the index λ is dummy. In the symbol $\alpha^{\lambda\mu}\beta_\lambda$ the index λ is a dummy, while μ is a free index. The symbol $u^{\lambda\mu\nu}v_\lambda w_{\mu\nu}$ does not have any free indices.

The Einstein convention enables us to write the formula for the partial derivatives of compound functions in a very simple form, namely

$$\frac{\partial \zeta^k}{\partial \xi^\mu} = \frac{\partial \zeta^k}{\partial \eta^Q} \cdot \frac{\partial \eta^Q}{\partial \xi^\mu}.$$

The index Q on the right-hand side of this formula is a dummy index; the summation extends over all values of the index Q. The indices k and μ are free. The letters of three different alphabets have been used in the formula, since the numbers of variables ζ^k, η^Q, ξ^μ may differ and be independent of each other.

Applying the above formula in the special case of transformation (1.20) and its inverse, together with the notations (1.23) and (1.25) we find the following two important relations which we shall frequently make use of in future

$$A_\lambda^{\lambda'} A_{\mu'}^{\lambda} = \frac{\partial \xi^{\lambda'}}{\partial \xi^{\mu'}} = \delta_{\mu'}^{\lambda'} = \begin{cases} 1 & \text{for} \quad \lambda' = \mu' \ ^{(1)}, \\ 0 & \text{for} \quad \lambda' \neq \mu', \end{cases} \tag{1.30}$$

$$A_{\lambda'}^{\lambda} A_{\mu}^{\lambda'} = \frac{\partial \xi^{\lambda}}{\partial \xi^{\mu}} = \delta_{\mu}^{\lambda} = \begin{cases} 1 & \text{for} \quad \lambda = \mu, \\ 0 & \text{for} \quad \lambda \neq \mu. \end{cases}$$

We proceed to generalize the above formulae in a certain sense.

By induction we establish the relationships

$$A_{\mu'}^{\lambda(p+1)} \prod_{j=1}^{p} A_{\lambda(j+1)}^{\lambda(j)} = \delta_{\mu'}^{\lambda'},$$

---

(1) Note: no summation is intended when $\lambda' = \mu'$ and $\lambda = \mu$!

where $\lambda^{(j)}$ denotes λ with (j) primes, and the even more general relationships

$$A_{\mu}^{\lambda(p+1)} \prod_{j=1}^{p} A_{\lambda(j+1)}^{\lambda(j)} = \frac{\partial \xi^{\lambda'}}{\partial \xi^{\mu}}. \tag{1.31}$$

Moreover, for the sake of brevity we introduce the convention

$$A_{\mu_1 \dots \mu_q}^{\lambda_1 \dots \lambda_q} \overset{\text{def}}{=} A_{\mu_1}^{\lambda_1} A_{\mu_2}^{\lambda_2} \dots A_{\mu_q}^{\lambda_q}, \tag{1.32}$$

where in the above formula indexes λ_i, μ_i may be indices belonging to the same alphabet, or to different alphabets.

We thus have, for example,

$$\begin{aligned} A_{k\lambda''\lambda'\lambda}^{\lambda\mu'k\nu''} &= A_k^\lambda A_{\lambda''}^{\mu'} A_{\lambda'}^{k} A_{\lambda}^{\nu''} = (A_k^\lambda A_{\lambda'}^{k})(A_{\lambda''}^{\mu'} A_{\lambda}^{\nu''}) = A_{\lambda'}^{\lambda}(A_{\lambda''}^{\mu'} A_{\lambda}^{\nu''}) \\ &= (A_{\lambda'}^{\lambda} A_{\lambda}^{\nu''}) A_{\lambda''}^{\mu'} = A_{\lambda'}^{\nu''} A_{\lambda''}^{\mu'} = A_{\lambda'\lambda''}^{\nu''\mu'}. \end{aligned}$$

From relations (1.30) we draw one important conclusion: in the relationships

$$A_{\lambda}^{\lambda'} A_{\mu'}^{\lambda} = \delta_{\mu'}^{\lambda'} \qquad (\lambda', \mu' = 1', 2', \dots, n')$$

we regard the index μ' as being fixed (not running). In this case

$$A_{\lambda}^{\lambda'} A_{\mu'}^{\lambda} = \delta_{\mu'}^{\lambda'} \qquad (\lambda' = 1', 2', \dots, n')$$

can be treated as a system of n linear equations with n unknowns $A_{\mu}^{\lambda'}$ $(\lambda = 1, 2, \dots, n)$ with given coefficients $A_{\lambda}^{\lambda'}$ and free terms $\delta_{\mu'}^{\lambda'}$. Since the determinant of the coefficients is in this case the Jacobian $J \neq 0$, we can use Cramer's formulae to calculate $A_{\mu'}^{\lambda}$ and write

$$A_{\mu'}^{\lambda} = W_{\lambda}/J, \tag{1.33}$$

where W_{λ} is a determinant obtained from J by replacing the λ-th column with a column of free terms, i.e.

$$W_{\lambda} = \begin{vmatrix} A_1^{1'} & \dots\ 0\ \dots & A_n^{1'} \\ A_1^{2'} & \dots\ 0\ \dots & A_n^{2'} \\ \dots & \dots & \dots \\ A_1^{(\mu-1)'} & \dots\ 0\ \dots & A_{\mu'}^{(\mu-1)'} \\ A_1^{\mu'} & \dots\ 1\ \dots & A_n^{n'} \\ \dots & \dots & \dots \\ A_1^{n'} & \dots\ 0\ \dots & A_n^{n'} \end{vmatrix}, \tag{1.34}$$

Expanding this determinant by the λ-th column, we write

$$A^{\lambda}_{\mu'} = \frac{\text{minor } A^{\mu'}_{\lambda}}{J}, \tag{1.35}$$

where the minor is taken in the algebraic sense. In this way we have expressed the partial derivatives of the inverse transformation in terms of the partial derivatives of the original transformation since both the minor $A^{\mu'}_{\lambda}$ and the determinant J contain only the elements $A^{\lambda'}_{\lambda}$. We put

$$\bar{J} = J^{-1} \tag{1.36}$$

and write the dual formulae

$$A^{\mu'}_{\lambda} = \frac{\text{minor } A^{\lambda}_{\mu'}}{\bar{J}}. \tag{1.37}$$

These formulae, will be used frequently in what follows.

To end this section, let us introduce one more simplified symbol for partial differentiation, one which is particularly applicable to the indicial method, viz.

$$\partial_{\lambda} \stackrel{\text{def}}{=} \partial/\partial\xi^{\lambda}. \tag{1.38}$$

There is one special case when this symbol will not be used, viz. in the case of the partial derivative

$$\partial\xi^{\lambda'}/\partial\xi^{\lambda} = \partial_{\lambda}\xi^{\lambda'}$$

when the even simpler symbol $A^{\lambda'}_{\lambda}$ will be employed. We shall also write more concisely

$$\partial_{\mu}A^{\lambda'}_{\lambda} = \frac{\partial A^{\lambda'}_{\lambda}}{\partial\xi^{\mu}} = \frac{\partial^2\xi^{\lambda'}}{\partial\xi^{\mu}\partial\xi^{\lambda}} = \frac{\partial^2\xi^{\lambda'}}{\partial\xi^{\lambda}\partial\xi^{\mu}} = \partial_{\lambda}A^{\lambda'}_{\mu} = A^{\lambda'}_{\lambda\mu} = A^{\lambda'}_{\mu\lambda}. \tag{1.39}$$

In connection with the above partial derivatives (1.39) we give one more important formula which will be helpful to us later. When we expand the determinant J by its λ'-th row, or λ-th column, we obtain

$$\partial J/\partial A^{\lambda'}_{\lambda} = \text{minor } A^{\lambda'}_{\lambda}.$$

Accordingly, by virtue of equation (1.35) we can write

$$\partial J/\partial A^{\lambda'}_{\lambda} = JA^{\lambda}_{\lambda'},$$

whence

$$A^{\lambda}_{\lambda'} = \frac{1}{J}\cdot\frac{\partial J}{\partial A^{\lambda'}_{\lambda}} = \frac{\partial \ln|J|}{\partial A^{\lambda'}_{\lambda}}.$$

Multiplying both sides of this relationship by $\partial_{\mu'} A_\lambda^{\lambda'}$ and taking the double sums over the indices λ, λ', we obtain

$$\frac{\partial \ln|J|}{\partial A_\lambda^{\lambda'}} \partial_{\mu'} A_\lambda^{\lambda'} = A_{\lambda'}^\lambda \partial_{\mu'} A_\lambda^{\lambda'}.$$

But, by the formula for the partial differentiation of compound functions, the left-hand side of the latter relationship is equal to $\partial_{\mu'} \ln|J|$. Thus we have

$$A_{\lambda'}^\lambda \partial_{\mu'} A_\lambda^{\lambda'} = \partial_{\mu'} \ln|J|. \tag{1.40}$$

This relationship will be used on many occasions. We now cast it into yet another form. Namely, we have

$$\partial_{\mu'}(A_{\lambda'}^\lambda A_\lambda^{\lambda'}) = A_{\lambda'}^\lambda \partial_{\mu'} A_{\lambda}^{\lambda'} + A_\lambda^{\lambda'} \partial_{\mu'} A_{\lambda'}^\lambda.$$

But

$$A_{\lambda'}^\lambda A_\lambda^{\lambda'} = A_\lambda^\lambda = \delta_\lambda^\lambda = n,$$

whence

$$\partial_{\mu'}(A_{\lambda'}^\lambda A_\lambda^{\lambda'}) = 0,$$

and, accordingly

$$A_{\lambda'}^\lambda \partial_{\mu'} A_\lambda^{\lambda'} = -A_\lambda^{\lambda'} \partial_{\mu'} A_{\lambda'}^\lambda,$$

or

$$A_\lambda^{\lambda'} \partial_{\mu'} A_{\lambda'}^\lambda = A_\lambda^{\lambda'} A_{\mu'\lambda'}^\lambda = -\partial_{\mu'} \ln|J|.$$

The left-hand side of this last relationship can be simplified still further. Note that

$$\partial_{\mu'} A_{\lambda'}^\lambda = \partial_{\lambda'} A_{\mu'}^\lambda,$$

whence

$$A_\lambda^{\lambda'} \partial_{\mu'} A_{\lambda'}^\lambda = A_\lambda^{\lambda'} \partial_{\lambda'} A_{\mu'}^\lambda.$$

Moreover, it follows from the definition of the symbol A that

$$A_\lambda^{\lambda'} \partial_{\lambda'} = \partial_\lambda,$$

so that we finally have

$$\partial_\lambda A_{\mu'}^\lambda = -\partial_{\mu'} \ln|J|. \tag{1.41}$$

Bearing in mind the definitive notation, we can rewrite equations (1.8) of an affine transformation as

$$\xi^{\lambda'} = \alpha_\lambda^{\lambda'} \xi^\lambda + \beta^{\lambda'} \quad (\lambda' = 1', 2', \ldots, n'). \tag{1.42}$$

But since in this case

$$A_\lambda^{\lambda'} = \frac{\partial \xi^{\lambda'}}{\partial \xi^\lambda} = \alpha_\lambda^{\lambda'},$$

equations (1.42) may also be written as

$$\xi^{\lambda'} = A^{\lambda'}_{\lambda}\xi^{\lambda}+\beta^{\lambda'}. \tag{1.43}$$

The equations of a centred affine transformation assume an extremely simple form in the new notation, namely

$$\xi^{\lambda'} = A^{\lambda'}_{\lambda}\xi^{\lambda} \qquad (\lambda' = 1', 2', \ldots, n'), \tag{1.44}$$

while the equations of the inverse transformation are also simple in form and can at once be written as

$$\xi^{\lambda} = A^{\lambda}_{\lambda'}\xi^{\lambda'} \qquad (\lambda = 1, 2, \ldots, n), \tag{1.45}$$

with the algebraic relationships (1.30) holding between $A^{\lambda'}_{\lambda}$ and $A^{\lambda}_{\lambda'}$ just as in the general case.

CHAPTER 2

GEOMETRIC OBJECTS

9. Vectors. Contravariant Vectors. Apart from the point, the simplest geometric concept is that of the vector. This concept can be introduced in many ways. Vectors can be introduced axiomatically as a set of objects (we shall denote them by the letters u, v, w) which can be: (i) multiplied by any real number [1], the result always being a uniquely defined vector; (ii) added to each other, to yield in all cases a uniquely defined vector.

These operations have the following properties:

I. $u+v \equiv v+u$,

II. $(u+v)+w \equiv u+(v+w)$,

III. $\alpha(\beta u) \equiv (\alpha\beta)u$,

IV. $\alpha(u+v) \equiv \alpha u+\alpha v$,

V. $(\alpha+\beta)u \equiv \alpha u+\beta u$,

VI. from $u+v = u+w$ it follows that $v = w$,

VII. $1 \cdot u \equiv u$.

These properties can be show to imply [2] the existence of a unique zero (or *neutral element*) for addition, i.e. a vector θ which when added to any vector u leaves that vector unaltered, i.e. such that for all u, $u+\theta \equiv u$. Further, we have $\alpha\theta \equiv \theta$ for every number α and $0 \cdot u \equiv \theta$ for every vector u.

---

[1] Multiplication of vectors by complex numbers is also allowable, but we shall confine ourselves to the set of real numbers.

[2] Spaces (sets of elements) in which both the inner operation (+) and multiplication of elements by numbers are defined, yielding elements which belong to that space, and in which the aforementioned postulates are satisfied, were called *vector spaces* by Banach. These spaces today are more frequently referred to as *linear spaces*. In this textbook, the term linear space is used to mean a space based on an *affine group*.

Subtraction may be defined either as the operation inverse to addition, or, in terms of addition and multiplication by a number, as follows:

$$u-v \overset{\text{def}}{=} u+(-1)v.$$

If the *scalar product* of two vectors is to be defined, it is necessary either to assume additional properties or to define the length of a vector and the angle between two non-zero vectors.

The introduction of vectors axiomatically leads to a theory of vectors in linear (affine) spaces. In such spaces a vector (*bound* or *localized*) can be defined as an ordered pair of points (p_1, p_2). From the concept of *bound vector* we go on to that to *free vector* by the familiar process of abstraction.

Even without the notion of distance, the concept of the *middle of a pair of points* (p_1, p_2) can be defined in space E_n. If the point p_1 has coordinates $\underset{1}{\xi}^{\lambda}$ and the point p_2 has coordinates $\underset{2}{\xi}^{\lambda}$ in the system (λ), the point c whose coordinates are $\eta^{\lambda} = \frac{1}{2}(\underset{1}{\xi}^{\lambda}+\underset{2}{\xi}^{\lambda})$ is given invariantly, i.e. independently of the cartesian coordinate system. Indeed, on the one hand we have

$$\eta^{\lambda'} = A_{\lambda}^{\lambda'}\eta^{\lambda}+\beta^{\lambda'} = A_{\lambda}^{\lambda'}\tfrac{1}{2}(\underset{1}{\xi}^{\lambda}+\underset{2}{\xi}^{\lambda})+\beta^{\lambda'},$$

while on the other hand,

$$\begin{aligned}\tfrac{1}{2}(\underset{1}{\xi}^{\lambda'}+\underset{2}{\xi}^{\lambda'}) &= \tfrac{1}{2}(A_{\lambda}^{\lambda'}\underset{1}{\xi}^{\lambda}+\beta^{\lambda'}+A_{\lambda}^{\lambda'}\underset{2}{\xi}^{\lambda}+\beta^{\lambda'})\\ &= \tfrac{1}{2}A_{\lambda}^{\lambda'}(\underset{1}{\xi}^{\lambda}+\underset{2}{\xi}^{\lambda})+\beta^{\lambda'} = A_{\lambda}^{\lambda'}\tfrac{1}{2}(\underset{1}{\xi}^{\lambda}+\underset{2}{\xi}^{\lambda})+\beta^{\lambda'} = \eta^{\lambda'}.\end{aligned}$$

Consider now a pair of points (p_1, p_2) and (q_1, q_2). These two pairs will be said to be *equivalent* if

the middle of the pair (p_1, q_2) = *the middle of the pair* (p_2, q_1).

It is easily shown that this relationship is an *equivalence*, i.e. that it has the three properties of reflexivity, symmetry, and transitivity. Accordingly, abstracts [(1)] of this relationship can be formed and a set of all bound vectors equivalent to each other may be called a *free vector*. The operation of multiplication by a number and of addition can now be defined in terms of the abstracts so introduced, and the entire algebra of vectors can be developed. As long as we remain in the space E_n, this is a *non-metric algebra* since neither the lengths of vectors, nor angles between vectors are intro-

[(1)] An abstract is a set of elements which are equivalent in a given relation (*Translator's note.*)

duced. Only when we go on to the subgroup G_m and the space R_n based on it can we develop the *metric part* of vector algebra as well.

For our purposes, however, the *analytic method* turns out to be the most appropriate for defining a bound or free vector. Let us recall from three-dimensional analytic geometry that a free vector was defined in a given *Cartesian coordinate system* by three so-called *coordinates*[1], that is, by the measures of the projections of any one of the vectors on to the axes of the coordinate system. These measures of projections are the differences between the coordinates of the terminal points of the vector, i.e. the numbers

$$\underset{2}{\xi}^{\lambda} - \underset{1}{\xi}^{\lambda} \qquad (\lambda = 1, 2, 3). \tag{2.1}$$

These coordinates plainly vary from one system of coordinates to another. In a new coordinate system (λ'), the coordinates of the vector $\overrightarrow{p_1 p_2}$ will be given by a triad of numbers

$$\underset{2}{\xi}^{\lambda'} - \underset{1}{\xi}^{\lambda'} \qquad (\lambda' = 1', 2', 3'),$$

which is in general, different from the triad (2.1). We therefore find that from the analytic point of view, a free vector in three-dimensional space E_3 is the mapping

$$\omega = f(\lambda), \tag{2.2}$$

which assigns a triad of real numbers to each allowable coordinate system (λ) in an unambiguous manner (the symbol (λ) here denotes the coordinate system, while the symbol (ω) denotes a triad of numbers).

If we take an n-dimensional space E_n, a mapping of a sequence of n real numbers (namely the vector coordinates, which are the measures of the projections of the vector on to the coordinate axes) to each coordinate system will be a *free vector*.

If we now want to go over to a more general space, based on the infinite group G_1, we note first that it will no longer be possible to speak of free vectors. This is because the free vector was defined by using the concept of the middle of a pair of points, and this concept cannot be defined under the group G_1. If we were now to define the free vector as a geometric object which, in any coordinate system (λ) is characterized by n coordinates defined by equation (2.1), then vectors so defined could not be multiplied by real numbers or added together; such a concept would thus be utterly

---

[1] Some authors use the term components of a vector instead of coordinates, even though the former may be ambiguous.

useless. Indeed, let us put

$$u^\lambda \overset{\text{def}}{=} \underset{2}{\xi}^\lambda - \underset{1}{\xi}^\lambda, \qquad v^\lambda \overset{\text{def}}{=} \alpha u^\lambda = \underset{2}{\eta}^\lambda - \underset{1}{\eta}^\lambda,$$

where $\underset{i}{\eta}^\lambda = \alpha \underset{1}{\xi}^\lambda$ $(i = 1, 2)$. If this is so, then in any other (allowable) coordinate system (λ') we must have $v^{\lambda'} = \alpha u^{\lambda'}$. But

$$v^{\lambda'} = \alpha u^{\lambda'} = \alpha(\underset{2}{\xi}^{\lambda'} - \underset{1}{\xi}^{\lambda'}) = \alpha[\varphi^{\lambda'}(\underset{2}{\xi}^\lambda) - \varphi^{\lambda'}(\underset{1}{\xi}^\lambda)].$$

On the other hand,

$$v^{\lambda'} = \underset{2}{\eta}^{\lambda'} - \underset{1}{\eta}^{\lambda'} = \varphi^{\lambda'}(\alpha \underset{2}{\xi}^\lambda) - \varphi^{\lambda'}(\alpha \underset{1}{\xi}^\lambda).$$

Consequently,

$$\varphi^{\lambda'}(\alpha \underset{2}{\xi}^\lambda) - \varphi^{\lambda'}(\alpha \underset{1}{\xi}^\lambda) \equiv \alpha[\varphi^{\lambda'}(\underset{1}{\xi}^\lambda) - \varphi^{\lambda'}(\underset{2}{\xi}^\lambda)]$$

would have to hold for all α, $\underset{1}{\xi}^\lambda$, $\underset{2}{\xi}^\lambda$. With the assumption that the functions $\varphi^{\lambda'}$ are of class C^1 it can easily be shown that the identity above implies that the functions $\varphi^{\lambda'}$ must be linear in the variables ξ^λ, that is, that the group G_1 must reduce to the subgroup G_a. Similarly, it can be shown that when a vector is defined as an object with coordinates (2.1), then under the group G_1 it is not possible to add vectors (in the sense that the corresponding coordinates of the two summands are added to each other).

For these reasons we are forced to abandon the definition of a free vector in spaces based on the group G_1 and instead we introduce only the concept of a vector bound at the point $\underset{0}{\xi}$.

DEFINITION. A *vector v bound at the point* $\underset{0}{\xi}$ is defined to be the assignment to each coordinate system (λ) of a sequence of n numbers

$$v^\lambda \qquad (\lambda = 1, 2, \ldots, n), \tag{2.3}$$

which, upon transition to another coordinate system (λ'), defined by means of the transformation

$$\xi^{\lambda'} = \varphi^{\lambda'}(\xi^\lambda) \tag{2.4}$$

transform in keeping with the rule

$$v^{\lambda'} = \underset{0}{A}{}_\lambda^{\lambda'} v^\lambda, \tag{2.5}$$

where $\underset{0}{A}{}_\lambda^{\lambda'}$ here denotes the partial derivatives of the transformation (2.4) at the point $\underset{0}{\xi}$, that is

$$\underset{0}{A}{}_{\lambda}^{\lambda'} \overset{\text{def}}{=} [A_{\lambda}^{\lambda'}]_{\xi^\lambda = \underset{0}{\xi}{}^\lambda}. \tag{2.6}$$

It is easily verified that this general definition of a vector coincides with the definition of a vector already given in the case of the special space E_n, for in this case

$$\underset{0}{A}{}_{\lambda}^{\lambda'} = A_{\lambda}^{\lambda'} = \alpha_{\lambda}^{\lambda'}$$

and, consequently, if we make the assumption

$$v^\lambda \overset{\text{def}}{=} \xi^\lambda - \underset{0}{\xi}{}^\lambda, \qquad v^{\lambda'} = \underset{1}{\xi}{}^{\lambda'} - \underset{0}{\xi}{}^{\lambda'}$$

(since the vector is to be bound at the point $\underset{0}{\xi}$), we have

$$v^{\lambda'} = \underset{1}{\xi}{}^{\lambda'} - \underset{0}{\xi}{}^{\lambda'} = \alpha_{\lambda}^{\lambda'} \underset{1}{\xi}{}^{\lambda} - \alpha_{\lambda}^{\lambda'} \underset{0}{\xi}{}^{\lambda} = \alpha_{\lambda}^{\lambda'} (\underset{1}{\xi}{}^{\lambda} - \underset{0}{\xi}{}^{\lambda}) = \alpha_{\lambda}^{\lambda'} v^\lambda = \underset{0}{A}{}_{\lambda}^{\lambda'} v^\lambda.$$

Thus, we see that the definition of a bound vector in a space based on the group G_1 is a *generalization* of the definition of a bound vector in the space E_n.

To prove that the rule for transforming the coordinates of a vector leads to a one-to-one assignment of the vector coordinates in every coordinate system, it is necessary to show that the transformation law possesses the so-called group property.

What this means is the following. Let any three coordinate systems (λ), (λ') and (λ''), which are allowable with respect to the group G_1 be given.

Let in general

$$\omega = f(\lambda), \qquad \omega' = f(\lambda'), \qquad \omega'' = f(\lambda'');$$

and let us denote the transformations

$$\xi^{\lambda'} = \varphi^{\lambda'}(\xi^\lambda), \qquad \xi^{\lambda''} = \psi^{\lambda''}(\xi^{\lambda'}), \qquad \xi^{\lambda''} = \chi^{\lambda''}(\xi^\lambda)$$

by T_1, T_2, T_3, respectively. Relation (2.5) will generally be abbreviated to

$$\omega' = F(\omega, T_1), \tag{2.7}$$

where T stands for a transformation which in the case of (2.5) is specified by $\underset{0}{A}{}_{\lambda}^{\lambda'}$. Thus, if firstly $F(\omega, I) = \omega$, where I denotes an identity transformation and, secondly, $F(\omega, T_3) = F(\omega', T_2)$, that is,

$$F(\omega, T_3) = F(\omega, T_2 T_1) = F[F(\omega, T_1), T_2]$$

for all transformations T_1 and T_2, where T_3 is the composition of the transformations T_1 and T_2, we say that the transformation law (2.7) possesses the group property.

Let v^λ, $v^{\lambda'}$, $v^{\lambda''}$, respectively, denote the coordinates of the vector in the above coordinate systems. By the rule (2.5) we have

$$v^{\lambda'} = \underset{0}{A}{}^{\lambda'}_{\lambda} v^\lambda, \qquad v^{\lambda''} = \underset{0}{A}{}^{\lambda''}_{\lambda} v^{\lambda'},$$

whence on substituting for $v^{\lambda'}$ from the left-hand equation into the right-hand one we obtain

$$v^{\lambda''} = \underset{0}{A}{}^{\lambda''}_{\lambda'} (\underset{0}{A}{}^{\lambda'}_{\lambda} v^\lambda) = \underset{0}{A}{}^{\lambda''}_{\lambda'} \underset{0}{A}{}^{\lambda'}_{\lambda} v^\lambda. \tag{2.8}$$

On the other hand because the assignment of the coordinates to coordinate systems is to be one-one, and since it is possible to go directly from the system (λ) to the system (λ'') (the reason we consider pseudogroups is just for this to be true), then, by virtue of the law (2.5) which holds generally, we must have

$$v^{\lambda''} = \underset{0}{A}{}^{\lambda''}_{\lambda} v^\lambda. \tag{2.9}$$

By equation (1.31), on the basis of which

$$\underset{0}{A}{}^{\lambda''}_{\lambda'} \underset{0}{A}{}^{\lambda'}_{\lambda} = \underset{0}{A}{}^{\lambda''}_{\lambda}$$

we find that the two formulae (2.8) and (2.9) are consistent with each other, which shows that the law (5) possesses the group property.

Note that if we take two points $\underset{0}{\xi}$ and $\underset{1}{\xi}$ and attach two vectors $\underset{0}{v}$ and $\underset{1}{v}$ at those points, the transformation laws for the coordinates of the vectors will be, respectively,

$$\underset{0}{v}{}^{\lambda'} = \underset{0}{A}{}^{\lambda'}_{\lambda} \underset{0}{v}{}^\lambda, \qquad \underset{1}{v}{}^{\lambda'} = \underset{1}{A}{}^{\lambda'}_{\lambda} \underset{1}{v}{}^\lambda.$$

Hence, the laws will differ, although they will be of the same type since

$$\underset{0}{A}{}^{\lambda'}_{\lambda} = \left[\frac{\partial \xi^{\lambda'}}{\partial \xi^\lambda}\right]_{\xi^\lambda = \underset{0}{\xi}{}^\lambda}, \qquad \underset{1}{A}{}^{\lambda'}_{\lambda} = \left[\frac{\partial \xi^{\lambda'}}{\partial \xi^\lambda}\right]_{\xi^\lambda = \underset{1}{\xi}{}^\lambda}.$$

Thus, in general $\underset{0}{A}{}^{\lambda'}_{\lambda} \neq \underset{1}{A}{}^{\lambda'}_{\lambda}$ unless the transformation of the coordinates is linear. Accordingly, even if in some coordinate system (λ) the coordinates of the two vectors were equal to each other $\underset{0}{v}{}^\lambda = \underset{1}{v}{}^\lambda$ $(\lambda = 1, 2, \ldots, n)$, this equality would not be preserved under passage to another coordinate system (λ'). It is meaningless therefore, to speak of the equality of vectors $\underset{0}{v}$ and $\underset{1}{v}$ assigned at two different points $\underset{0}{\xi}$ and $\underset{1}{\xi}$.

It may be said, on the other hand, that the transformation laws for vectors "bound" at two different points are similar. This is because the right-hand sides of equations (2.5) are linear homogeneous functions, whose coefficients are always the partial derivatives of functions defining the transformation of the coordinate system calculated at the point at which the vector is bound.

Vectors bound at the same point $\underset{0}{\xi}$ can be multiplied by a real number α (all the coordinates of the vector are multiplied by α) and can be added together: one simply adds together the corresponding coordinates. It is easily shown that these operations are invariant under transformations of the group G_1. *A null vector* is a vector all of whose coordinates are zero. On the other hand, it would not be meaningful to say that a unit vector is a vector v whose coordinates v satisfy the condition

$$\sum_{\lambda} (v^{\lambda})^2 = 1,$$

since this relationship is not invariant, i.e. is not preserved under passage from the system (λ) to the system (λ') with the use of any transformation of C^1.

10. The Concept of Geometric Object. After this introduction concerning the concept of a vector in spaces based on the group G_1, we can now proceed to a general definition of the concept of geometric object.

This concept was first formulated by O. Veblen in 1928 in the paper *Differential Invariants and Geometry* which he read at a plenary session of the International Congress of Mathematics in Bologna, where he described this concept as *invariant*. The term *geometric object* was first introduced in 1930 by J. A. Schouten and E. R. van Kampen [146]. However, the definitions they gave were not precise. A precise definition of this important concept was presented in 1934 by A. Wundheiler.

In a paper delivered to an international conference organized in Moscow by Kagan and Schouten on the applications of tensor calculus in geometry and physics in 1934, Wundheiler laid the foundations for the theory of geometric objects [159]. A major paper devoted to the theory of geometric objects and the classification of objects was published by J. A. Schouten and

J. Haantjes in 1936–1937 [149]. From that time onwards, this theory has developed vigorously (in the Netherlands, Poland, the Soviet Union, Japan, and in recent years also in Hungary and Rumania). In 1952, A. Nijenhuis published the first monograph on gemetric objects [55]. Nijenhuis based his theory on two basic concepts; the concept of groupoid in the sense of Brandt, and the concept of *functional element*. In turn, the theory of geometric objects was placed in the setting of fibre spaces by J. Haantjes and Laman [124]. In 1960, the present author and J. Aczél jointly published a monograph [1] on the functional equations of the theory of geometric objects. More recently, M. Kucharzewski and M. Kuczma introduced the concept of abstract object. Their monograph [42] on the fundamental concepts of the theory of geometric objects is the most recent major treatment of the subject. Nijenhuis' present feeling is that it will be most advantageous to develop (and complete) the theory of geometric objects on the basis of the modern theory of Lie groups [140].

DEFINITION 1. Suppose that we are given an n-dimensional space based on the pseudogroup G_r of transformations, where r is a natural number independent of n, and suppose we consider a fixed point $\underset{0}{\xi}$. An *object* is said to be defined at point $\underset{0}{\xi}$ if to each allowable coordinate system (λ) a unique sequence of numbers

$$\omega_\alpha \qquad (\alpha = 1, 2, \ldots, m), \tag{2.10}$$

is assigned. These number are called the *coordinates of the object* in the system (λ). The coordinates of the object ω in the system (λ') will be denoted by

$$\omega_{\alpha'} \qquad (\alpha' = 1', 2', \ldots, m').$$

This can be expressed symbolically by writing

$$\omega_\alpha = f_\alpha(\lambda) \qquad (\alpha = 1, 2, \ldots, m)$$

or, more concisely,

$$\omega = f(\lambda). \tag{2.11}$$

The number m of coordinates of the object is in general independent of n, but in many special cases, indeed in several important cases, m will correspond exactly to the dimension n of the space.

The *point* ξ is thus an object with n coordinates, just as a vector is an object with n coordinates. A *scalar* is the simplest object with just one coordinate; it is the simplest because the function f in this case is a constant (independent of (λ)).

If an object is given at every point ξ of a domain D, we say that we have a *field of objects* defined in the domain D.

DEFINITION 2. An object ω is said to be a *geometric object* if for any two allowable coordinate systems (λ) and (λ') we have the correspondence

$$\omega_{\alpha'} = F_{\alpha'}[\omega_{\alpha};\ T(\lambda \to \lambda')], \tag{2.12}$$

i.e. the coordinates of the object ω in the system (λ') can be calculated from a knowledge of the coordinates of that object in (λ) and the transformation T which takes us from (λ) to (λ'). Relation (2.12) will be referred to as the *transformation law* for the given geometric object.

REMARK. In the light of these definitions, both the point and the vector are geometric objects [1].

On the basis of equation (2.12) we can say that the right-hand side of the transformation law for a geometric object is a function of the coordinates of the object and at the same time a function of the transformation T.

In special cases, the functions F may depend on a finite number of numerical parameters defined by the transformation T. This is so, for instance, in the case of a vector in which the functions F depend on n^2 parameters $A_{\lambda}^{\lambda'}$. This therefore suggests a further subclassification of geometric objects.

DEFINITION 3. If the transformation law of a geometric object can be written as

$$\omega_{\alpha'} = F_{\alpha'}\left[\omega_{\alpha};\ \underset{0}{\xi^{\lambda}},\ \underset{0}{\xi^{\lambda'}},\ \underset{0}{A_{\lambda}^{\lambda'}},\ [\partial_{\mu}A_{\lambda}^{\lambda'}]_0,\ \ldots,\ \left[\frac{\partial^p \xi^{\lambda'}}{\partial\xi^{\lambda_1}\ldots\partial\xi^{\lambda_p}}\right]_0\right], \tag{2.13}$$

that is, in a form in which the functions F depend on the coordinates of the object in the system (λ) and also on the coordinates of the point $\underset{0}{\xi}$ in the original and new systems and on the partial derivatives of the transformation T calculated at the point $\underset{0}{\xi}$ (indicated by the suffix 0 present in all the symbols) up to derivatives of the order p $(p \leqslant r)$, the object is called, following Schouten and Haantjes, a *geometric object of class p*.

---

(1) Examples of non-geometric objects were first given by Schouten and Haantjes [149].

According to this definition a *point is a geometric object of class zero*, and a *vector is a geometric object of class one.*

DEFINITION 4. If in a particular case the transformation law (2.13) is such that the functions F depend not on the variables $\underset{0}{\xi}^{\lambda}$ and $\underset{0}{\xi}^{\lambda'}$ but only on the partial derivatives $\dfrac{\partial^{p}\xi^{\varkappa'}}{\partial\xi^{\lambda_1}\ldots\partial\xi^{\lambda_p}}$ the object is known as a *purely differential geometric object* [1].

DEFINITION 5. If in a particular case the functions F do not depend on the partial derivatives of the coordinate transformation, that is, if the transformation law is of the form

$$\omega_{\alpha'} = F_{\alpha'}(\omega_{\alpha};\ \xi^{\lambda},\ \xi^{\lambda'}), \tag{2.14}$$

the object is said to be *non-differential.* Such objects do not as yet play any role in differential geometry. They, have, however, found interesting applications elsewhere [1], [115].

Thus, *vectors,* for instance, *are purely differential objects of class one.* It may also be said that they are objects of this type with the simplest possible form since in this case the functions F are linear and homogeneous in the variables ω_{α}, and in the variables $\underset{0}{A}{}_{\lambda}^{\lambda'}$. *Scalars,* on the other hand, *are non-differential objects.*

It should be carefully noted that the functions F on the right-hand sides of the transformation laws for geometric objects of class p cannot be arbitrary functions of the independent variables $\omega_{\alpha}, \ldots, \partial^{p}\xi$ since these functions must satisfy the group condition inasmuch as the coordinates of the object ω are uniquely associated with the given coordinate system. By reasoning just as in the case of vectors, we arrive at this condition in the following manner. We consider three arbitrary coordinate systems (λ), (λ') and (λ'') and write down formulae (2.13) for the following three transformations of the coordinate system

$$T(\lambda \to \lambda'), \qquad T(\lambda' \to \lambda''), \qquad T(\lambda \to \lambda'').$$

We obtain

$$\begin{aligned} \omega_{\alpha'} &= F_{\alpha'}(\omega_{\alpha};\ \xi^{\lambda},\ \xi^{\lambda'},\ \partial^{j}_{\lambda}\xi^{\lambda'}), \\ \omega_{\alpha''} &= F_{\alpha''}(\omega_{\alpha'};\ \xi^{\lambda'},\ \xi^{\lambda''},\ \partial^{j}_{\lambda'}\xi^{\lambda''}), \\ \omega_{\alpha''} &= F_{\alpha''}(\omega_{\alpha};\ \xi^{\lambda},\ \xi^{\lambda''},\ \partial^{j}_{\lambda}\xi^{\lambda''})\ ^{(1)}. \end{aligned} \tag{2.15}$$

(1) This notation is symbolic and highly condensed. Moreover, for simplicity we omit

Substituting from the first of these relations into the second, we get

$$\omega_{\alpha''} = F_{\alpha''}[F_{\alpha'}(\omega_\alpha;\ \xi^\lambda, \xi^{\lambda'}, \partial^j_\lambda \xi^{\lambda'});\ \xi^{\lambda'}, \xi^{\lambda''}, \partial^j_{\lambda'} \xi^{\lambda''}].$$

When we compare this relation with the third of equation (2.15), we arrive at the condition

$$F_{\alpha''}[F_{\alpha'}(\omega_\alpha;\ \xi^\lambda, \xi^{\lambda'}, \partial^j_\lambda \xi^{\lambda'});\ \xi^{\lambda'}, \xi^{\lambda''}, \partial^j_{\lambda'} \xi^{\lambda''}] \equiv F_{\alpha''}[\omega_\alpha;\ \xi^\lambda, \xi^{\lambda'}, \partial^j_\lambda \xi^{\lambda''}]. \qquad (2.16)$$

Now the values of the partial derivatives $\partial^j_\lambda \xi^{\lambda''}$ can be expressed as rational functions of the partial derivatives $\partial^j_\lambda \xi^{\mu'}$ and $\partial_{\lambda'}{}^j \xi^{\lambda''}$. This is indicated by the symbol

$$[\partial^j_\lambda \xi^{\lambda'}, \partial^j_{\lambda'} \xi^\lambda].$$

Equation (2.16) (or, rather, the set of equations) now becomes

$$F_{\alpha''}\{F_{\alpha''}[\omega_\alpha;\ \xi^\lambda, \xi^{\lambda'}, \partial^j_\lambda \xi^{\lambda'}];\ \xi^{\lambda'}, \xi^{\lambda''}, \partial^j_{\lambda'} \xi^{\lambda''}\} \equiv F_{\alpha''}\{\omega_\alpha;\ \xi^\lambda, \xi^{\lambda''}, [\partial^j_\lambda \xi^{\lambda'}, \partial^j_{\lambda'} \xi^{\lambda''}]\}^{(1)} \quad (\alpha = 1, 2, \ldots, m). \qquad (2.17)$$

The functions F_α must satisfy this equation identically, i.e. for all values of the variables

$$\omega_\alpha;\ \xi^\lambda, \xi^{\lambda'}, \xi^{\lambda''}, \partial^j_\lambda \xi^{\lambda'}, \partial^{j'}_\lambda \xi^{\lambda''}. \qquad (2.18)$$

This is a set of functional equations which are not differential but are merely iterated equations, such as occur in the theory of Lie groups.

Obviously, not every set of functions F_α will satisfy relations (2.17) and indeed, only certain exceptional sets will be solutions of (2.17).

The problem thus arises of determining (for given values of the natural numbers n, m, p) all the solutions of the set of equations (2.17). This problem has not yet been solved in its full generality. For certain values of n, m, p (i.e. for spaces of particular dimension n, for objects with a particular number of coordinates m, and for objects of a certain class p) solutions have been found by E. Cartan, W. W. Wagner and Yu. Pentsov under the assumption that the functions are analytic. With relatively weak assumptions as to the functions F, the set of equations (2.17) has been solved for some

---

the suffix 0: this should be tacitly-understood since the object in question is one defined (bound, attached) at the point $\underset{0}{\xi}$.

(1) The functions F_α $F_{\alpha'}$, or $F_{\alpha''}$ are, of course, identical. The addition of an appropriate index in formulae (2.17) which define the transformation law of the object is intended to indicate which of the m functions corresponds to the left-hand side, where the kind of index (and not just its number) is important.

n, m, p by Gołąb [111], A. Woźniacki, Pidek-Łopuszańska [142] and J. Aczél [1].

The objective in this textbook, however, is not to present an exhaustive list of geometric objects of various classes, but to describe the most important ones, including tensors, and to construct an algebra for them, to discuss their basic analytic properties and to give some applications.

11. Covariant Vectors. Suppose we are given a scalar field $\sigma = \sigma(\xi)$. We thus assume that in a domain D each point ξ is uniquely assigned a number σ which is independent of the choice of coordinate system. If we write this scalar field as a function of the coordinates ξ^λ in any coordinate system

$$\sigma = \sigma(\xi^1, \xi^2, \ldots, \xi^n) \tag{2.19}$$

we assume that the function σ is of class C^1. If we admit the group of transformations G_1, the function σ will also be a function of class C^1 in any other coordinate system.

As on p. 27, we introduce the abbreviated notation

$$\sigma_\lambda \overset{\text{def}}{=} \partial_\lambda \sigma = \frac{\partial \sigma}{\partial \xi^\lambda} \qquad (\lambda = 1, 2, \ldots, n) \tag{2.20}$$

and we calculate $\sigma_{\lambda'} = \partial_{\lambda'} \sigma$. Accordingly, we have

$$\partial_{\lambda'} \sigma = \frac{\partial \sigma}{\partial \xi^{\lambda'}} = \frac{\partial \sigma}{\partial \xi^\lambda} \cdot \frac{\partial \xi^\lambda}{\partial \xi^{\lambda'}} = \sigma_\lambda A_\lambda^{\lambda'},$$

or

$$\sigma_{\lambda'} = A_{\lambda'}^\lambda \sigma_\lambda. \tag{2.21}$$

We thus find that the partial derivatives of the scalar field transform under passage to a new system in the same way as the coordinates of a vector, except that the transformation law for the vector has coefficients $A_\lambda^{\lambda'}$ whereas the transformation law (2.21) has coefficients $A_{\lambda'}^\lambda$, i.e. the partial derivatives of the inverse transformation.

It can be shown (just as for the transformation law for vector coordinates) that the transformation law (2.21) possesses the group property. It therefore defines a certain geometric object, or more precisely a whole field of objects, defined in the domain D.

DEFINITION. A geometric object obeying the transformation law

$$u_{\lambda'} = \underset{0}{A_{\lambda'}^\lambda} \cdot u_\lambda \tag{2.22}$$

is called a *covariant vector* attached at the point ξ.

In keeping with this terminology, objects hitherto referred to briefly as vectors will henceforth be known as contravariant vectors. The traditional usage of the terms contravariant and covariant vector is wrong and in fact the meanings should be interchanged, i.e. contravariant vectors should be called *covariant* and vice versa. Adopting a group of centred affine transformations

$$\xi^{\lambda'} = \alpha^{\lambda'}_{\lambda} \xi^{\lambda}$$

we have for contravariant vectors the transformation law

$$v^{\lambda'} = A^{\lambda'}_{\lambda} v^{\lambda} = \alpha^{\lambda'}_{\lambda} v^{\lambda},$$

which is the same as the coordinate transformation rule; it would thus be more appropriate to call these vectors covariant.

REMARK. If we wanted to write the coordinates of a covariant vector by adding a superscript i.e. u^{λ}, the transformation law would then be of the form

$$u^{\lambda'} = \sum_{\lambda} A^{\lambda}_{\lambda'} u^{\lambda}$$

and we would not be entitled (according to the convention adopted) to omit the summation sign. This would be inconvenient.

We therefore see that the use of an inferior running index for covariant vectors enables the transformation law to be written down not only simply but also purely mechanically, just as in the case of the transformation law for contravariant vectors. Henceforth we have only to remember that the running index must be superior for contravariant vectors, and inferior for covariant vectors. Moreover, this rule is consistent with the rule for writing the differentiation index at the bottom and the fact that $\partial_{\lambda}\sigma$ is a covariant vector.

The question we now ask is for what transformations of coordinates will the transformation law for covariant vectors coincide with that for contravariant vectors. This will happen when

$$A^{\lambda'}_{\lambda} = A^{\lambda}_{\lambda'} \quad (\lambda = 1, 2, \ldots, n;\ \lambda' = 1', 2', \ldots, n'). \tag{2.23}$$

The first of relations (1.30) together with the equations above yields

$$A^{\lambda'}_{\lambda} A^{\lambda}_{\mu'} = \sum_{\lambda} A^{\lambda}_{\lambda'} A^{\lambda}_{\mu'} = \delta^{\lambda'}_{\mu'} \quad (\lambda', \mu' = 1', 2', \ldots, n'). \tag{2.24}$$

Equations (2.24) are merely relations (1.9), which define the orthogonal group of transformations, rewritten in fresh notation. Thus, under an orthogonal group of transformations the difference between covariant and contravariant vectors disappears since the two transformation laws are identical.

Covariant vectors (attached at the same point), as well as contravariant vectors, can be multiplied by numbers (by multiplying the coordinates) and added together (by adding corresponding coordinates). To show that this is true, suppose $\overset{1}{u}$ and $\overset{2}{u}$ are two covariant vectors attached at the same point ξ, so that the transformation laws will be

$$\overset{1}{u}_{\lambda'} = \underset{0}{A}{}^{\lambda}_{\lambda'} \overset{1}{u}_{\lambda}, \qquad \overset{2}{u}_{\lambda'} = \underset{0}{A}{}^{\lambda}_{\lambda'} \overset{2}{u}_{\lambda}.$$

Suppose that

$$v_{\lambda} \overset{\text{def}}{=} \varrho \overset{1}{u}_{\lambda}, \qquad w_{\lambda} = \overset{1}{u}_{\lambda} + \overset{2}{u}_{\lambda}.$$

We now calculate

$$v_{\lambda'} = \varrho \overset{1}{u}_{\lambda'} = \varrho \underset{0}{A}{}^{\lambda}_{\lambda'} \overset{1}{u}_{\lambda} = \underset{0}{A}{}^{\lambda}_{\lambda'} \varrho \overset{1}{u}_{\lambda} = \underset{0}{A}{}^{\lambda}_{\lambda'} v_{\lambda},$$

$$w_{\lambda'} = \overset{1}{u}_{\lambda'} + \overset{2}{u}_{\lambda'} = \underset{0}{A}{}^{\lambda}_{\lambda'} \overset{1}{u}_{\lambda} + \underset{0}{A}{}^{\lambda}_{\lambda'} \overset{2}{u}_{\lambda} = \underset{0}{A}{}^{\lambda}_{\lambda'} (\overset{1}{u}_{\lambda} + \overset{2}{u}_{\lambda}) = \underset{0}{A}{}^{\lambda}_{\lambda'} w_{\lambda}.$$

Thus, v_{λ} and w_{λ} do indeed transform as required under the transformation law for the coördinates of a covariant vector. It is easily shown that just as for contravariant vectors, addition of covariant vectors is commutative and associative and that the (unique) neutral element for addition is a vector with all of its coordinates zero. This latter property is, of course, invariant, which means that it holds in every coordinate system, if it holds in at least one.

A covariant vector can not, however, be added to a contravariant vector even if both are bound at the same point ξ. How is this negative statement to be interpreted? It means that the expression

$$v^{\lambda} + u_{\lambda}, \tag{2.25}$$

where v^{λ} and u_{λ} are, respectively, the coordinates of contravariant and covariant vector, attached at the point $\underset{0}{\xi}$, does not transform either according to the law for contravariant vectors or to the law for covariant vectors. This is not difficult to prove. However, there is the possibility that expression (2.25) transforms under passage to a new system (λ') in keeping with some other transformation law for geometric objects of class one.

Assume for the moment that $v^{\lambda}+u_{\lambda}$ constitute a geometric object of class one. This means that we can write

$$v^{\lambda'}+u_{\lambda'} = F_{\lambda'}[v^{\lambda}+u_{\lambda}, \underset{0}{\xi^{\lambda}}, \underset{0}{\xi^{\lambda'}}, \underset{0}{A_{\lambda}^{\lambda'}}],$$

or

$$\underset{0}{A_{\lambda}^{\lambda'}}v^{\lambda}+\underset{0}{A_{\lambda'}^{\lambda}}u_{\lambda} = F_{\lambda'}[v^{\lambda}+u_{\lambda}; \underset{0}{\xi^{\lambda}}, \underset{0}{\xi^{\lambda'}}, \underset{0}{A_{\lambda}^{\lambda'}}]$$

$$(\lambda = 1, 2, \ldots, n;\ \lambda' = 1', 2', \ldots, n'). \qquad (2.26)$$

These relationships are to be satisfied identically for the variables v^{λ}, u_{λ}. It can be shown, although the proof is omitted, that this condition leads to the conclusion that the group is an orthogonal subgroup and then, as we already know, the difference between covariant and contravariant vectors disappears. It follows that only in this case that can contravariant vectors be added to covariant ones.

12. The Geometric Interpretation of the Covariant Vector. Following Schouten, we give the geometric meaning of the covariant vector in a space E_n based on the affine group G_a. In E_n the dual to a point is a hyperplane ($(n-1)$-dimensional). Just as a (contravariant) bound vector is defined in E_n as a pair of points, so can we define a (bound) covariant vector as a pair of parallel hyperplanes $\overset{1}{E}_{n-1}$ and $\overset{2}{E}_{n-1}$.

Next, we define to be equivalent any two such pairs which are parallel to each other, have equal distances (this concept can be defined in E_n despite the lack of a metric, since the purpose here is to compare distances in the same direction) and have the same sense, which means that if we cut both pairs with a straight line l intersecting the pairs at the points q_1, q_2, q_3, q_4, respectively, the then senses of the vectors $\overrightarrow{q_1q_2}$ and $\overrightarrow{q_3q_4}$ must be the same. Finally, we construct the classes of this equivalence relation, and call it a covariant vector. The aim now is to attach a geometric interpretation to the same coordinates u_{λ}. For this purpose, we take a pair of parallel hyperplanes (not both the same). Assume now that in the coordinate system (λ) the equations of these planes have the intercept form

$$a_{\lambda}\xi^{\lambda} = 1, \qquad b_{\lambda}\xi^{\lambda} = 1,$$

where

$$b_{\lambda} = \varrho a_{\lambda} \qquad (\varrho \neq 0,\ 1)$$

and this slight loss of generality is offset by the simplified calculation to which it leads ($\varrho \neq 1$ means that the two planes are not the same).

When (λ) is transformed into (λ') by means of the equations

$$\xi^\lambda = A^\lambda_{\lambda'}\xi^{\lambda'} + \beta^\lambda \qquad (A^\lambda_{\lambda'}, \beta^\lambda \text{ constant})$$

the equations of the hyperplanes become, respectively,

$$a_\lambda A^\lambda_{\lambda'}\xi^{\lambda'} = 1 - a_\lambda \beta^\lambda, \qquad b_\lambda A^\lambda_{\lambda'}\xi^{\lambda'} = 1 - b_\lambda \beta^\lambda,$$

or, in intercept form,

$$\frac{a_\lambda A^\lambda_{\lambda'}}{1-\tau}\xi^{\lambda'} = 1, \qquad \frac{b_\lambda A^\lambda_{\lambda'}}{1-\varrho\tau}\xi^{\lambda'} = 1 \qquad (\tau = a_\mu \beta^\mu).$$

If we denote the coefficients on the left-hand sides of these equations by $a_{\lambda'}$ and $b_{\lambda'}$, respectively, we obtain the formulae

$$a_{\lambda'} = \frac{a_\lambda A^\lambda_{\lambda'}}{1-\tau}, \qquad b_{\lambda'} = \frac{b_\lambda A^\lambda_{\lambda'}}{1-\varrho\tau}.$$

The pair of hyperplanes in question cuts out a vector on the axis ξ^λ and we now calculate the measure of the vector on the axis. The intercept of the origin of the vector will be (we take $\xi^\mu = 0$ for $\mu \neq \lambda$) $\underset{1}{\xi^\lambda} = 1/a_\lambda$, whereas the intercept of the terminal point will be $\underset{2}{\xi^\lambda} = 1/b_\lambda$; hence the measure of this vector will be

$$\underset{2}{\xi^\lambda} - \underset{1}{\xi^\lambda} = \frac{1}{b_\lambda} - \frac{1}{a_\lambda} = \frac{1-\varrho}{\varrho a_\lambda}.$$

Let us now take the reciprocal of this measure and denote it by u_λ

$$u_\lambda \overset{\text{def}}{=} \frac{1}{\underset{2}{\xi^\lambda} - \underset{1}{\xi^\lambda}} \frac{\varrho}{1-\varrho} a_\lambda. \tag{2.27}$$

We now calculate $u_{\lambda'}$ in the new coordinate system (λ'):

$$u_{\lambda'} = \frac{\varrho'}{1-\varrho'} a_{\lambda'}.$$

Since $a_{\lambda'}$ has already been calculated, we must find ϱ', which turns out to be

$$\varrho' = \frac{b_{\lambda'}}{a_{\lambda'}} = \frac{\varrho a_\lambda A^\lambda_{\lambda'}}{1-\varrho\tau} \cdot \frac{1-\tau}{a_\lambda A^\lambda_{\lambda'}} = \frac{\varrho(1-\tau)}{1-\varrho\tau}.$$

Thus

$$1-\varrho' = \frac{1-\varrho\tau-\varrho+\varrho\tau}{1-\varrho\tau} = \frac{1-\varrho}{1-\varrho\tau},$$

whence

$$\frac{\varrho'}{1-\varrho'} = \frac{\varrho(1-\tau)}{1-\varrho\tau} \cdot \frac{1-\varrho\tau}{1-\varrho} = \frac{\varrho(1-\tau)}{1-\varrho}$$

and consequently

$$u_{\lambda'} = \frac{\varrho(1-\tau)}{1-\varrho} \cdot \frac{a_\lambda A^\lambda_{\lambda'}}{1-\tau} = \frac{\varrho}{1-\varrho} a_\lambda A^\lambda_{\lambda'} = u_\lambda A^\lambda_{\lambda'}.$$

We thus find that the quantities (2.27) transform under change of coordinate system like the coordinates of a covariant vector. The foregoing geometric interpretation of the covariant vector is due to Schouten.

Multiplication of a covariant vector by a number α signifies an $|\alpha|$-*fold reduction* of the spacing between the hyperplanes of the vector with a possible change of sense if $\alpha < 0$. The geometric interpretation of the sum of two covariant vectors will be given for the special case $n = 2$. In this case a covariant vector is a pair of parallel lines. Now take two vectors u_λ, v_λ. Assume that they are not linearly dependent, i.e. that the straight lines of the vector v_λ are not parallel to those of the vector u_λ (the case of linear dependence is much more easily disposed of). For simplicity of calculation assume that the first straight line of each pair pass through the origin of the system, and denote the straight lines of the vector u, or vector v, by L_1, L_2 or L_3, L_4 (Fig. 1).

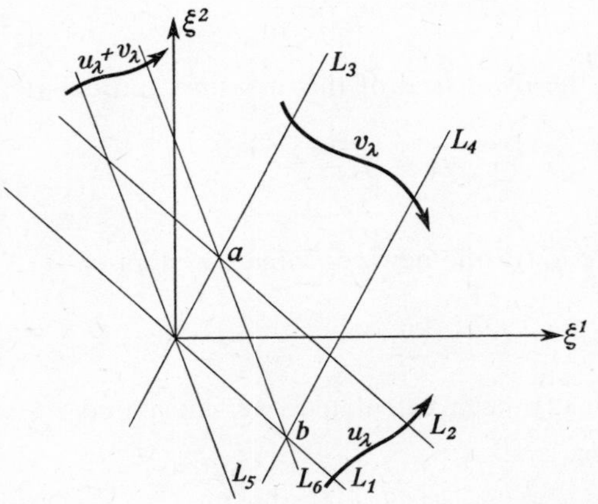

Fig. 1.

In the light of what has been said above about the geometric significance of a covariant vector, the straight line L_2 will intersect the axes of the system at the points $1/u_1, 1/u_2$; the straight line L_4 will intersect the axes of the system at the points $1/v_1$, $1/v_2$. Accordingly, the equations of the straight lines L_1, L_2, L_3, L_4 will be, respectively,

$$L_1: u_1\xi^1 + u_2\xi^2 = 0,$$
$$L_2: u_1\xi^1 + u_2\xi^2 = 1,$$
$$L_3: v_1\xi^1 + v_2\xi^2 = 0,$$
$$L_4: v_1\xi^1 + v_2\xi^2 = 1.$$

The straight lines L_i $(i = 1, 2, 3, 4)$ form a parallelogram with one vertex at the origin. In this parallelogram take the diagonal which does not pass through the origin and denote by L_6 the straight line which contains that diagonal. This is the straight line running from a to b, where a is the point of intersection of L_2 and L_3, while b is the point of intersection of L_1 and L_4. Writing

$$W \overset{\text{def}}{=} \begin{vmatrix} u_1 & u_2 \\ v_1 & v_2 \end{vmatrix},$$

we obtain the coordinates of the point a

$$a\left[\frac{v_2}{W}, \; -\frac{v_1}{W}\right]$$

and the coordinates of the point b

$$b\left[-\frac{u_2}{W}, \; \frac{u_1}{W}\right].$$

Let L_5 denote the straight line which is parallel to L_6 and which passes through the origin and, finally, let w_λ denote the covariant vector given by the pair of straight lines (L_5, L_6). The coordinates of w_λ will be the reciprocals of the intercepts of L_6 on the axes ξ^1, ξ^2.

The equation of the straight line L_6 can easily be shown to be

$$(u_1+v_1)\xi^1 + (u_2+v_2)\xi^2 = 1,$$

whence

$$w_1 = u_1+v_1, \qquad w_2 = u_2+v_2,$$

or

$$w_\lambda = u_\lambda + v_\lambda.$$

We thus have the geometric construction of the vector $u_\lambda+v_\lambda$ when u_λ and v_λ are given.

13. The Mutual Relationship Between Contra- and Co-variant Vectors.

DEFINITION. A contravariant (free) vector v^λ in the space E_n is said *to lie on a covariant (free) vector* u_λ if the direction of the vector v^λ is parallel to the hyperplanes determining the vector u_λ.

Let us now study the analytic expression of the geometric fact that v^λ lies on u_λ. Attach both vectors v and u at the origin of the coordinate system. Then v^λ are the coordinates of the terminal point of v while u_λ are the reciprocals of the intercepts of the other hyperplane of u with the axes of the system. Its equation is

$$u_\lambda \xi^\lambda = 1. \tag{2.28}$$

The equation of the first hyperplane of the vector u is

$$u_\lambda \xi^\lambda = 0. \tag{2.29}$$

Since v is to lie on the vector u, the terminal point of v must lie in the hyperplane (2.29), that is, the relation

$$u_\lambda v^\lambda = 0 \tag{2.30}$$

must hold. Since this relation expresses the geometric fact that v lies on u, it must be invariant, i.e. it must also hold in every other coordinate system (λ'). We now verify this analytically.

Application of the transformation laws

$$u_{\lambda'} = A^\lambda_{\lambda'} u_\lambda, \qquad v^{\lambda'} = A^{\lambda'}_\mu v^\mu$$

(the change of the dummy index in the second transformation law is necessary because the same index letter cannot be used repeatedly in any one formula for several independent summations) yields

$$u_{\lambda'} v^{\lambda'} = A^\lambda_{\lambda'} u_\lambda A^{\lambda'}_\mu v^\mu = A^\lambda_{\lambda'} A^{\lambda'}_\mu u_\lambda v^\mu = A^{\lambda\lambda'}_{\lambda'\mu} u_\lambda v^\mu = \delta^\lambda_\mu u_\lambda v^\mu = u_\lambda v^\lambda.$$

In this way we have shown that

$$u_{\lambda'} v^{\lambda'} = u_\lambda v^\lambda, \tag{2.31}$$

i.e. that the expression

$$\sigma \overset{\text{def}}{=} u_\lambda v^\lambda \tag{2.32}$$

is a *scalar*, or an *invariant* with respect to transformations of coordinate systems.

This implies in particular that equation (2.30) is invariant in character, as indeed we could have foreseen from a knowledge of the geometric interpretation of this relationship.

It can be shown that the equality $u_\lambda v^\lambda = 1$, which is also invariant, means geometrically speaking that if the origin of the vector v is situated in the first hyperplane of the vector u, then its terminal point will fall in the second hyperplane of u. The operation of constructing a scalar σ from two given vectors u, v by means of equation (2.32) is called the *transvection* of vectors on to each other.

Note that whereas the discussion concerning the geometric significance of vectors applied to a space based on the group G_a, the derivation of relation (2.31) and the proof that expression (2.32) is an invariant were of a more general nature and apply to a space based on the group G_1.

On the other hand, the expressions

$$\sum_\lambda u^\lambda v^\lambda \quad \text{and} \quad \sum_\lambda u_\lambda v_\lambda, \tag{2.33}$$

consisting of sums of products of the coordinates of two contravariant or two covariant vectors are not invariants — as can be easily seen — with respect to the general group G_1, or even with respect to the subgroup G_a of coordinate transformations.

It is true that in analytic geometry one is taught that the expression $\sum_\lambda u^\lambda v^\lambda$ is the so-called scalar product of the vectors u and v, and is hence an invariant. It should be remembered, however, that the analytic geometry learnt in schools or used for classical mechanics concerns Euclidean spaces, i.e. spaces based on the metric group G_m, and with respect to this subgroup it can be shown that the expressions (2.33) are indeed invariants. Conversely it is not difficult to show that if the expressions (2.33) are to be invariants for any pair of vectors u, v the transformations must belong to the orthogonal subgroup G_m.

14. Vector Algebra with the Pseudogroup G_1. Once the point ξ has been fixed in a space based on the pseudogroup G_1, it is possible to construct for vectors attached at that point an algebra embracing the following operations and properties:

I. Addition of contravariant and covariant vectors according to the formulae

$$w = u+v \text{ equivalent to } \begin{Bmatrix} w^\lambda = u^\lambda + v^\lambda \\ w_\lambda = u_\lambda + v_\lambda \end{Bmatrix} \text{ for } \begin{Bmatrix} \text{contravariant} \\ \text{covariant} \end{Bmatrix} \text{ vectors.}$$

II. Multiplication of vectors by real numbers

$$v = \alpha \cdot u = u \cdot \alpha \text{ equivalent to } \left\{ \begin{array}{l} v^\lambda = \alpha u^\lambda \\ v_\lambda = \alpha u_\lambda \end{array} \right\} \text{ for } \left\{ \begin{array}{l} \text{contravariant} \\ \text{covariant} \end{array} \right\} \text{ vectors.}$$

III. Subtraction may be defined either as the operation inverse to addition, or in terms of the first two operations:

$$u - v = u + (-1)v.$$

IV. The addition of contravariant vectors has a unique neutral element, namely the zero vector θ with coordinates

$$v^\lambda = 0 \qquad (\lambda = 1, 2, \ldots, n).$$

Similarly, the addition of covariant vectors has a unique neutral element, the zero vector $\bar{\theta}$ (which may also be denoted by the symbol θ without risk of misunderstanding), with coordinates

$$u_\lambda = 0 \qquad (\lambda = 1, 2, \ldots, n).$$

V. Addition is commutative

$$u + v = v + u$$

and is associative

$$(u + v) + w = u + (v + w).$$

VI. The law

$$\alpha(\beta u) = (\alpha\beta)u$$

holds.

VII. The first law of distributivity of multiplication with respect to addition holds

$$\alpha(u + v) = \alpha u + \alpha v$$

and so does the second law of distributivity of multiplication with respect to addition

$$(\alpha + \beta)u = \alpha u + \beta u.$$

VIII. The relation

$$\alpha\theta = \theta \qquad \text{holds for all choices of } \alpha$$

and the relation

$$0 \cdot u = \theta \qquad \text{holds for all choices of } u.$$

IX. Every pair consisting of a contravariant vector v^λ and a covariant vector u_λ can be transvected (cf. p. 48), and we thus obtain the scalar

$$\sigma = v^\lambda u_\lambda.$$

With the group G_1, no further algebraic operations are introduced. The operation of scalar multiplication can be introdced, as we shall see, either by replacing the group G_1 with the subgroup G_m or by defining at the point ξ an additional geometric object — the *metric tensor*.

Most of the theorems of vector and tensor algebra, and later, analysis, will take the form of equalities which can be written either in symbolic form, e.g.

$$u+v = v+u,$$

or in terms of coordinates, e.g.

$$u_\lambda + v_\lambda = v_\lambda + u_\lambda \qquad (\lambda = 1, 2, \dots, n) \tag{2.34}$$

or

$$\sigma = u_\lambda v^\lambda . \tag{2.35}$$

Equalities such as (2.34) or (2.35) are invariant, i.e. hold in every allowable coordinate system.

Accordingly, we adopt the convention that all relationships in the form of equalities are taken to be invariant; otherwise, we must use the sign $\overset{*}{=}$ to indicate that the equality holds in the given coordinate system (indicated by the presence of an appropriate index), although it is not necessarily valid in other coordinate systems. This convention will enable us to avoid possible misunderstandings in future.

15. The Linear Dependence of Vectors. Let p be a natural number independent of the dimension n of the space and suppose that we are given p vectors (contravariant vectors) attached at the same point

$$\underset{i}{v} \qquad (i = 1, 2, \dots, p) \tag{2.36}$$

with coordinates $\underset{i}{v}^\lambda$ $(\lambda = 1, 2, \dots, n)$ in the coordinate system (λ). Letters of another alphabet are used as indices to distinguish between different vectors since the vectors (2.36) need not be related to the system (λ) in any way.

DEFINITION. The vectors (2.36) are said to be *linearly dependent* if there exists a set of numbers $\overset{i}{\alpha}$ $(i = 1, 2, \dots, p)$ such that

$$\sum_i \overset{i}{\alpha}{}^2 > 0 \tag{2.37}$$

and such that $\overset{i}{\alpha}\underset{i}{v} = \theta$, or, writing in terms of coordinates, $\overset{i}{\alpha}\underset{i}{v}^\lambda = 0$ for

$\lambda = 1, 2, \ldots, n$. Vectors (2.36) are said to be *linearly independent* if they are not linearly dependent.

THEOREM. *At every point ξ there exists a set of n vectors ($p = n$) which are linearly independent.*

Indeed, it suffices to consider the (square) matrix of numbers

$$\varrho_\mu^\lambda \quad (\lambda, \mu = 1, 2, \ldots, n) \tag{2.38}$$

with the property that the determinant

$$|\varrho_\mu^\lambda| \neq 0, \tag{2.39}$$

and to set

$$\underset{\mu}{v}^\lambda \overset{*}{=} \varrho_\mu^\lambda, \tag{2.40}$$

to obtain a set of linearly independent vectors. The assumption that the set is linearly dependent leads to the relations $\overset{\mu}{\alpha}\underset{\mu}{\varrho}^\lambda = 0$ ($\lambda = 1, 2, \ldots, n$). If regarded as a system of equations in the unknowns $\overset{\mu}{\alpha}$ with coefficients $\underset{\mu}{\varrho}^\lambda$, Cramer's formulae tell us that $\overset{\mu}{\alpha} = 0$ ($\mu = 1, 2, \ldots, n$), which contradicts inequality (2.37).

THEOREM. *Every set of p vectors, where $p \geqslant n+1$, constitutes a set of linearly dependent vectors.*

Indeed, if the set $\overset{i}{\alpha}\underset{i}{v}^\lambda = 0$ ($\lambda = 1, 2, \ldots, n$) is regarded as a system of equations in the p unknowns $\overset{i}{\alpha}$ and if the rank of the matrix $[\underset{i}{v}^\lambda]$ is denoted by r, we can assign arbitrary, and hence, non-zero values to $(p-r)$ of the unknowns $\overset{i}{\alpha}$ and since $r \leqslant n$ and $p \geqslant n+1$, we do have $p-r \geqslant 1$.

16. Basis Vectors. If in particular we set

$$\underset{\mu}{v}^\lambda = \underset{\mu}{e}^\lambda \overset{*}{=} \delta_\mu^\lambda, \tag{2.41}$$

the set of n vectors

$$\underset{\mu}{e}^\lambda, \tag{2.42}$$

defined at a given point p of space will be called a set of *basis vectors bound to the system* (λ).

The fact that the equalities are written

$$\underset{\mu}{e}^{\lambda} \overset{*}{=} \delta^{\lambda}_{\mu}$$

mean that in another system (λ') these equalities will in general no longer be valid, i.e. that

$$\underset{\mu}{e}^{\lambda'} \neq \delta^{\lambda'}_{\mu},$$

which is understandable.

REMARK 1. The use of the inferior index μ indicates that our vectors are closely related to the coordinate system (λ). The index μ is here a dead index. It does not transform under passage to another coordinate system.

REMARK 2. If in particular the space is an affine space E_n, the set of basis vectors is a set of vectors which are parallel to the ξ^{λ}-axis of the system and have a measure (1) (not length!) of 1.

The set of basis vectors (2.42) naturally constitutes a set of linearly independent vectors since the value of the determinant (2.39) is 1:

$$|\underset{\mu}{e}^{\lambda}| = 1.$$

The definition of *linearly dependent covariant vectors* is exactly analogous and the theorems for covariant vectors are similar. The set of vectors

$$\overset{\mu}{e}_{\lambda} \overset{*}{=} \delta^{\lambda}_{\mu} \quad (\lambda, \mu = 1, 2, \ldots, n) \tag{2.43}$$

is called the *basis system associated with the coordinate system* (λ). In the space E_n these vectors are such that their first hyperplanes contain $n-1$ axes of the system, while their second hyperplanes have an intercept of 1 with the other axis.

THEOREM. *For each system of n linearly independent vectors* $\underset{\mu}{v}^{\lambda}$ *there exists a unique system of linearly independent vectors* $\overset{\lambda}{\bar{v}}_{\mu}$ *which satisfy the relations*

$$\underset{\mu}{v}^{\lambda} \overset{\nu}{\bar{v}}_{\lambda} = \delta^{\nu}_{\mu} \quad (\mu, \nu = 1, 2, \ldots, n). \tag{2.44}$$

Indeed, upon fixing the index ν, we can regard (2.44) as a system of n linear

(1) The *measure of a vector on an axis* is, as is well known, the difference between the coordinates of the terminal point and of the origin of the vector.

equations in the n unknowns $\overset{\nu}{\bar{v}}_\lambda$. Since the determinant $|\underset{\mu}{v}^\lambda| \neq 0$, the system has the unique solution

$$\overset{\nu}{\bar{v}}_\lambda = \frac{\text{minor}\ \overset{\nu}{v}^\lambda}{|\underset{\mu}{v}^\lambda|} \qquad (\lambda = 1, 2, \ldots, n). \tag{2.45}$$

System (2.45) is a system of linearly independent vectors since $\det \overset{\nu}{\bar{v}}_\lambda = (\det \underset{\mu}{v}^\lambda)^{-1} \neq 0$. The system of vectors $\overset{\nu}{\bar{v}}_\lambda$ satisfying relations (2.44) is called the *inverse of the system* $\underset{\nu}{v}^\lambda$. Figure 2 shows in the case $n = 2$, how to obtain the system $\bar{v}$ geometrically for any given v's.

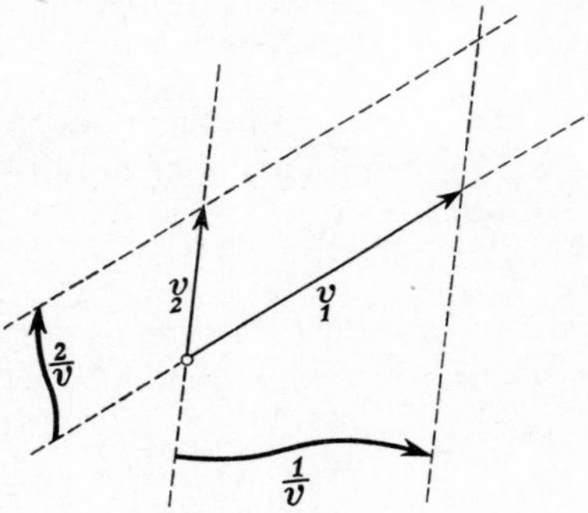

Fig. 2.

THEOREM. *For each system of n linearly independent vectors* $\overset{\nu}{v}_\lambda$ *there exists a unique (inverse) system of vectors* $\underset{\nu}{\bar{v}}^\lambda$ *which satisfy the relationships*

$$\overset{\nu}{v}_\lambda \underset{\mu}{\bar{v}}^\lambda = \delta^\nu_\mu \qquad (\nu, \mu = 1, 2, \ldots, n). \tag{2.46}$$

The proof is analogous to that of the preceding theorem.

THEOREM. *The inverse of an inverse system coincides with the original system.*

PROOF. Suppose we are first given a system

$$\underset{\mu}{v}^\lambda \tag{2.47}$$

of linearly independent vectors. Let

$$\overset{\nu}{\bar{v}}_{\lambda} \tag{2.48}$$

denote the inverse and

$$\underset{\mu}{\underline{\underline{v}}}{}^{\lambda} \tag{2.49}$$

the inverse of (2.48). We thus have the relations

$$\underset{\mu}{\underline{\underline{v}}}{}^{\lambda}\overset{\nu}{\bar{v}}_{\lambda} = \delta^{\nu}_{\mu}. \tag{2.50}$$

Relations (2.44) and (2.50) taken together yield

$$(\underset{\mu}{\underline{\underline{v}}}{}^{\lambda} - \underset{\mu}{v}{}^{\lambda})\overset{\nu}{\bar{v}}_{\lambda} = 0 \qquad (\mu, \nu = 1, 2, \ldots, n).$$

Since $|\overset{\nu}{\bar{v}}_{\lambda}| \neq 0$, it follows that for every fixed μ we have

$$\underset{\mu}{\underline{\underline{v}}}{}^{\lambda} - \underset{\mu}{v}{}^{\lambda} = 0 \qquad (\lambda = 1, 2, \ldots, n),$$

and the theorem is thus proved.

In particular, the systems of basis vectors of every coordinate system are the inverses of each other.

DEFINITION. A vector w [1] is said to be a *linear combination* of a system of linearly independent vectors $\underset{i}{v}$ $(i = 1, 2, \ldots, p)$ if there exists a set of (scalar) coefficients $\overset{i}{\alpha}$ with the property that

$$w = \overset{i}{\alpha}\underset{i}{v}. \tag{2.51}$$

The coefficients $\overset{i}{\alpha}$ are called the *coordinates of the vector w in the system* $\underset{i}{v}$.

REMARK .The coefficients $\overset{i}{\alpha}$ are scalars. Indeed, assume that

$$w^{\lambda} \overset{*}{=} \overset{i}{\alpha}\underset{i}{v} \qquad (\lambda = 1, 2, \ldots, n)$$

and consider any other allowable coordinate system (λ'). Then on the one hand we have

$$w^{\lambda'} = A^{\lambda'}_{\lambda}\, w^{\lambda}.$$

[1] If we use the term vector without specifying whether it is contra- or co-variant, it is to be tacitly understood that the statement holds for both contravariant and covariant vectors.

On the other hand, however,

$$\underset{i}{v}^{\lambda'} = A_{\lambda}^{\lambda'}\underset{i}{v}^{\lambda} \qquad (i = 1, 2, \ldots, p;\ \lambda = 1, 2, \ldots, n).$$

Thus,

$$\overset{i}{\alpha}\underset{i}{v}^{\lambda'} = \overset{i}{\alpha}A_{\lambda}^{\lambda'}\underset{i}{v}^{\lambda} = A_{\lambda}^{\lambda'}\overset{i}{\alpha}\underset{i}{v}^{\lambda} = A_{\lambda}^{\lambda'}w^{\lambda},$$

whence

$$w^{\lambda'} = \overset{i}{\alpha}\underset{i}{v}^{\lambda'},$$

which means that the coefficients $\overset{i}{\alpha}$ do not alter under passage to the system (λ').

THEOREM. *If the vectors*

$$\underset{i}{v} \qquad (i = 1, 2, \ldots, n) \tag{2.52}$$

constitute a system of n linearly independent vectors, each vector w can be expressed in just one way as a linear combination of the vectors (2.52), *i.e. there exists a uniquely determined set of constants* $\overset{i}{\alpha}$ *with properties* (2.51).

The proof of this theorem follows from the statement that the system of equations

$$\underset{i}{v}^{\lambda}\overset{i}{\alpha} = w^{\lambda} \qquad (\lambda = 1, 2, \ldots, n),$$

regarded as a system of n equations in the n unknowns $\overset{i}{\alpha}$, has a unique solution since the determinant of the coefficients is non-zero.

If in particular we assume that

$$\underset{\mu}{v}^{\lambda} = \underset{\mu}{e}^{\lambda},$$

then it follows from the relations

$$w^{\lambda} = \overset{\mu}{\alpha}\underset{\mu}{e}^{\lambda} \tag{2.53}$$

that

$$\overset{\lambda}{\alpha} \overset{*}{=} w^{\lambda}.$$

An asterisk must be added here because the preceding summation involved dead indices and caused the living index on the right-hand side to be moved to the dead position. Moreover, an equality which has a living index on one side and a dead index on the other cannot be valid in all coordinate systems. If we pass to a new coordinate system (λ'), equality (2.53) becomes

$$w^{\lambda'} = \overset{\mu}{\alpha} \underset{\mu}{e}^{\lambda'},$$

since the dead index is not transformed. And since

$$\underset{\mu}{e}^{\lambda'} = A^{\lambda'}_{\lambda} \underset{\mu}{e} \overset{*}{=} A^{\lambda'}_{\lambda} \delta^{\lambda}_{\mu} = A^{\lambda'}_{\mu},$$

we have

$$w^{\lambda'} \overset{*}{=} \overset{\mu}{\alpha} A^{\lambda'}_{\mu},$$

or

$$w^{\lambda'} \neq \overset{\mu}{\alpha} \delta^{\lambda'}_{\mu}.$$

Instead of $\overset{\lambda}{\alpha}$ we can write $\overset{\lambda}{w}$, and the formula

$$w^{\lambda} = \overset{\mu}{w} e^{\lambda} \tag{2.54}$$

is now in a form which is easy to remember.

If the vector w^{λ} is transvected on to any of the covariant basis vectors $\overset{\nu}{e}_{\lambda}$, we obtain

$$w^{\lambda} \overset{\nu}{e}_{\lambda} = \overset{\mu}{w} \underset{\mu}{e}^{\lambda} \overset{\nu}{e} = \overset{\mu}{w} \delta^{\nu}_{\mu} = \overset{\nu}{w},$$

or

$$\overset{\mu}{w} = w^{\lambda} \overset{\mu}{e}_{\lambda}.$$

Scalars $\overset{\mu}{w}$ can thus be obtained from the coordinates of the vector w by transvecting the latter onto the basis covariant vectors in turn. Relation (2.54), conversely, generates the coordinates of a vector from the scalars $\overset{\mu}{w}$ in terms of basis contravariant vectors. The operation of transition from w^{λ} to $\overset{\lambda}{w}$ has been called "strangling of the index" by Schouten. The inverse operation could be called fluxing of the index [69].

CHAPTER III

TENSORS AND TENSOR ALGEBRA

17. The Concept of Tensor. Consider two contravariant vectors u^λ and v^λ, attached at the same point ξ. We now form the square matrix of numbers

$$w^{\lambda\mu} \stackrel{\text{def}}{=} u^\lambda v^\mu \qquad (\lambda, \mu = 1, 2, \ldots, n).$$

In a new system of coordinates (λ') we assume by analogy that

$$w^{\lambda'\mu'} \stackrel{\text{def}}{=} u^{\lambda'} v^{\mu'},$$

and consider whether $w^{\lambda'\mu'}$ can be calculated from a knowledge of $w^{\lambda\mu}$ and the transformations $T(\lambda \to \lambda')$. Accordingly we have

$$u^{\lambda'} = A^{\lambda'}_{\lambda} u^\lambda, \qquad v^{\mu'} = A^{\mu'}_{\mu} u^\mu,$$

whence

$$w^{\lambda'\mu'} = u^{\lambda'} v^{\mu'} = A^{\lambda'}_{\lambda} u^\lambda A^{\mu'}_{\mu} v^\mu = A^{\lambda'\mu'}_{\lambda\,\mu} v^\lambda v^\mu = A^{\lambda'\mu'}_{\lambda\,\mu} w^{\lambda\mu}.$$

We thus find that $w^{\lambda'\mu'}$ can indeed be calculated if $w^{\lambda\mu}$ and $A^{\lambda'}_{\lambda}$ are known. It can be seen that the transformation law

$$w^{\lambda'\mu'} = A^{\lambda'\mu'}_{\lambda\,\mu} w^{\lambda\mu} \tag{3.1}$$

possesses the group property, i.e. that it defines a geometric object of class one. This law is more complicated than the corresponding one for vectors. This is because even though the $w^{\lambda'\mu'}$ are linear homogeneous functions of the variables $w^{\lambda\mu}$, with respect to $A^{\lambda'}_{\lambda}$, they are functions of the second degree rather than the first as in the case of the transformation law for vectors.

A geometric object with the transformation law (3.1) is called a *tensor*. This is the name given to a broader class of geometric objects. We may say in this case that the object $w^{\lambda\mu}$ was generated by multiplying two vectors. But more generally, in the system (λ) we can take an arbitrary (square) array, i.e. a matrix of coordinates

$$w^{\lambda\mu} \tag{3.2}$$

and assign to any other coordinate system (λ') the array $w^{\lambda'\mu'}$ obtained from the array (3.2) by means of formulae (3.1). This assignment will be one-to-one because of the group property. In this way, definition (3.1) of an object can be completely separated from the vectors u, v originally used to define w. It is possible, in fact, to display an array (3.2) for which no pair of vectors u, v exists with the property that

$$u^{\lambda}v^{\mu} = w^{\lambda\mu}. \tag{3.3}$$

Indeed for such an array to exist it suffices that the determinant W constructed from this array be non-zero, for then the set of equations (3.3) cannot have any solution with respect to the unknowns u^{λ}, v^{μ}, as the reader can easily verify.

Generalizing this example, we can say that the transformation law

$$w^{\lambda'\mu'\nu'} = A^{\lambda'\mu'\nu'}_{\lambda\ \mu\ \nu}w^{\lambda\mu\nu}$$

possesses the group property, and therefore defines a certain geometric object of class one, which we shall also call a *tensor* (1). A particular class of such objects consists of those obtained by multiplying three vectors $\underset{1}{v}^{\lambda}, \underset{2}{v}^{\lambda}, \underset{3}{v}^{\lambda}$:

$$w^{\lambda\mu\nu} \overset{\text{def}}{=} \underset{1}{v}^{\lambda}\underset{2}{v}^{\mu}\underset{3}{v}^{\nu} \qquad (\lambda, \mu, \nu = 1, 2, \ldots, n).$$

The object $w^{\lambda\mu}$ has n^2 coordinates while the object $w^{\lambda\mu\nu}$ has n^3.

Tensors with any number p of indices can be constructed in this way. In this general case, we can no longer use different letters to denote consecutive indices, but must use indices with (sub-)indices, $\lambda_1, \lambda_2, \ldots, \lambda_p$. Such a tensor will have n^p coordinates since each one can take any value from 1 to n; these coordinates will be written briefly as $w^{\lambda_1\ldots\lambda_p}$. The transformation law for the coordinates of this tensor, known as a *contravariant* tensor, can be written briefly as

$$w^{\lambda'_1\ldots\lambda'_p} = A^{\lambda'_1\ldots\lambda'_p}_{\lambda_1\ldots\lambda_p}w^{\lambda_1\ldots\lambda_p}. \tag{3.4}$$

The natural number p is in general independent of the dimension n of the space.

---

(1) Tensors are also sometimes called *affinors*, although this term is being used less and less.

In similar fashion we arrive at the concept of *covariant tensor* with coordinates $w_{\lambda_1\ldots\lambda_p}$, for which the transformation law is

$$w_{\lambda'_1\ldots\lambda'_p} = A^{\lambda_1\ldots\lambda_p}_{\lambda'_1\ldots\lambda'_p} w_{\lambda_1\ldots\lambda_p}. \tag{3.5}$$

In addition to contravariant tensors with the transformation law (3.4) and covariant tensors with the transformation law (3.5), it is also possible to introduce *mixed tensors*

$$w^{\lambda_1\ldots\lambda_p}{}_{\mu_1\ldots\mu_q} \tag{3.6}$$

governed by the transformation law

$$w^{\lambda'_1\ldots\lambda'_p}{}_{\mu'_1\ldots\mu'_q} = A^{\lambda'_1\ldots\lambda'_p\mu_1\ldots\mu_q}_{\lambda_1\ldots\lambda_p\mu'_1\ldots\mu'_q} w^{\lambda_1\ldots\lambda_p}{}_{\mu_1\ldots\mu_q}. \tag{3.7}$$

Such a tensor is said to be *p-fold contravariant* and *q-fold covariant*, or, more briefly, a tensor with valence (p, q), where the first number denotes the number of contravariant indices and the second, the number of covariant indices. If $p = 0$ or $q = 0$, the tensor becomes either purely covariant or purely contravariant, as the case may be.

In this terminology, contravariant tensors are tensors with valence (1, 0) while covariant tensors have a valence of (0, 1). Scalars can be regarded as tensors with valence (0, 0). Tensors with valence (p, q), where $p+q = 2$ will be referred to as *tensors of rank two*. The array of coordinates of such a tensor in any coordinate system is a square matrix. A second-rank tensor should not, however, be identified with a square matrix since the concept of the matrix does not embody any law governing the transformation of its elements upon transition from one coordinate system to another. A tensor of rank two is not a matrix but rather a matrix together with an associated coordinate system.

As its definition implies, an object is a function which sets up a unique correspondence between a sequence of numbers w_a, called the *coordinates of the object*, and any allowable system of coordinates. The set of allowable systems depends on the pseudogroup of transformations adopted. Given a geometric object together with a particular law for the transformation of the coordinates of the object, we in fact obtain a different object by restricting the pseudogroup while retaining the transformation law. It can happen that the restriction of a pseudogroup may itself cause a change (simplication) in the transformation law. Such examples will be dealt with later on. In particular, when we restrict the group G_1 to be an orthogonal group, the transformation law for tensors will remain formally unchanged, it is true, nevertheless the set of allowable values $A^{\lambda'}_{\lambda}$ and $A^{\lambda}_{\lambda'}$, appearing in the transformation law is restricted. Some authors use the term "Cartesian tensors" for tensors when the group is restricted to be an orthogonal group, although this name is not a very good choice because Cartesian systems are, after all, arbitrary rectilinear systems, and not necessarily orthogonal with equal units on each of the axes.

18. The Unit Tensor. Consider the mixed tensor $w^{\lambda}{}_{\mu}$ defined by the relations

$$w^{\lambda}{}_{\mu} = \underset{\nu}{e^{\lambda}}\overset{\nu}{e}_{\mu}. \tag{3.8}$$

We first of all verify that this is indeed a tensor; in fact we see that

$$w^{\lambda'}{}_{\mu'} = \underset{\nu}{e^{\lambda'}}\overset{\nu}{e}_{\mu'} = A^{\lambda'}_{\lambda}\underset{\nu}{e^{\lambda}}A^{\mu}_{\mu'}\overset{\nu}{e}_{\mu} = A^{\lambda'\mu}_{\lambda\ \mu'}\underset{\nu}{e^{\lambda}}\overset{\nu}{e}_{\mu} = A^{\lambda'\mu}_{\lambda\ \mu'}w^{\lambda}{}_{\mu},$$

which shows that it is a tensor. Furthermore, we have

$$w^{\lambda}{}_{\mu} \overset{*}{=} \delta^{\lambda}_{\nu}\delta^{\nu}_{\mu} = \delta^{\lambda}_{\mu}, \tag{3.9}$$

which tells us that the array of coordinates of this tensor has ones on the main diagonal and zeros everywhere else in the system (λ) [(1)]. What will the coordinates of this tensor be in another coordinate system (λ')? Using equation (3.9), we may write

$$w^{\lambda'}{}_{\mu'} = A^{\lambda'\mu}_{\lambda\ \mu'}\delta^{\lambda}_{\mu} = A^{\lambda'\lambda}_{\lambda\ \mu'} = A^{\lambda'}_{\mu'} = \delta^{\lambda'}_{\mu'},$$

which means that the array of this tensor looks the same in any other coordinate system (λ'). This could therefore be called a *constant tensor,* since for a tensor—just as for every other geometric object—the coordinate system (λ) is an independent variable while the array of coordinates is a dependent variable.

This tensor has been called a *unit tensor* (this term will be explained later) and denoted by the symbol

$$A^{\lambda}_{\mu}\ ^{(2)}. \tag{3.10}$$

This notation does not conflict with the symbols $A^{\lambda'}_{\mu}$, $A^{\lambda}_{\mu'}$: in fact, it fits in with them extremely well. The transformation law can now be written

$$A^{\lambda'}_{\mu'} = A^{\lambda'\mu}_{\lambda\ \mu'}A^{\lambda}_{\mu}, \tag{3.11}$$

---

(1) It is thus a *unit matrix.*

(2) We write A^{λ}_{μ} instead of $A^{\lambda}{}_{\mu}$. The reason for writing $w^{\lambda_1\ldots\lambda_p}{}_{\mu_1\ldots\mu_g}$ rather than the more abbreviated symbol $w^{\lambda_1\ldots\lambda_p}_{\mu_1\ldots\mu_g}$ will make itself clear further on.

which leads to a tautology when A^{λ}_{μ}, $A^{\lambda'}_{\mu'}$ are taken into account, the symbols on the right-hand side of relations (3.11) are identified and the laws (1.31) taken into consideration. The advantage of using the kernel letter A for the unit tensor as well as for the partial derivatives of the transformation of the coordinate system will become evident later when we introduce intermediate coordinates for objects.

Now take an arbitrary contravariant vector v^{λ} and transvect the unit tensor A onto it, i.e. put

$$\bar{v}^{\lambda} \stackrel{\text{def}}{=} A^{\lambda}_{\mu} v^{\mu}.$$

This leads to

$$\bar{v}^{\lambda'} = A^{\lambda'}_{\mu'} v^{\mu'} = A^{\lambda'}_{\mu'} A^{\mu'}_{\mu} v^{\mu} = A^{\lambda' \mu'}_{\mu' \mu} v^{\mu} = A^{\lambda'}_{\mu} v^{\mu} = A^{\lambda'}_{\mu} A^{\mu}_{\lambda} v^{\lambda} = A^{\lambda'}_{\mu} \bar{v}^{\mu},$$

which implies that $\bar{v}^{\lambda}$ is a vector. On the other hand

$$\bar{v}^{\lambda} \stackrel{*}{=} \delta^{\lambda}_{\mu} v^{\mu} = v^{\lambda},$$

and the vector $\bar{v}^{\lambda}$ is thus identical with the vector v^{λ}, i.e. transvection of a unit tensor onto the vector v produces the original vector v. Hence the name "unit tensor".

It can be shown more generally that if we have an arbitrary tensor of arbitrary valence $w^{\lambda_1 \ldots \lambda_p}{}_{\mu_1 \ldots \mu_q}$ and if we fix, say the index $\lambda_i(\mu_j)$ provided that $p \geqslant 1$ $(q \geqslant 1)$ and transvect the tensor A^{λ}_{μ} onto the tensor w with respect to the fixed index (when we are transvecting a vector it is not necessary to add with respect to which index this is done since there is only one index), or in other words if we form the expression

$$w^{\lambda_1 \ldots \lambda_{i-1} \varrho \lambda_{i+1} \ldots \lambda_p}{}_{\mu_1 \ldots \mu_q} A^{\lambda_i}_{\varrho} \quad {}^{(1)}$$

we obtain a tensor, which is in fact the original tensor w.

Now take a mixed tensor

$$w^{\lambda}{}_{\mu}. \tag{3.12}$$

Transvecting this onto an arbitrary contravariant vector v^{λ}, we obtain

$$\bar{v}^{\lambda} = w^{\lambda}{}_{\mu} v^{\mu},$$

i.e. a new contravariant vector $\bar{v}$. We can thus make use of the tensor w to assign to each vector v a new vector $\bar{v}$ in a unique fashion; in other words, the tensor w establishes a *vector-vector transformation*. This transformation is linear in the sense that it takes the vector $\alpha \cdot v$ over into

(1) Or $w^{\lambda_1 \ldots \lambda_p}{}_{\mu_1 \ldots \mu_{j-1} \varrho \mu_{j+1} \ldots \mu_q} A^{\varrho}_{\mu_j}$.

the vector $\alpha \cdot \bar{v}$ and the sum $v_1 + v_2$ over into the sum $\underset{1}{\bar{v}} + \underset{2}{\bar{v}}$. For some authors(1) the linear transformation of vectors into vectors forms a basis for defining a two-index tensor. It is here that we find the historical beginnings of the concept of tensor introduced by A. Einstein, while the term itself had been introduced in elastomechanics back in 1898 by Voigt [85].

In a similar way the mixed tensor (3.12) can be used to transform covariant vectors into covariant vectors

$$\bar{u}_\lambda = w^\mu{}_\lambda u_\mu$$

and tensors of valence (2,0) (or (0,2)) can be used to transform covariant (contravariant) vectors into contravariant (covariant) vectors

$$\bar{v}^\mu = w^{\lambda\mu} v_\lambda, \qquad \bar{u}_\mu = w_{\lambda\mu} v^\lambda.$$

Transformations of this type using the unit tensor A^λ_μ determine identity transformations which explains the origin of the name "unit tensor".

Algebraic operations which produce new tensors from given ones are the subject of tensor algebra. As we shall see, there are more operations of this kind for tensors than there are for vectors.

19. The Product of a Tensor with a Number α. Suppose that we are given some tensor w of valence (p, q). Now put

$$\bar{w} \overset{\text{def}}{=} \alpha w, \tag{3.13}$$

to indicate that each coordinate of the tensor w is obtained by multiplying the corresponding coordinate of the tensor w by the same scalar factor α. It is very easily shown that this yields a new tensor $\bar{w}$ of the same valence as w. *Tensors can* thus *be multiplied by numbers* just as vectors can be so multiplied.

Multiplication by zero always yields a tensor with all of its coordinates equal to zero and this property is invariant, as follows directly from the transformation law. Such tensors will be referred to as *null* or *zero tensors*. Since there is no danger of misunderstanding we shall use the symbol 0 to denote zero tensors of arbitrary valence. Accordingly, the equation

$$v^\lambda = 0 \qquad (\lambda = 1, 2, \ldots, n)$$

will be written briefly as

$$v = 0 \tag{3.14}$$

(1) Rubinowicz [64].

and, similarly, the equation

$$w^{\lambda_1 \dots \lambda_p}{}_{\mu_1 \dots \mu_q} = 0 \qquad (\lambda_i, \mu_j = 1, 2, \dots, n)$$

will also be abbreviated to

$$w = 0 \tag{3.15}$$

although the symbol 0 on the right-hand sides of equations (3.14) and (3.15) stands for different things in the two cases.

It follows immediately from the definition of multiplication of tensors by scalars that the product of a null tensor with any number is always a null tensor.

20. The Sum of Tensors. Only tensors of the same valence can be added to each other. Given two tensors of the same valence

$$\underset{1}{w}^{\lambda_1 \dots \lambda_p}{}_{\mu_1 \dots \mu_q}, \qquad \underset{2}{w}^{\lambda_1 \dots \lambda_p}{}_{\mu_1 \dots \mu_q}$$

and put

$$\underset{3}{w}^{\lambda_1 \dots \lambda_p}{}_{\mu_1 \dots \mu_q} \overset{\text{def}}{=} \underset{1}{w}^{\lambda_1 \dots \lambda_p}{}_{\mu_1 \dots \mu_q} + \underset{2}{w}^{\lambda_1 \dots \lambda_p}{}_{\mu_1 \dots \mu_q}.$$

It is our contention that $\underset{3}{w}$ is again a tensor of the same valence as w_1 and w_2.

Indeed, we have

$$\begin{aligned}
\underset{3}{w}^{\lambda'_1 \dots \lambda'_p}{}_{\mu'_1 \dots \mu'_q} &= \underset{1}{w}^{\lambda'_1 \dots \lambda'_p}{}_{\mu'_1 \dots \mu'_q} + \underset{2}{w}^{\lambda'_1 \dots \lambda'_p}{}_{\mu'_1 \dots \mu'_q} \\
&= A^{\lambda'_1 \dots \lambda'_p \mu_1 \dots \mu_q}_{\lambda_1 \dots \lambda_p \mu'_1 \dots \mu'_q} \underset{1}{w}^{\lambda_1 \dots \lambda_p}{}_{\mu_1 \dots \mu_q} + A^{\lambda'_1 \dots \lambda'_p \mu_1 \dots \mu_q}_{\lambda_1 \dots \lambda_p \mu'_1 \dots \mu'_p} \underset{2}{w}^{\lambda_1 \dots \lambda_p}{}_{\mu_1 \dots \mu_q} \\
&= A^{\lambda'_1 \dots \lambda'_p \mu_1 \dots \mu_q}_{\lambda_1 \dots \lambda_p \mu'_1 \dots \mu'_q} \left(\underset{1}{w}^{\lambda_1 \dots \lambda_p}{}_{\mu_1 \dots \mu_q} + \underset{2}{w}^{\lambda_1 \dots \lambda_p}{}_{\mu_1 \dots \mu_q} \right) \\
&= A^{\lambda'_1 \dots \lambda'_p \mu_1 \dots \mu_q}_{\lambda_1 \dots \lambda_p \mu'_1 \dots \mu'_q} \underset{3}{w}^{\lambda_1 \dots \lambda_p}{}_{\mu_1 \dots \mu_q}.
\end{aligned}$$

Tensors of different valences cannot be added together for reasons similar to those for which covariant vectors cannot be added to contravariant vectors. It could be said, in short, that tensors are added by adding their corresponding coordinates.

21. Linear Dependence of Tensors. If

$$\underset{1}{w}, \underset{2}{w}, \dots, \underset{r}{w} \tag{3.16}$$

denotes a sequence of tensors of the same valence, to say that they form a *system of linearly dependent tensors* means that there exist r constants $\overset{1}{a}, \overset{2}{a}, \ldots, \overset{r}{a}$ with the property

$$\sum_{i=1}^{r} (\overset{i}{a})^2 > 0$$

and such that

$$\overset{i}{a} \underset{i}{w} = 0. \tag{3.17}$$

This relationship is, of course, invariant in character. If no such constants exist the system is said to be a *system of linearly independent tensors.*

The maximum number of linearly independent vectors is n, or in other words is equal to the dimension of the space. The maximum number of linearly independent tensors of a particular valence depends not only on n but also on the numbers p, q which define the valence of the tensor, and in general is n^{p+q}. For certain distinguished tensors with a given valence the maximum number of independent tensors may be less than n^{p+q}, e.g. the number of independent antisymmetric tensors, is, as we shall see below, less than both n^p and n^q.

22. The Tensor Product. Just as the multiplication of vectors resulted in tensors (of rank two), so tensors of any valence, not necessarily equal, can be multiplied together. The result of the multiplication will always be a tensor of a certain valence; we shall therefore not arrive at any new quantities in this way.

Suppose we are given two tensors

$$\underset{1}{w}^{\lambda_1 \ldots \lambda_p}{}_{\mu_1 \ldots \mu_q} \quad \text{and} \quad \underset{2}{w}^{\varrho_1 \ldots \varrho_r}{}_{\sigma_1 \ldots \sigma_s},$$

and let us write

$$\underset{3}{w}^{\nu_1 \ldots \nu_{p+r}}{}_{\tau_1 \ldots \tau_{q+s}} \overset{\text{def}}{=} \underset{1}{w}^{\nu_1 \ldots \nu_p}{}_{\tau_1 \ldots \tau_q} \underset{2}{w}^{\nu_{p+1} \ldots \nu_{p+r}}{}_{\tau_{q+1} \ldots \tau_{q+s}}. \tag{3.18}$$

It is easily shown that the object (3.18) so defined is a tensor of valence $(p+r, q+s)$. Thus, in multiplication the corresponding components of the valence are added together.

Note that the product of tensors depends on the ordering of the factors. The tensor $\underset{4}{w}$ defined by the formula

$$\underset{4}{w}^{\nu_1 \ldots \nu_{p+r}}{}_{\tau_1 \ldots \tau_{q+s}} \overset{\text{def}}{=} \underset{2}{w}^{\nu_1 \ldots \nu_r}{}_{\tau_1 \ldots \tau_s} \underset{1}{w}^{\nu_{r+1} \ldots \nu_{r+p}}{}_{\tau_{s+1} \ldots \tau_{s+q}} \tag{3.19}$$

will in general differ from the tensor $\underset{3}{w}$, as can be easily verified by considering the simplest example of multiplication of two (different) tensors. In fact, if we set

$$\underset{3}{w}^{\lambda\mu} = \underset{1}{v}^{\lambda}\underset{2}{v}^{\mu}, \qquad \underset{4}{w}^{\lambda\mu} = \underset{2}{v}^{\lambda}\underset{1}{v}^{\mu},$$

then in general we have

$$\underset{1}{v}^{\lambda}\underset{2}{v}^{\mu} \neq \underset{2}{v}^{\lambda}\underset{1}{v}^{\mu} \qquad (\lambda \neq \mu),$$

from which it follows that $\underset{4}{w} \neq \underset{3}{w}$.

23. The Contraction of Tensors. Suppose that we are given a tensor w of valence (p, q) where $p \geqslant 1$, $q \geqslant 1$:

$$w^{\lambda_1 \ldots \lambda_p}{}_{\mu_1 \ldots \mu_q}.$$

Now fix one of the superior indices, e.g. λ_i, and one of the inferior indices, e.g. μ_j. Next, we put

$$\overset{*}{w}{}^{\lambda_1 \ldots \lambda_{i-1}\lambda_{i+1} \ldots \lambda_p}{}_{\mu_1 \ldots \mu_{j-1}\mu_{j+1} \ldots \mu_q} \overset{\text{def}}{=} w^{\lambda_1 \ldots \lambda_{i-1}\varrho\lambda_{i+1} \ldots \lambda_p}{}_{\mu_1 \ldots \mu_{j-1}\varrho\mu_{j+1} \ldots \mu_q}. \tag{3.20}$$

Explicitly, we sum the coordinates of the tensor w with respect to the i-th superior and the j-th inferior indices. We claim that the *object* $\overset{*}{w}$ *so defined is a tensor of valence* $(p-1, q-1)$.

PROOF. We have

$$\overset{*}{w}{}^{\lambda'_1 \ldots \lambda'_{i-1}\lambda'_{i+1} \ldots \lambda'_p}{}_{\mu'_1 \ldots \mu'_{j-1}\mu'_{j+1} \ldots \mu'_q} = w^{\lambda'_1 \ldots \lambda'_{i-1}\nu'\lambda'_{+1} \ldots \lambda'_p}{}_{\mu'_1 \ldots \mu'_{j-1}\nu'\mu'_{j+1} \ldots \mu'_q}.$$

But

$$\begin{aligned} & w^{\lambda'_1 \ldots \lambda'_{i-1}\nu'\lambda'_{i+1} \ldots \lambda'_p}{}_{\mu'_1 \ldots \mu'_{j-1}\varrho'\mu'_{j+1} \ldots \mu'_p} \\ &\quad = A^{\lambda'_1 \ldots \lambda'_{i-1}\nu'\lambda'_{i+1} \ldots \lambda'_p\mu_1 \ldots \mu_{j-1}\varrho\mu_{j+1} \ldots \mu_q}_{\lambda_1 \ldots \lambda_{i-1}\nu\lambda_{i+1} \ldots \lambda_p\mu'_1 \ldots \mu'_{j-1}\varrho'\mu'_{j+1} \ldots \mu'_q} w^{\lambda_1 \ldots \lambda_{i-1}\nu\lambda_{i+1}\lambda_p}{}_{\mu_1 \ldots \mu_{j-1}\varrho\mu_{j+1} \ldots \mu_q}. \end{aligned}$$

We therefore obtain

$$\begin{aligned} & w^{\lambda'_1 \ldots \lambda'_{i-1}\nu'\lambda'_{i+1} \ldots \lambda'_p}{}_{\mu'_1 \ldots \mu'_{j-1}\nu'\mu'_{j+1} \ldots \mu'_q} \\ &\quad = A^{\lambda'_1 \ldots \lambda'_{i-1}\nu'\lambda'_{i+1} \ldots \lambda'_p\mu_1 \ldots \mu_{j-1}\varrho\mu_{j+1} \ldots \mu_q}_{\lambda_1 \ldots \lambda_{i-1}\nu\lambda_{i+1} \ldots \lambda_p\mu'_1 \ldots \mu'_{j-1}\nu'\mu'_{j+1} \ldots \mu'_q} w^{\lambda_1 \ldots \lambda_{i-1}\nu\lambda_{i+1} \ldots \lambda_p}{}_{\mu_1 \ldots \mu_{j-1}\varrho\mu_{j+1} \ldots \mu_q} \\ &\quad = A^{\nu'\varrho}_{\nu\nu'} A^{\lambda'_1 \ldots \lambda'_{i-1}\lambda'_{i+1} \ldots \lambda'_p\mu_1 \ldots \mu_{j-1} \ldots \mu_{j+1} \ldots \mu_q}_{\lambda_1 \ldots \lambda_{i-1}\lambda_{i+1} \ldots \lambda_p\mu'_1 \ldots \mu'_{j-1}\mu'_{j+1} \ldots \mu^1_q} w^{\lambda_1 \ldots \lambda_{i-1}\nu\lambda_{i+1} \ldots \lambda_p}{}_{\mu_1 \ldots \mu_{i-1}\varrho\mu_{j+1} \ldots \mu_q} \\ &\quad = A^{\varrho}_{\nu} A^{\cdots}_{\cdots} w^{\cdots}{}_{\cdots} = \delta^{\varrho}_{\nu} A^{\cdots}_{\cdots} w^{\cdots}{}_{\cdots} \\ &\quad = A^{\lambda'_1 \ldots \lambda_{i-1} \ldots \lambda'_{i+1} \ldots \lambda'_p\mu_1 \ldots \mu_{j-1}\mu_{j+1} \ldots \mu_q}_{\lambda_1 \ldots \lambda_{i-1} \ldots \lambda_{i+1} \ldots \lambda_p\mu'_1 \ldots \mu'_{j-1}\mu'_{j+1} \ldots \mu'_q} \overset{*}{w}{}^{\lambda_1 \ldots \lambda_{i-1}\lambda_{i+1} \ldots \lambda_p}{}_{\mu_1 \ldots \mu_{j-1}\mu_{j+1} \ldots \mu_q}. \end{aligned}$$

which is just what we had to prove.

The tensor $\overset{*}{w}$ obtained from the tensor w by the operation defined by equation (3.20) is said to be a *tensor obtained by contraction with respect to the indices* λ_i *and* μ_j. In particular, if we have a mixed tensor w of rank two and contract it, we obtain a tensor of valence (0, 0), in fact a *scalar*

$$w^{\lambda}{}_{\lambda} = \sigma. \tag{3.21}$$

For the unit tensor A^{λ}_{μ} contraction yields a scalar equal to

$$A^{\lambda}_{\lambda} = n. \tag{3.22}$$

Contraction with respect to the indices λ, ν in the tensor $w^{\lambda\mu}{}_{\nu}$ leads to a contravariant vector.

If a tensor w is of valence (p, q), various tensors of valence $(p-1, q-1)$ can in general be obtained from it by contraction with respect to various pairs of indices.

Contraction of a tensor can be repeated until the indices are "exhausted", i.e. until we obtain a tensor with the number zero as its only component of valence. A sequence of contractions can be replaced by a single operation. In this way it turns out that the result does not depend on the order of contraction with respect to the various pairs of indices. Thus, for example, let us perform on the tensor

$$w^{\lambda_1\lambda_2\lambda_3}{}_{\mu_1\mu_2\mu_3\mu_4\mu_5}$$

three contractions, say with respect to the pairs of indices (λ_1, μ_4) (λ_3, μ_2) and (λ_2, μ_1). The result of this is a covariant tensor of rank two (two-index tensor) which could also be obtained by means of a single, collective contraction, or triple summation

$$\overset{*}{w}_{\mu_3\mu_5} = w^{\varrho\tau\sigma}{}_{\tau\sigma\mu_3\varrho\mu_5}. \tag{3.23}$$

Recalling the rule that each independent summation must be indicated by the use of a separate dummy index, formula (3.23) clearly shows with respect to which pairs of indices the contraction operations have been carried out.

24. Transvection of Tensors. We have already defined the transvection of a covariant vector onto some given contravariant vector. Mention has also been made of the transvection of the unit tensor onto tensors, an operation which, as we know, does not change tensors into any other entities.

We are now going to introduce a generalization of these concepts. Suppose that we are given two arbitrary tensors

$$u^{\lambda_1\ldots\lambda_p}{}_{\mu_1\ldots\mu_q} \quad \text{and} \quad w^{\lambda_1\ldots\lambda_r}{}_{\mu_1\ldots\mu_s},$$

where the numbers p, q, r, s are independent of one another. Assume that $p \geqslant 1$, $s \geqslant 1$ and fix the index λ_i from the set $\lambda_1, \lambda_2, \ldots, \lambda_p$ and index μ_j from the set $\mu_1, \mu_2, \ldots, \mu_s$. Next, let us put

$$v^{\lambda_1\ldots\lambda_{i-1}\lambda_{i+1}\ldots\lambda_p\lambda_{p+1}\ldots\lambda_{p+r}}{}_{\mu_1\ldots\mu_q\mu_{q+1}\ldots\mu_{q+j-1}\mu_{q+j+1}\ldots\mu_{q+s}}$$
$$\stackrel{\text{def}}{=} u^{\lambda_1\ldots\lambda_{i-1}\varrho\lambda_{i+1}\ldots\lambda_p}{}_{\mu_1\ldots\mu_q} w^{\lambda_{p+1}\ldots\lambda_{p+r}}{}_{\mu_{q+1}\ldots\mu_{q+j-1}\varrho\mu_{q+j+1}\ldots\mu_{q+s}}.$$

It can be shown (the proof, which although simple, is tedious to write out in full, is omitted) that an object v so defined is a tensor of valence $(p+r-1, q+s-1)$. We say that it is the tensor generated by transvecting the tensor u onto the tensor w with respect to the indices λ_i, μ_j (λ_i for u and μ_j for w). If $r \geqslant 1$ and $q \geqslant 1$, then by fixing the index λ_i in the tensor w and the index μ_j in the tensor u, we can similarly effect transvection with respect to these indices.

Multiple transvection can also be effected by fixing several pairs of indices simultaneously (an inferior index in one tensor and a superior index in the other) and transvecting with respect to the fixed pairs. For example

$$u^{\lambda_1\varrho}{}_{\sigma\mu_2\tau} w^{\tau\lambda_4\lambda_5\tau}{}_{\varrho\mu_5} = v^{\lambda_1\lambda_4\lambda_5}{}_{\mu_2\mu_5}.$$

If a tensor u of valence (p, q) is transvected t-fold onto a tensor w of valence (r, s), the result will be a tensor of valence $(p+r-t, q+s-t)$. In the example above, we have $p = 2$, $q = 3$, $r = 4$, $s = 2$, $t = 3$. It is easily seen that the operation of transvection can be regarded as a compound operation consisting of a multiplication followed by a contraction.

25. Mixing, or Symmetrization of Indices. If we generate a tensor by multiplying a vector v^λ by itself

$$w^{\lambda\mu} = v^\lambda v^\mu,$$

we find that

$$w^{\lambda\mu} = w^{\mu\lambda} \quad (\text{for all } \lambda, \mu). \tag{3.24}$$

This property is invariant, i.e. it holds in every allowable coordinate system (λ'). However, if the tensor is generated by multiplying two different vectors

$$w^{\lambda\mu} = u^\lambda v^\mu,$$

equality (3.24) will in general not hold.

Suppose that we are given some tensor $w^{\lambda\mu}$ such that relations (3.24) hold. We now ask whether or not analogous relationships will also hold in any arbitrary coordinate system (λ'). To begin with, we have

$$w^{\lambda'\mu'} = A^{\lambda'\mu'}_{\lambda\ \mu} w^{\lambda\mu}.$$

Similarly,

$$w^{\mu'\lambda'} = A^{\mu'\lambda'}_{\varrho\ \sigma} w^{\varrho\sigma}.$$

Making use of the equality $w^{\varrho\sigma} = w^{\sigma\varrho}$, we can write further

$$w^{\mu'\lambda'} = A^{\mu'\lambda'}_{\varrho\ \sigma} w^{\sigma\varrho}.$$

The law by which any letter may be used for the dummy index leads to the following further transformation

$$w^{\mu'\lambda'} = A^{\mu'\lambda'}_{\varrho\ \sigma} w^{\sigma\varrho} = A^{\mu'\lambda'}_{\mu\ \lambda} w^{\lambda\mu}.$$

But

$$A^{\mu'\lambda'}_{\mu\ \lambda} = A^{\mu'}_{\mu} A^{\lambda'}_{\lambda} = A^{\lambda'}_{\lambda} A^{\mu'}_{\mu} = A^{\lambda'\mu'}_{\lambda\ \mu},$$

and hence

$$w^{\mu'\lambda'} = A^{\lambda'\mu'}_{\lambda\ \mu} w^{\lambda\mu} = w^{\lambda'\mu'}.$$

We have thus shown that equality (3.24) implies the equality $w^{\lambda'\mu'} = w^{\mu'\lambda'}$ for all λ', μ', i.e. that the property expressed by relations (3.24) is invariant, and hence geometric in nature. If a tensor has property (3.24) we call it a *symmetric tensor*.

We emphasize here that symmetry means that the coordinates of the tensor are equal when each pair of fixed indices is interchanged i.e. the matrix of coordinates of the tensor $w^{\lambda\mu}$ is symmetric about the main diagonal.

The equality of the coordinates under the interchange of a pair of constant indices is by no means an invariant property. For example, the equality $w^{13} = w^{31}$ by no means implies the equality $w^{1'3'} = w^{3'1'}$.

Now take an arbitrary tensor w of valence (p, q) where $p \geqslant 2$, and fix in it two contravariant indices λ_i and λ_j $(i \neq j)$. It turns out, just as in the special case that the property expressed by the formula

$$w^{\lambda_1\cdots\lambda_{i-1}\lambda_i\lambda_{i+1}\cdots\lambda_{j-1}\lambda_j\lambda_{j+1}\cdots\lambda_p}{}_{\mu_1\cdots\mu_q} = w^{\lambda_1\cdots\lambda_{i-1}\lambda_j\lambda_{i+1}\cdots\lambda_{j-1}\lambda_i\lambda_{j+1}\cdots\lambda_p}{}_{\mu_1\cdots\mu_q}$$

$$(\text{for all } \lambda_i, \lambda_j = 1, 2, \ldots, n) \tag{3.25}$$

is invariant and the tensor is then said to be *symmetric in the indices* λ_i, λ_j. Similarly, if $q \geqslant 2$, and we fix two inferior indices μ_i, μ_j (i.e. we fix i and j), the property expressed by the formula

$$w^{\lambda_1\ldots\lambda_p}{}_{\mu_1\ldots\mu_i\ldots\mu_j\ldots\mu_q} = w^{\lambda_1\ldots\lambda_p}{}_{\mu_1\ldots\mu_j\ldots\mu_i\ldots\mu_q} \qquad (\text{for all } \mu_i, \mu_j = 1, 2, \ldots, n) \tag{3.26}$$

is also invariant and we then say that the tensor is symmetric in the indices μ_i and μ_j.

A tensor may also possess ramified symmetry. Let r be a number such that $2 \leqslant r \leqslant p$ and let

$$\alpha_1, \alpha_2, \ldots, \alpha_r \tag{3.27}$$

be a permutation of r different elements taken from the numbers of the sequence $1, 2, \ldots, p$. Further, take two coordinates of a tensor with identical covariant indices and identical contravariant indices not located in any of the positions given by numbers (3.27). On the other hand, indices that are located in positions given by numbers (3.27) are only permutations for both coordinates. If every two such coordinates are equal to each other (there must thus be $r!$ equal coordinates), it can be shown that this property is once again invariant and we then say that the given tensor is symmetric in those r indices which are in positions (3.27). Thus, for instance, the statement that the tensor $w^{\lambda_1\lambda_2\lambda_3\lambda_4\lambda_5}$ is symmetric in the second, fourth, and fifth indices means that

$$w^{\lambda\varrho\mu\sigma\tau} = w^{\lambda\sigma\mu\tau\varrho} = w^{\lambda\tau\mu\varrho\sigma} = w^{\lambda\varrho\mu\tau\sigma} = w^{\lambda\tau\mu\sigma\varrho} = w^{\lambda\sigma\mu\varrho\tau}$$

for all $\lambda, \mu, \varrho, \sigma, \tau$'s.

If in particular $r = p$, that is if the value of the coordinate $w^{\lambda_1\ldots\lambda_p}{}_{\mu_1\ldots\mu_p}$ of the tensor remains unaltered under any arbitrary permutation of the indices $\lambda_1, \lambda_2, \ldots, \lambda_p$, the tensor is referred to briefly as a tensor symmetric in the superior indices.

All of this can mutatis mutandis be said about symmetry with respect to a given set of s inferior indices of the tensor where $2 \leqslant s \leqslant q$.

Symmetry with respect to a pair of different kinds of indices (one superior, one inferior), on the other hand, is not an invariant property, as can be shown by considering the simplest example of a mixed tensor of rank two.

Namely, let $n = 2$ and assume that

$$w^1{}_2 = w^2{}_1. \tag{3.28}$$

We now calculate $w^{1'}{}_{2'}$ and $w^{2'}{}_{1'}$. We have

$$\begin{aligned} w^{1'}{}_{2'} &= A^{1'}_{\lambda} A^{\mu}_{2'} w^{\lambda}{}_{\mu} = A^{1'}_{1} A^{1}_{2'} w^1{}_1 + A^{1'}_{2} A^{1}_{2'} w^2{}_1 + A^{1'}_{1} A^{2}_{2'} w^1{}_2 + A^{1'}_{2} A^{2}_{2'} w^2{}_2, \\ w^{2'}{}_{1'} &= A^{2'}_{\lambda} A^{\mu}_{1'} w^{\lambda}{}_{\mu} = A^{2'}_{1} A^{1}_{1'} w^1{}_1 + A^{2'}_{2} A^{1}_{1'} w^2{}_1 + A^{2'}_{1} A^{2}_{1'} w^1{}_2 + A^{2'}_{2} A^{2}_{1'} w^2{}_2. \end{aligned} \tag{3.29}$$

Recalling relations (1.35), i.e. that

$$A^{1}_{1'} = \frac{A^{2'}_{2}}{J}, \qquad A^{1}_{2'} = -\frac{A^{1'}_{2}}{J}, \qquad A^{2}_{1'} = -\frac{A^{2'}_{1}}{J}, \qquad A^{2}_{2'} = \frac{A^{1'}_{1}}{J},$$

we can rewrite relations (3.29) in the following form, using equality (3.28):

$$w^{1'}{}_{2'} = \frac{1}{J}[-A^{1'}_{1}A^{1'}_{2}w^{1}{}_{1} + (-A^{1'}_{2}A^{1'}_{2} + A^{1'}_{1}A^{1'}_{1})w^{1}{}_{2} + A^{1'}_{2}A^{1'}_{1}w^{2}{}_{2}],$$

$$w^{2'}{}_{1'} = \frac{1}{J}[A^{2'}_{1}A^{2'}_{2}w^{1}{}_{1} + (A^{2'}_{2}A^{2'}_{2} - A^{2'}_{1}A^{2'}_{1})w^{1}{}_{2} - A^{2'}_{2}A^{2'}_{1}w^{2}{}_{2}].$$

Thus, if for all w's the equality

$$w^{1'}{}_{2'} = w^{2'}{}_{1'}$$

was to hold, as a consequence of equation (3.28), we would then have to have

$$-A^{1'}_{1}A^{1'}_{2} = A^{2'}_{1}A^{2'}_{2}, \qquad -A^{1'}_{2}A^{1'}_{2} + A^{1'}_{1}A^{1'}_{1} = A^{2'}_{2}A^{1'}_{2} - A^{2'}_{1}A^{2'}_{1},$$
$$A^{1'}_{2}A^{1'}_{1} = -A^{2'}_{2}A^{2'}_{1},$$

or

$$A^{1'}_{1}A^{1'}_{2} + A^{2'}_{1}A^{2'}_{2} = 0, \qquad A^{1'}_{1}A^{1'}_{1} + A^{2'}_{1}A^{2'}_{1} = A^{1'}_{2}A^{1'}_{2} + A^{2'}_{2}A^{2'}_{2}. \tag{3.30}$$

These relationships lead to a restriction on the group and are in general not satisfied under the general pseudogroup G_1.

A subgroup defined by means of relations (3.30) is called a *similarity group* and in the case $n = 2$ is a two-parameter group which admits a representation in terms of two independent parameters α, β in the form

$$A^{1'}_{1} = \beta\cos\alpha, \qquad A^{1'}_{2} = -\beta\sin\alpha, \qquad A^{2'}_{1} = \beta\sin\alpha, \qquad A^{2'}_{2} = \beta\cos\alpha. \tag{3.31}$$

The name of the subgroup derives from the fact that using it, the *angle between two non-zero vectors* can be defined, even though the length of the vectors cannot be determined.

Assume now that for any arbitrary n we are going to confine ourselves to a subgroup defined by the requirement that the relations

$$\sum_{\lambda'} A^{\lambda'}_{\lambda}A^{\lambda'}_{\mu} = \omega\delta_{\lambda\mu} \qquad (\omega > 0;\ \lambda, \mu = 1, 2, \ldots, n) \tag{3.32}$$

be satisfied (of course, it should be proved — although we do not do here — that the set of transformations satisfying relations (3.32) does in fact constitute a group).

Suppose further that we are given two non-zero vectors $\underset{1}{v}^{\lambda}$ and $\underset{2}{v}^{\lambda}$, i.e.

such that the inequalities

$$\sum_{\lambda} (\underset{i}{v^{\lambda}})^2 > 0 \qquad (i = 1, 2) \tag{3.33}$$

are satisfied. We now define the angle φ between these vectors by

$$\cos\varphi \overset{\text{def}}{=} \frac{\sum_{\lambda} \underset{1}{v^{\lambda}} \underset{2}{v^{\lambda}}}{[\sum_{\lambda} (\underset{1}{v^{\lambda}})^2]^{1/2} [\sum_{\lambda} (\underset{2}{v^{\lambda}})^2]^{1/2}} . \tag{3.34}$$

It should be emphasized that the right-hand side of (3.34) is invariant under the transformations of the subgroup (3.32). Indeed, we have

$$\sum_{\lambda'} (\underset{i}{v^{\lambda'}})^2 = \sum_{\lambda'} \underset{i}{v^{\lambda'}} \underset{i}{v^{\lambda'}} = \sum_{\lambda'} A_{\lambda}^{\lambda'} \underset{i}{v^{\lambda}} A_{\mu}^{\lambda'} \underset{i}{v^{\mu}} = \sum_{\lambda'} A_{\lambda}^{\lambda'} A_{\mu}^{\lambda'} \underset{i}{v^{\lambda}} \underset{i}{v^{\mu}}$$

$$= \omega \delta_{\lambda\mu} \underset{i}{v^{\lambda}} \underset{i}{v^{\mu}} = \omega \sum_{\mu} \underset{i}{v^{\mu}} \underset{i}{v^{\mu}} = \omega \sum_{\mu} (\underset{i}{v^{\mu}})^2 .$$

Consequently,

$$\sum_{\lambda'} \underset{1}{v^{\lambda'}} \underset{2}{v^{\lambda'}} = \sum_{\lambda'} A_{\mu}^{\lambda'} \underset{1}{v^{\mu}} A_{\lambda}^{\lambda'} \underset{2}{v^{\lambda}} = \sum_{\lambda'} A_{\mu}^{\lambda'} A_{\lambda}^{\lambda'} \underset{1}{v^{\mu}} \underset{2}{v^{\lambda}} = \omega \delta_{\mu\lambda} \underset{1}{v^{\mu}} \underset{2}{v^{\lambda}} = \omega \sum_{\lambda} \underset{1}{v^{\lambda}} \underset{2}{v^{\lambda}},$$

whence

$$\frac{\sum_{\lambda'} \underset{1}{v^{\lambda'}} \underset{2}{v^{\lambda'}}}{[\sum_{\lambda'} (\underset{1}{v^{\lambda'}})^2]^{1/2} [\sum_{\lambda'} (\underset{2}{v^{\lambda'}})^2]^{1/2}} = \frac{\omega \sum_{\lambda} \underset{1}{v^{\lambda}} \underset{2}{v^{\lambda}}}{[\omega^2 \sum_{\lambda} (\underset{1}{v^{\lambda}})^2 \sum_{\mu} (\underset{2}{v^{\mu}})^2]^{1/2}}$$

$$= \frac{\sum_{\lambda} \underset{1}{v^{\lambda}} \underset{2}{v^{\lambda}}}{[\sum_{\lambda} (\underset{1}{v^{\lambda}})^2]^{1/2} [\sum_{\lambda} (\underset{2}{v^{\lambda}})^2]^{1/2}} .$$

The Lagrange identity can be used to show that the right-hand side of equation (3.34) has an absolute value less than or equal to 1 and thus that the angle φ is real.

The *similarity subgroup* (3.30) G_s contains the metric subgroup G_m and hence is itself in turn a subgroup of the affine group G_a.

A symmetric contravariant tensor $g^{\lambda\mu}$ or covariant tensor $g_{\lambda\mu}$ of rank two has at most

$$n^2 - \binom{n}{2} = \tfrac{1}{2} n(n+1) = \binom{n+1}{2}$$

different coordinates, for the equations $g_{\lambda\mu} = g_{\mu\lambda}$ or $g^{\lambda\mu} = g^{\mu\lambda}$ yield $\binom{n}{2}$

relations. It follows that in two-dimensional space a tensor of rank two has in general $\binom{3}{2} = 3$ different coordinates. In three-dimensional space, it has $\binom{4}{2} = 6$ different coordinates, and so on.

This is the number that is meant when it is said briefly that a tensor has a certain number of coordinates. This abbreviation is taken to mean that zero coordinates are not counted while repetitions of coordinates which are equal to each other are only counted once.

Given a tensor a two-index tensor $w_{\mu\lambda}$, say, set $\overline{w}_{\lambda\mu} = w_{\lambda\mu}$. Since

$$\overline{w}_{\lambda'\mu'} = w_{\mu'\lambda'} = A^{\mu\ \lambda}_{\mu'\lambda'} w_{\mu\lambda} = A^{\mu\ \lambda}_{\mu'\lambda'} \overline{w}_{\lambda\mu} = A^{\lambda\ \mu}_{\lambda'\mu'} \overline{w}_{\lambda\mu},$$

$\overline{w}$ is also a tensor (of the same valence). From a given tensor $\overline{w}_{\lambda\mu}$ we have thus generated a new tensor $\overline{w}_{\lambda\mu}$ by interchanging indices. The only difference is that this new tensor has interchanged coordinates, as we say. But since a geometric object, in this case a two-index covariant tensor, is the association of a sequence, and not just a set, of numbers with a coordinate system the tensor $\overline{w}$ is different from the tensor w. Only in the special case when the tensor w is symmetric is $\overline{w}$ the same as the original tensor w.

The tensor $\overline{w}$ is called the *isomer* of the tensor w. In the special case of our two-index tensor, the operation of transition from w to the isomer $\overline{w}$ consists in transposing the matrix $w_{\lambda\mu}$, i.e. changing the rows into columns and conversely.

We shall now give a general definition of isomers.

26. The Isomers of a Tensor.

DEFINITION. From a group of indices of one kind, e.g. contravariant (the same will hold for covariant indices) $\lambda_1, \lambda_2, \ldots, \lambda_p$ let us select r indices $(2 \leqslant r \leqslant p)$, namely

$$\alpha_1, \alpha_2, \ldots, \alpha_r, \tag{3.35}$$

where sequence (3.35) is a permutation of r different numbers taken from the numbers of the sequence $1, 2, \ldots, p$. Furthermore, let $\beta_1, \beta_2, \ldots, \beta_r$ denote an arbitrary permutation of the sequence (3.35). If we set

$$\overline{\overline{w}}^{\lambda_1\ldots\lambda_p}{}_{\nu_1\ldots\nu_q} \overset{\text{def}}{=} w^{\mu_1\ldots\mu_p}{}_{\nu_1\ldots\nu_q},$$

where $\lambda_j = \mu_j$ for $j \neq \alpha_k$ $(k = 1, 2, \ldots, r)$, while $\lambda_{\alpha k} = \mu_{\beta_k}$ for $k = 1, 2, \ldots, r$, we obtain a tensor of the same valence (p, q) as has been shown. This tensor is called an *isomer* (with respect to the indices $\alpha_1, \alpha_2, \ldots, \alpha_r$) of the given tensor w, and the operation of transition from w to the isomer $\overline{\overline{w}}$, is known as *isomerization*.

The total number of all isomers (in indices (3.35)) will, of course, be $r!$ since this is the number of different permutations $\beta_1, \beta_2, \ldots, \beta_r$ of the sequence $\alpha_1, \alpha_2, \ldots, \alpha_r$.

If we now construct the arithmetic mean of all the isomers generated from a given tensor w by permuting the indices in the positions (3.35), including the original given tensor in the sum, then, remembering that tensors of the same valence can be summed and multiplied or divided by numbers, we obtain a new tensor $\overset{*}{w}$ of the same valence which will be symmetric in the indices in the selected positions. This new tensor will no longer only have coordinates which have been transposed with respect to the given tensor. For example, if $n = 3$, $r = 3$ and if the tensor $w^{\lambda\mu}$ in the system (λ) can be represented by means of the matrix

$$\begin{bmatrix} 0 & -1 & 2 \\ 3 & 1 & 0 \\ -4 & -4 & 1 \end{bmatrix},$$

then the tensor $\overset{*}{w}{}^{\lambda\mu}$ will be represented by the matrix

$$\begin{bmatrix} 0 & 1 & -1 \\ 1 & 1 & -2 \\ -1 & -2 & 1 \end{bmatrix}.$$

In the general case, if (3.35) is an arbitrary combination of r numbers from $1, 2, \ldots, p$, no abbreviated symbol is possible for the tensor $\overset{*}{w}$. However, Schouten has introduced a brief and lucid symbol for $\overset{*}{w}$ which can be used in those cases when the operation involves indices lying together in a compact group. For instance, if we are given a tensor $w^{\lambda_1\lambda_2\lambda_3\lambda_4\lambda_5}$ then the symbols $w^{\lambda_1(\lambda_2\lambda_3)\lambda_4\lambda_5}$, or $w^{\lambda_1\lambda_2(\lambda_3\lambda_4\lambda_5)}$, denote tensors $\overset{*}{w}$ obtained by mixing isomers formed by permuting the second and third, or third, fourth and fifth indices. We then have

$$w^{\lambda_1(\lambda_2\lambda_3)\lambda_4\lambda_5} = \frac{1}{2!}(w^{\lambda_1\lambda_2\lambda_3\lambda_4\lambda_5} + w^{\lambda_1\lambda_3\lambda_2\lambda_4\lambda_5})$$

and similarly

$$w^{\lambda_1\lambda_2(\lambda_3\lambda_4\lambda_5)} = \frac{1}{3!}(w^{\lambda_1\lambda_2\lambda_3\lambda_4\lambda_5} + w^{\lambda_1\lambda_2\lambda_4\lambda_5\lambda_3} + w^{\lambda_1\lambda_2\lambda_5\lambda_3\lambda_4} + w^{\lambda_1\lambda_2\lambda_5\lambda_4\lambda_3} + w^{\lambda_1\lambda_2\lambda_3\lambda_5\lambda_4}, + w^{\lambda_1\lambda_2\lambda_4\lambda_3\lambda_5}).$$

In the simplest case we have

$$g_{(\lambda\mu)} = \tfrac{1}{2}(g_{\lambda\mu} + g_{\mu\lambda}).$$

When all the indices of one kind are mixed, by this peration we obtain the tensor $w^{(\lambda_1 \dots \lambda_p)}$. We thus see that the operation of mixing (symmetrization) is the result of operations of isomerization, addition, and multiplication by a number.

27. Skew (Antisymmetric) Tensors. We prove similarly that the property

$$w_{\lambda\mu} = -w_{\mu\lambda} \tag{3.36}$$

is invariant for all λ, μ's; whereby in particular we have the relations

$$w_{\lambda\lambda} = 0 \quad (\text{for } \lambda = 1, 2, \dots, n). \tag{3.37}$$

If we are given a tensor $a_{\lambda\mu}$ and we construct from it the tensor $a_{(\lambda\mu)}$ and $a_{[\lambda\mu]}$ (where $a_{[\lambda\mu]} = \frac{1}{2}(a_{\lambda\mu} - a_{\mu\lambda})$, cf. p. 77), we quickly find that

$$a_{\lambda\mu} = a_{(\lambda\mu)} + a_{[\lambda\mu]}. \tag{3.38}$$

In the matrix representation, for $n = 3$ this equality assumes the form

$$\begin{bmatrix} a_{11} & a_{12} & a_{13} \\ a_{21} & a_{22} & a_{23} \\ a_{31} & a_{32} & a_{33} \end{bmatrix}$$

$$= \begin{bmatrix} a_{11} & \dfrac{a_{12}+a_{21}}{2} & \dfrac{a_{13}+a_{31}}{2} \\ \dfrac{a_{21}+a_{12}}{2} & a_{22} & \dfrac{a_{23}+a_{32}}{2} \\ \dfrac{a_{31}+a_{13}}{2} & \dfrac{a_{32}+a_{23}}{2} & a_{33} \end{bmatrix} + \begin{bmatrix} 0 & \dfrac{a_{12}-a_{21}}{2} & \dfrac{a_{13}-a_{31}}{2} \\ \dfrac{a_{21}-a_{12}}{2} & 0 & \dfrac{a_{23}-a_{32}}{2} \\ \dfrac{a_{31}-a_{13}}{2} & \dfrac{a_{32}-a_{23}}{2} & 0 \end{bmatrix}.$$

Equation (3.38) represents the decomposition of the given tensor $a_{\lambda\mu}$ into the sum of two tensors, viz. a symmetric part $a_{(\lambda\mu)}$ and an antisymmetric part $a_{[\lambda\mu]}$. This decomposition is easily shown to be unique. Similarly, a tensor of valence $(2, 0)$ can be decomposed into a symmetric and an antisymmetric part. On the other hand, for tensors of valences $(p, 0)$ and $(0, p)$, where $p \geqslant 3$, such a decomposition is in general no longer possible; the reader is urged to check this for the case $p = 3$.

More generally, if in a tensor the interchange of two indices of the same kind located in particular places always (i.e. for all constant values of these running indices) results in a change of sign, the property is invariant and the tensor is then said to be *antisymmetric* or *skew-symmetric* with respect to these two indices. For instance, if we set

$$w^{\lambda\mu} \overset{\text{def}}{=} \underset{1}{v}^{\lambda}\underset{2}{v}^{\mu} - \underset{1}{v}^{\mu}\underset{2}{v}^{\lambda},$$

we find that the tensor $w^{\lambda\mu}$ is antisymmetric (note that here it is not necessary to state with respect to which indices).

Turning now to the general case, we take a group of r indices of one kind $\lambda_{\alpha_1}, \lambda_{\alpha_2}, \ldots, \lambda_{\alpha_r}$ where $r \geqslant 2$ and $\alpha_1, \alpha_2, \ldots, \alpha_r$ constitute a permutation of certain numbers from the sequence $1, 2, \ldots, p$ (or $1, 2, \ldots, q$, since the same applies to both species of indices). Let $\beta_1, \beta_2, \ldots, \beta_r$ be an arbitrary permutation of the sequence $\alpha_1, \alpha_2, \ldots, \alpha_r$. If the interchange of any two indices λ_i and λ_j, where $i = \alpha_k, j = \alpha_l$, causes the coordinate of the tensor to change sign, the tensor is said to be *antisymmetric* (*skew-symmetric*) with respect to the given group of r indices.

This definition can also be recast in the following equivalent form. Any two coordinates, one of indices $\lambda_1, \lambda_2, \ldots, \lambda_p$, and the other of indices $\mu_1, \mu_2, \ldots, \mu_p$ are either equal when

$$\mu_j = \lambda_j \quad \text{for} \quad j \neq \alpha_k \quad (k = 1, 2, \ldots, r)$$

and

$$\mu_{\alpha_k} = \lambda_{\beta_k} \quad \text{for} \quad k = 1, 2, \ldots, r,$$

where $\beta_1, \beta_2, \ldots, \beta_r$ is an even permutation of the sequence $\alpha_1, \alpha_2, \ldots, \alpha_r$, or are of opposite sign when

$$\mu_j = \lambda_j \quad \text{for } j \neq \alpha_k,$$

while

$$\mu_{\alpha_k} = \lambda_{\beta_k} \quad \text{for } k = 1, 2, \ldots, r,$$

where $\beta_1, \beta_2, \ldots, \beta_r$ is an odd permutation of the sequence $\alpha_1, \alpha_2, \ldots, \alpha_r$.

In the special case, when $r = p$ (or $r = q$), i.e. when there is antisymmetry between all the indices of one kind, the tensor is referred to briefly as an antisymmetric tensor.

28. Polyvectors (Multivectors). A pure antisymmetric p-index tensor (of valence $(p, 0)$ or $(0, p)$) is called a *p-vector*. The reason for this will become apparent further on. Some authors also use the name "*antisymmetric tensors*" for *polyvectors* (*multivectors*), i.e. p-vectors when $p \geqslant 2$.

If in any tensor we take a group of r ($r \geqslant 2$) indices of one kind, operations resembling mixing can be performed to yield a tensor which is antisymmetric with respect to the chosen group of indices. To this end we take all the isomers of a tensor with respect to a given group of indices; including the given tensor they number $r!$ Next, the plus sign is assigned to every isomer in which the indices of the chosen group constitute an even permutation of the indices of that group in the original tensor, and the minus sign is taken for those isomers in which the indices of the aforementioned groups form an odd permutation. We then take the algebraic sum of isomers so obtained. This sum is divided by $r!$ to yield the desired tensor which is antisymmetric with respect to the chosen group of indices.

If all the indices of the group subjected to "shuffling" and antisymmetrization are adjacent, the skewed tensor is denoted by brackets around the given group of indices in the original tensor. Thus, with this notational convention we have

$$w^{[\lambda\mu]} = \tfrac{1}{2}(w^{\lambda\mu} - w^{\mu\lambda}),$$

$$w_{[\lambda\mu\nu]} = \tfrac{1}{6}(w_{\lambda\mu\nu} + w_{\mu\nu\lambda} + w_{\nu\lambda\mu} - w_{\lambda\nu\mu} - w_{\nu\mu\lambda} - w_{\mu\lambda\nu})$$

and so on.

Note that with these notations the following simple relationships hold:

1. If the tensor $w^{\lambda_1 \cdots \lambda_p}$ is symmetric, then

$$w^{(\lambda_1 \cdots \lambda_p)} = w^{\lambda_1 \cdots \lambda_p}, \qquad w^{[\lambda_1 \cdots \lambda_p]} = 0.$$

2. If the tensor $w^{\lambda_1 \cdots \lambda_p}$ is antisymmetric (is a p-vector), then

$$w^{[\lambda_1 \cdots \lambda_p]} = w^{\lambda_1 \cdots \lambda_p}, \qquad w^{(\lambda_1 \cdots \lambda_p)} = 0.$$

Note that if the tensor possesses p contravariant (covariant) indices, where $p > n$, and if we skew it with respect to all these indices, the result will be a null vector. Indeed, since $p > n$, then for every set of constant values for the indices $\lambda_1, \lambda_2, \ldots, \lambda_p$ at least two must have the same value. Since an interchange of two indices in an antisymmetric tensor causes a change in the sign of the coordinate and since, on the other hand, the transposition of these two identical indices does not change the coordinate, every coordinate must be zero and so the tensor is a null tensor (cf. p. 63).

When the tensor $g_{\lambda\mu} = g_{\mu\lambda}$, which has $\frac{1}{2}n(n+1)$ coordinates, is multiplied by itself it is possible that the result is a four-index tensor $g_{\lambda\mu}g_{\varrho\sigma}$ which is not symmetric (e.g. $g_{\lambda\mu}g_{\varrho\sigma}$ will in general be different from $g_{\lambda\varrho}g_{\mu\sigma}$). Nevertheless, if all the indices in it are skewed, i.e. if we construct the tensor $g_{[\lambda\mu}g_{\varrho\sigma]}$ the result is a null tensor. The antisymmetrization of three indices also leads to a null tensor. Skewing of two indices of the same factor plainly leads to a null tensor. However, the skewing of two indices, one in each factor, leads to a tensor which in general is not a null tensor.

For instance, let us set

$$G_{\lambda\mu\varrho\sigma} \overset{\text{def}}{=} g_{[\lambda|\varrho|}g_{\mu]\sigma}\ ^{(1)}.$$

REMARK. It can be checked that the skewing of another pair of indices leads only to an isomer.

We obtain the tensor

$$G_{\lambda\mu\varrho\sigma} = \tfrac{1}{2}[g_{\lambda\varrho}g_{\mu\sigma} - g_{\mu\varrho}g_{\lambda\sigma}]$$

which is symmetric with respect to the pairs (λ, μ), (ϱ, σ) and skew-symmetric with respect to the pairs (λ, ϱ), (μ, σ), and which no longer has $[\frac{1}{2}n(n+1)]^2$ coordinates, but only $\frac{1}{12}n^2(n^2-1)$ coordinates [2]. Thus, for $n = 2$ for example we get only one essential coordinate

$$G_{1212} = \tfrac{1}{2}(g_{12}^2 - g_{11}g_{22}) = \tfrac{1}{2}\mathfrak{g},$$

where $\mathfrak{g}$ denote the value of the determinant $|g_{\mu\lambda}|$. For $n = 3$ the tensor G has six essential coordinates

$$G_{1212}, G_{1213}, G_{1223}, G_{1313}, G_{1323}, G_{2323};$$

while for $n = 4$, there are 20.

In Part Two, we shall become familiar with the four-index tensor R which has important applications and which also has the property of antisymmetry for the first and for the second pair of indices as well as the property of symmetry under simultaneous interchange of both pairs since, as is easily verified, in addition to the relationships

$$R_{(\lambda\mu)\varrho\sigma} = R_{\lambda\mu(\varrho\sigma)} = 0$$

the relationships $R_{\lambda\mu\varrho\sigma} = R_{\varrho\sigma\lambda\mu}$ are also satisfied.

The special case of the antisymmetric contravariant (covariant) p-vector is obtained if we take p contravariant (covariant) vectors

$$\underset{i}{v}^{\lambda} \quad (i = 1, 2, \ldots, p;\ \lambda = 1, 2, \ldots, n), \tag{3.39}$$

form their product

$$\underset{1}{v}^{\lambda_1}\underset{2}{v}^{\lambda_2} \ldots \underset{p}{v}^{\lambda_p}$$

and then skew this product

$$w^{\lambda_1 \ldots \lambda_p} \overset{\text{def}}{=} \underset{1}{v}^{[\lambda_1} \ldots \underset{p}{v}^{\lambda_p]}.$$

---

(1) The bars around the index ϱ indicate that this index ϱ does not participate in the skewing operation.

(2) Different in the sense established on p. 73, and independent.

Since they are skewed products of vectors, we shall refer to p-vectors of this kind as simple p-vectors.

If $p \leqslant n$, then of the n^p coordinates of the simple p-vector only $\binom{n}{p}$ are different as can easily be proved by induction. In particular, a simple n-vector will have essentially only one coordinate.

Not every p-vector is simple, however, and this can be shown by means of a straightforward example. Take, in four-dimensional space ($n = 4$), a bivector $F^{\lambda\mu}$ such that

$$F^{12}F^{34}+F^{23}F^{14}+F^{31}F^{24} \neq 0. \tag{3.40}$$

It can easily be shown that this inequality is invariant. Now write out the product of the bivector F with itself,

$$F^{\lambda\mu}F^{\nu\varkappa},$$

then antisymmetrize with respect to the group consisting of the first three indices, and denote the tensor so obtained briefly by the symbol G for brevity

$$G^{\lambda\mu\nu\varkappa} \overset{\text{def}}{=} F^{[\lambda\mu}F^{\nu]\varkappa}. \tag{3.41}$$

It is plain that the only non-zero coordinates of the tensor G will be those in which all four indices are different. Indeed, if any two of the indices λ, μ, ν are equal to each other, the corresponding coordinate is zero. Suppose now that the fourth index $\varkappa$ is equal to one of the three λ, μ, ν. When $\varkappa = \lambda$, we have

$$F^{[\lambda\mu}F^{\nu]\lambda} = \tfrac{1}{6}(F^{\lambda\mu}F^{\nu\lambda}+F^{\mu\nu}F^{\lambda\lambda}+F^{\nu\lambda}F^{\mu\lambda}-F^{\lambda\nu}F^{\mu\lambda}-F^{\nu\mu}F^{\lambda\lambda}-F^{\mu\lambda}F^{\nu\lambda}).$$

The second and fifth terms in the parentheses are zero because $F^{\lambda\lambda} = 0$; the third and sixth terms obviously cancel; and the fourth term

$$-F^{\lambda\nu}F^{\mu\lambda} = F^{\nu\lambda}F^{\mu\lambda} = -F^{\nu\lambda}F^{\lambda\mu} = -F^{\lambda\mu}F^{\nu\lambda}$$

cancels with the first. Thus, the entire expression within the parentheses is equal to zero. Similarly, it turns out that when $\varkappa = \mu$ or $\varkappa = \nu$, we again have $G = 0$.

It can be proved that the tensor G is a four-vector, and that it has only one non-zero coordinate, provided it is itself different from zero. Accordingly,

$$\begin{aligned} G^{1234} &= F^{[12}F^{3]4} \\ &= \tfrac{1}{6}(F^{12}F^{34}+F^{23}F^{14}+F^{31}F^{24}-F^{13}F^{24}-F^{32}F^{14}-F^{21}F^{34}) \\ &= \tfrac{1}{3}(F^{12}F^{34}+F^{23}F^{14}+F^{31}F^{24}). \end{aligned}$$

If we had $G^{1234} \overset{*}{=} 0$, then since the other coordinates can only differ in sign at most, we would have $G^{\lambda\mu\varkappa\nu} \overset{*}{=} 0$, and then since G is a tensor, we would have

$$G^{\lambda\mu\nu\varkappa} = 0$$

in every coordinate system and the invariant character of inequality (3.40) is thus established.

We now choose six coordinates

$$F^{12}, F^{13}, F^{14}, F^{23}, F^{24}, F^{34},$$

of the bivector $F^{\lambda\mu}$ so that inequality (3.40) is satisfied.

The system of six equations in the eight unknowns $\underset{1}{v}^{\lambda}, \underset{2}{v}^{\lambda}$ $(\lambda = 1, 2, 3, 4)$, viz.

$$\underset{1}{v}^{[\lambda}\underset{2}{v}^{\mu]} = F^{\lambda\mu}$$

then has no solution: this can be shown by means of an elementary, but rather long calculation, which we omit but which we advise the reader to carry out. We thus have proof that the bivector $F^{\lambda\mu}$, for which the four-vector G defined by equation (3.41) is non-zero, is not a simple bivector.

In three-dimensional space, by contrast, every bivector is simple. We also have a more general theorem which states that in n-dimensional space every $(n-1)$-vector is simple [(1)].

If a p-vector is simple, there exist an infinite number of sets (3.39) of p vectors from which it can be "built". Take, for instance, the contravariant bivector

$$b^{\lambda\mu} = \underset{1}{v}^{[\lambda}\underset{2}{v}^{\mu]}.$$

In order for this bivector to be non-zero, the vectors $\underset{1}{v}$ and $\underset{2}{v}$ must be linearly independent; in general, a simple p-vector defined by means of the vectors $\underset{i}{v}$ $(i = 1, 2, \ldots, p)$ is non-zero if and only if vectors (3.39) are linearly independent.

If we set

$$\underset{1}{u} = \alpha\underset{1}{v} + \beta\underset{2}{v}, \qquad \underset{2}{u} = \gamma\underset{1}{v} + \delta\underset{2}{v}$$

and if we choose the coefficients α, β, γ, δ so that

$$\alpha\delta - \beta\gamma = 1,$$

---

(1) Schouten [69], Vol. I, p. 16.

then a simple calculation yields

$$\underset{1}{u}^{[\lambda}\underset{2}{u}^{\mu]} = \underset{1}{v}^{[\lambda}\underset{2}{v}^{\mu]}.$$

In general: if we are given a simple p-vector

$$\underset{1}{v}^{[\lambda_1} \dots \underset{p}{v}^{\lambda_p]}$$

and if we put

$$\underset{i}{u}^{\lambda} = \underset{i\,k}{\overset{k}{\alpha}}\, v^{\lambda} \qquad (i, k = 1, 2, \dots, p),$$

then we have

$$\underset{1}{u}^{[\lambda_1} \dots \underset{p}{u}^{\lambda_p]} = |\underset{i}{\overset{k}{\alpha}}|\, \underset{1}{v}^{[\lambda_1} \dots \underset{p}{v}^{\lambda_p]},$$

where $|\underset{i}{\overset{k}{\alpha}}|$ denotes the value of the determinant (of order p) constructed from the coordinates $\underset{i}{\overset{k}{\alpha}}$.

As stated above, the n-vector

$$w^{\lambda_1 \dots \lambda_n} = w^{[\lambda_1 \dots \lambda_n]} \tag{3.42}$$

has essentially only one coordinate $\left[\binom{n}{p} = \binom{n}{n} = 1\right]$, which means that all the coordinates are either zero (if at least two indices λ_i and λ_j have identical values), or have a common absolute value. This value is

$$|w^{1 \dots n}|. \tag{3.43}$$

Each n-vector is simple. When the value (3.43) is zero, this is trivially the case; it then suffices to set

$$\underset{i}{v}^{\lambda} = 0 \qquad (i = 1, 2, \dots, n;\ \lambda = 1, 2, \dots, n)$$

for the relationship

$$w^{[\lambda_1 \dots \lambda_n]} = \underset{1}{v}^{[\lambda_1} \dots \underset{n}{v}^{\lambda_n]}.$$

to be satisfied. If, on the other hand, the value (3.43) is non-zero it is sufficient to write

$$\underset{\mu}{v}^{\lambda} = \underset{\mu}{e}^{\lambda} \qquad (\mu = 1, 2, \dots, n-1),$$

while

$$\underset{n}{v}^{\lambda} = \varrho \underset{n}{e}^{\lambda},$$

where

$$\varrho \overset{*}{=} n!\; w^{1\ldots n}\,{}^{(1)}.$$

We then have

$$\underset{1}{v}^{[\lambda_1} \ldots \underset{n}{v}^{\lambda_n[} = \varrho \underset{1}{e}^{[\lambda_1} \ldots \underset{n}{e}^{\lambda_n]}.$$

But

$$\underset{1}{v}^{1} \ldots \underset{n}{v}^{n} \overset{*}{=} \varrho \underset{1}{e}^{[1} \ldots \underset{n}{e}^{n]} \overset{*}{=} \varrho \frac{1}{n!} = w^{1\ldots n},$$

and thus

$$\underset{1}{v}^{[\lambda_1} \ldots \underset{n}{v}^{\lambda_n]} = w^{[\lambda_1\ldots\lambda_n]}.$$

The simplicity of the n-vector follows directly from a certain general theorem [(2)]. By this theorem the condition that the tensor

$$w^{[\lambda_1\ldots\lambda_p} w^{\mu_1]\mu_2\ldots\mu_p}$$

be zero is necessary and sufficient for the p-vector $w^{[\lambda_1\ldots\lambda_p]}$ to be simple. An even more general theorem [(3)], giving the condition

$$w^{[\lambda_1\ldots\lambda_p} w^{\mu_1\mu_2]\ldots\mu_p} = 0$$

as necessary and sufficient for the simplicity of the p-vector w, immediately implies that every $(n-1)$-vector is simple (p. 80). However, not every $(n-2)$-vector need be simple, as we have already seen in the example $n = 4$.

29. Densities. We return now to the n-vector

$$w^{\lambda_1\ldots\lambda_n} = w^{[\lambda_1\ldots\lambda_n]}. \tag{3.44}$$

If at least two indices λ_i and λ_j have the same value, the coordinate is zero. If all the λ_i's are different, the sequence $\lambda_1, \lambda_2, \ldots, \lambda_n$ is a permutation of the sequence $1, 2, \ldots, n$ and then

$$w^{\lambda_1\ldots\lambda_n} = \pm w^{1\ldots n}, \tag{3.45}$$

where the plus sign should be taken when the permutation $\lambda_1, \ldots, \lambda_n$ is even, and the minus sign when it is odd.

Following Ricci we introduce the very convenient symbol

---

(1) We have to insert the symbol $*$ since ϱ is a scalar, whereas $w^{1\ldots n}$ is not a scalar, as we shall see presently.

(2) Schouten [69], p. 53.

(3) Weitzenböck [88].

$$\varepsilon^{\lambda_1 \ldots \lambda_n} = \varepsilon_{\lambda_1 \ldots \lambda_n} \stackrel{\text{def}}{=} \begin{cases} 0, & \text{when at least two indices } \lambda_i,\ \lambda_j \text{ are equal,} \\ 1, & \text{when all the } \lambda_i\text{'s differ and the permutation } \lambda_1, \ldots, \lambda_n \text{ is even,} \\ -1, & \text{when all the } \lambda_i\text{'s differ and the permutation } \lambda_1, \ldots, \lambda_n \text{ is odd.} \end{cases} \tag{3.46}$$

Equation (3.45) can then be rewritten in the more precise form

$$w^{\lambda_1 \ldots \lambda_n} = \varepsilon^{\lambda_1 \ldots \lambda_n} w^{1 \ldots n}. \tag{3.47}$$

The Ricci symbol can also be used when writing the determinant

$$D = |\overset{\lambda}{\underset{\mu}{a}}|$$

in the expanded form

$$D = \varepsilon^{\lambda_1 \ldots \lambda_n} \overset{1}{\underset{\lambda_1}{a}} \ldots \overset{n}{\underset{\lambda_n}{a}} = \varepsilon_{\lambda_1 \ldots \lambda_n} \overset{\lambda_1}{\underset{1}{a}} \ldots \overset{\lambda_n}{\underset{n}{a}}. \tag{3.48}$$

The question now is how the coordinate

$$\mathfrak{w} \stackrel{\text{def}}{=} w^{1 \ldots n} \tag{3.49}$$

of the n-vector transforms under transition to a new system (λ'). We have

$$\mathfrak{w}' = w^{1' \ldots n'} = A^{1'}_{\lambda_1} \ldots A^{n'}_{\lambda_n} w^{\lambda_1 \ldots \lambda_n} = A^{1'}_{\lambda_1} \ldots A^{n'}_{\lambda_n} \varepsilon^{\lambda_1 \ldots \lambda_n} w^{1 \ldots n} = J\mathfrak{w},$$

or, in other words, the transformation law for $\mathfrak{w}$ is

$$\mathfrak{w}' = J\mathfrak{w}. \tag{3.50}$$

This law, as might be expected, possesses the group property. Indeed, consider a third coordinate system (λ'') and write $J_1 = |A^{\lambda''}_{\lambda'}|$ and $J_2 = |A^{\lambda''}_{\lambda}|$. Then, by equation (3.50) we have

$$\mathfrak{w}'' = J_1 \mathfrak{w}', \qquad \mathfrak{w}'' = J_2 \mathfrak{w}.$$

But

$$J_1 \mathfrak{w}' = J_1 J \mathfrak{w},$$

which furthermore is equal to $J_2 \mathfrak{w}$ by virtue of the well-known theorem of analysis concerning the Jacobian of a compound transformation. Thus, an object which has one coordinate $\mathfrak{w}$ and transforms according to the law (3.50) is a geometric object of class one. It belongs to the subclass of objects of class J, i.e. to objects for which the transformation law is of the form

$$\omega' = f(\omega, J). \tag{3.51}$$

The set of all objects of this class has been determined by Gołąb [109], under the weakest possible assumptions as to the regularity of the function f: each function f specifying an object of this class is expressed by one of two formulae

$$f(x, y) = \Phi[y\varphi(x)] \quad \text{or} \quad f(x, y) = \Phi[|y| \cdot \varphi(x)], \tag{3.52}$$

where the function φ is arbitrary, except for the condition that there exist a function Φ with the property

$$\Phi[\varphi(x)] \equiv x. \tag{3.53}$$

Objects obeying the transformation law (3.50) are called *densities* [1]. More generally, if r stands for an integer, the law expressed by the formula

$$\mathfrak{g}' = J^r \mathfrak{g} \tag{3.54}$$

possesses the group property, i.e. it defines a special geometric object of class one, since the first derivatives $A_{\lambda}^{\lambda'}$ appear in the Jacobian J. This object is known as a density of weight $-r$. The concept of density can be further generalized to include arbitrary weights which are not necessarily integers. Here, however, we must distinguish between two kinds of densities: ordinary densities (in the terminology of Schouten), or G-densities, and Weyl-densities or W-densities. The transformation law for the latter is

$$\mathfrak{g}' = |J|^r \cdot \mathfrak{g}, \tag{3.55}$$

while that of the former is

$$\mathfrak{g}' = \operatorname{sgn} J \cdot |J|^r \cdot \mathfrak{g}. \tag{3.56}$$

Both laws can be written by using the single formula

$$\mathfrak{g}' = \varepsilon |J|^r \cdot \mathfrak{g}, \tag{3.57}$$

where

$$\varepsilon = \begin{cases} 1 & \text{for } W\text{-densities}, \\ \operatorname{sgn} J & \text{for } G\text{-densities}. \end{cases} \tag{3.58}$$

Let us verify that the transformation law (3.57) does indeed have the group property. We therefore write

$$\mathfrak{g}'' = \varepsilon_1 |J_1|^r \cdot \mathfrak{g}'. \tag{3.59}$$

Substituting from equation (3.57) into the last formula, we obtain

$$\mathfrak{g}'' = \varepsilon_1 |J_1|^r \cdot \varepsilon |J|^r \mathfrak{g} = \varepsilon_1 \varepsilon |J_1 J|^r \cdot \mathfrak{g}.$$

It is seen that $J_1 J$ is simply the Jacobian of the transformation $(\lambda) \to (\lambda'')$. On the other hand,

$$\varepsilon_1 \varepsilon = 1 \quad \text{when} \quad \varepsilon_1 = 1 \quad \text{and} \quad \varepsilon = 1$$

---

[1] Brillouin [6] retains the name of density for the case $r = 1$, and introduces the term "capacity" when $r = -1$.

and

$$\varepsilon_1 \varepsilon = \operatorname{sgn}(J_1 J) \quad \text{when} \quad \varepsilon_1 = \operatorname{sgn} J_1 \quad \text{and} \quad \varepsilon = \operatorname{sgn} J.$$

The group property is thus established.

A density obeying the law (3.50) is a G-density of weight -1. Formula (3.57) is a special instance of the general formula for J-objects (3.51). In this case, we have

$$\varphi(x) = \varepsilon \sqrt[r]{|x|}, \qquad \varepsilon = \operatorname{sgn} x,$$

where for G-densities the function φ is defined everywhere, and for W-densities, only on a half-line (i.e. either just for $x > 0$, or just for $x < 0$).

Note that if the group under consideration is specialized to a subgroup, for which $J > 0$, then the difference between G-densities and W-densities disappears. Note further that the p-th power, for an arbitrary real number, of a positive W-density of weight 1 is a W-density of weight p. Indeed, let

$$\mathfrak{g}' = |J|^{-1} \cdot \mathfrak{g}$$

and set

$$\mathfrak{G} \overset{\text{def}}{=} \mathfrak{g}^p.$$

In this case we have

$$\mathfrak{G}' = (\mathfrak{g}')^p = [|J|^{-1} \cdot \mathfrak{g}]^p = |J|^{-p} \cdot \mathfrak{g}^p = |J|^{-p} \cdot \mathfrak{G}.$$

For G-densities p can no longer be an arbitrary number since the sign of $\mathfrak{g}$ then depends on the coordinate system. While J-densities can be defined only for integer exponents r, G-densities and W-densities can be defined for arbitrary real exponents [1].

30. The Geometric Interpretation of n-Vectors. Volume Measure. Suppose we are given in the space E_n a set of n linearly independent vectors

$$\underset{\lambda}{v}^{\mu} \qquad (\lambda, \mu = 1, 2, \ldots, n). \tag{3.60}$$

If we fix the origins of these vectors at one point and through the terminal point of each we draw a hyperplane parallel to the hyperplane generated by the remaining vectors, the space will be divided into a finite number of

---

[1] This view seems to be shared by W. W. Wagner, though he does not state it explicity, in his paper *Klassifikatsiya lineinykh svyaznostei v sostavnom mnogoobrazii* $X_{n+(1)}$ *po ikh gruppakh golonomi* (*Classification of linear relations in multi-image sets* $X_{n+(1)}$ *according to their holonomy groups*), Trudy Semin. po vekt. i tenz. anal. 7 (1949), p. 205–226, Wagner calls a W-density a proper W-density, while referring to a G-density as an improper W-density. Wagner's terminology seems inappropriate to us.

parts, only one of which will be bounded. This bounded part will be called a *parallelepiped spanned by the system of vectors* (3.60).

It is proved in three-dimensional Euclidean geometry R_3, that for rectangular Cartesian systems the absolute value of the determinant constructed from the coordinates of three vectors $\underset{1}{v}, \underset{2}{v}, \underset{3}{v}$, i.e. the quantity

$$\operatorname{mod}\begin{vmatrix} \underset{1}{v}^1 & \underset{1}{v}^2 & \underset{1}{v}^3 \\ \underset{2}{v}^1 & \underset{2}{v}^2 & \underset{2}{v}^3 \\ \underset{3}{v}^1 & \underset{3}{v}^2 & \underset{3}{v}^3 \end{vmatrix}$$

represents the volume of the parallelepiped spanned by these vectors.

We now attempt to generalize this result to the space E_n. From the sequence of vectors (3.60) we construct the n-vector

$$w^{[\lambda_1 \ldots \lambda_n]} \overset{\text{def}}{=} \underset{1}{v}^{[\lambda_1} \ldots \underset{n}{v}^{\lambda_n]}. \tag{3.61}$$

The n-vector (3.61) is known to have a single non-zero coordinate; suppose, for example, that it is

$$w^{1 \ldots n} = \underset{1}{v}^{[1} \ldots \underset{n}{v}^{n]}.$$

Now set

$$\mathfrak{v} \overset{\text{def}}{=} |\underset{\lambda}{v}^{\nu}| = n! w^{1 \ldots n}. \tag{3.62}$$

As a possible way to generalize the definition from three-dimensional Euclidean space, we could take the value

$$|\mathfrak{v}| = n! |w^{1 \ldots n}| \tag{3.63}$$

to be the measure of the volume of the parallelepiped spanned by vectors (3.60). Such a definition would, however, be incorrect as we want the volume measure to be an *absolute invariant* (and thus a scalar) under transformations of coordinates, and since it obeys the transformation law

$$w^{1' \ldots n'} = J w^{1 \ldots n},$$

$w^{1 \ldots n}$ is a density (called a *relative invariant* by some authors).

In Euclidean spaces R_n, as we know, $J^2 = 1$; hence, (3.63) becomes an absolute invariant and can, therefore, serve as a basis for determining the volume measure of the parallelepiped. Even in more general spaces, based on the unimodular group G_u the formula

$$V = |\mathfrak{v}|$$

can be used to introduce the volume measure of the parallelepiped; this is not possible, however, in the space E_n based on the group G_a.

It is true that the coordinate of an n-vector constructed from the basis vectors $\underset{\mu}{e}^{\lambda}$, i.e.

$$e^{[\lambda_1 \ldots \lambda_n]} \overset{\text{def}}{=} \underset{1}{e}^{[\lambda_1} \ldots \underset{n}{e}^{\lambda_n]}. \tag{3.64}$$

is also a density, inasmuch as

$$e^{[1' \ldots n']} = J e^{[1 \ldots n]},$$

and hence, the quotient

$$I = \frac{w^{1 \ldots n}}{e^{[1 \ldots n]}}$$

is subject to the transformation law

$$I' = \frac{w^{1' \ldots n'}}{e^{[1' \ldots n']}} = \frac{J w^{1 \ldots n}}{J e^{[1 \ldots n]}} = I,$$

i.e., is an invariant. However, the definition of the volume measure of a parallelepiped spanned by vectors (3.60) by means of the formula

$$V = I$$

would also not be invariant since the n-vector $e^{[\lambda'_1 \ldots \lambda'_n]}$ defined in terms of the set of basis vectors $\underset{\mu'}{e}^{\lambda'}$ belonging to another coordinate system (λ') is not compatible with the n-vector (3.64).

Following Schouten all that can be said is that the number I represents the volume measure of the parallelepiped spanned by the vectors (3.60) if the parallelopiped spanned by the basis vectors of the system $\underset{\mu}{e}^{\lambda}$ is defined to have unit volume measure.

However, suppose that in the space E_n we are given a priori a certain non-zero W-density, i.e. a geometric object with one coordinate $\mathfrak{g}$ which transforms according to the law

$$\mathfrak{g}' = |J| \mathfrak{g} \quad (\mathfrak{g} \neq 0). \tag{3.65}$$

Then, each parallelepiped spanned by the sequence of vectors (3.60) can be uniquely and invariantly assigned a particular volume measure by taking

$$V \overset{\text{def}}{=} \left| \frac{\mathfrak{v}}{\mathfrak{g}} \right|. \tag{3.66}$$

31. Space Orientation.

DEFINITION. A system (sequence) of linearly independent vectors $\underset{\mu}{v}^\lambda$ i.e. a system such that

$$\mathfrak{v} = |\underset{\mu}{v}^\lambda| \neq 0$$

defines an *orientation* in the space E_n.

We now show that a non-zero G-density, of non-zero weight, uniquely determines one of two possible orientations in space E_n. To this end suppose that a certain non-zero G-density of non-zero weight, e.g. of weight -1, obeys the transformation law

$$\mathfrak{g}' = J \cdot \mathfrak{g}$$

and note that each coordinate system (λ) defines an orientation in E_n, viz. the one which is defined in it by the set of basis vectors associated with (λ).

The set of all coordinate systems (λ') for which

$$\operatorname{sgn} \underset{1}{e}^{[1'} \ldots \underset{n}{e}^{n']} = \operatorname{sgn} \mathfrak{g},$$

is such that the transition from one to another gives rise to a transformation Jacobian $J > 0$; all such system thus determine the same orientation of space. It is therefore possible to distinguish the orientation which is determined by those coordinate systems for which $\mathfrak{g} > 0$.

We now pass on to spaces based on the group G_1. We assume that a density, or rather that a field of non-zero G-densities $\mathfrak{g} = \mathfrak{g}(\xi) \neq 0$, has been defined in a certain domain D of this space. (Note that while the density itself is not an invariant the fact that it is not equal to zero at some point is an invariant property.)

By assuming that $\mathfrak{g}(\xi) \neq 0$ in D, we are of course making the tacit assumption that $\mathfrak{g}(\xi)$ is a continuous function and hence that it is of constant sign. If D is a connected domain in the space X_n, then the functions $g(\xi)$ of constant sign will, of course, always exist. However, if the space X_n is replaced by a manifold (cf. p. 344), it is possible that density fields over the entire manifold may no longer exist. If such fields do exist, the manifold is said to be orientable for the following reason. Having fixed a point p_0, and a local coordinate system at this point we can then set

$$\mathfrak{g}(\underset{0}{\xi}) = \underset{1}{e}^{[1} \ldots \underset{n}{e}^{n]},$$

where the $\underset{\lambda}{e}^\nu$ are the basis vectors at the point p_0. If the system $\underset{\lambda}{e}^\nu(\xi)$ can then be extended continuously over the entire manifold in a unique manner, the manifold is called *orientable* and it is then possible to extend the density $\mathfrak{g}$ to the field $\mathfrak{g}(\xi)$ which is defined over the entire manifold and which is of constant sign over the entire manifold. Conversely, the orientability of a manifold follows from the fact that a field of G-densities $\mathfrak{g}(\xi)$ of constant sign exists over the entire manifold. Non-orientable two-dimensional manifolds are known to exist in three-dimensional Euclidean space.

We shall now show that if a non-trivial G density exists in a domain D, a volume metric can be introduced in that domain. Take a domain D_0 that is bounded and measurable in the Jordan sense, and which is contained in the domain D.

Now consider the integral

$$I \overset{\text{def}}{=} \int_{D(\lambda)} |\mathfrak{g}|^{-1} d\xi^1 \dots d\xi^n, \tag{3.67}$$

where $D(\lambda)$ is the domain of the variables ξ^λ and is the image of the domain D_0 in the analytic space X_n.

Analogously we construct

$$I' \overset{\text{def}}{=} \int_{D(\lambda')} |\mathfrak{g}'|^{-1} d\xi^{1'} \dots d\xi^{n'}$$

and study the relation of I' to I. For this purpose we employ a change of variables in the latter integral to obtain

$$d\xi^{1'} \dots d\xi^{n'} = |J| \, d\xi^1 \dots d\xi^n,$$

and since

$$\mathfrak{g}' = J\mathfrak{g},$$

we therefore have

$$I' = \int_{D(\lambda)} |J\mathfrak{g}|^{-1} |J| \, d\xi^1 \dots d\xi^n = \int_{D(\lambda)} |\mathfrak{g}|^{-1} d\xi^1 \dots d\xi^n = I.$$

In other words, the integral I is an invariant and may consequently be adopted as a volume measure of the domain D_0.

We therefore see that the existence of a field of non-zero densities enables a volume metric to be introduced in spaces based on the group G_1.

REMARK 1. In addition to an absolute volume measure, we can introduce a relative volume measure which depends on the orientation of the domain D_0; we do this by setting

$$I^* \overset{\text{def}}{=} \int_{D(\lambda)} \mathfrak{g}^{-1} d\xi^1 \dots d\xi^n.$$

Then I^* will be positive if the orientation of D_0 agrees with the orientation specified by the density $\mathfrak{g}$, and negative otherwise.

REMARK 2. The term density introduced on p. 84 derives from the fact that just as (in Euclidean space) the total mass distributed over a domain is obtained by integrating the density over this domain (element of mass = density times element of volume), so the volume of a domain (in an arbitrary space, in general non-Euclidean) is obtained by integrating a certain "density". The analogy is not very close, of course, since physical density is a scalar, and hence an absolute invariant, whereas our density is a relative invariant, which becomes absolute only in spaces based on a group of transformations with the property

$$J^2 = 1.$$

32. The Algebra of Densities. Among the geometric objects of class J there exist the so-called *biscalars* [1], i.e. objects whose coordinate $\tilde{\mathfrak{g}}$ can only take two values. When we pass from the coordinate system (λ) to (λ'), the coordinate is unaltered if $J > 0$, but is changed if $J < 0$.

So-called W-scalars, i.e. objects with the transformation law

$$\tilde{\mathfrak{g}}' = \operatorname{sgn} J \cdot \tilde{\mathfrak{g}},$$

are a special case of biscalars. The term W-scalars, which was introduced by Schouten [68], p. 31, is unfortunate. It would be better to refer to the concept of G-densities introduced above and speak of G-scalars rather than W-scalars.

These densities posses certain algebraic properties viz.

1. If a W-density is multiplied by a scalar, the result is a W-density.
2. If a G-density is multiplied by a scalar, the result is a G-density.
3. If a W-density is multiplied by a W-scalar, the result is a G-density.
4. If a G-density is multiplied by a W-scalar, the result is a W-density.
5. The product of two W-densities is a W-density.
6. The product of two G-densities is a W-density.
7. The product of a W-density and a G-density is a G-density.

When densities are multiplied, the weights are always added. The proofs of the theorems given above are routine: as a typical example we give the proof of the last theorem (No. 7).

---

(1) A term introduced by Gołąb [109], p. 110.

Let $\mathfrak{g}_1$ be a W-density of weight $-r$, and $\mathfrak{g}_2$ a G-density of weight $-s$. The transformation laws are then:

$$\mathfrak{g}_1' = |J|^r \mathfrak{g}_1,$$
$$\mathfrak{g}_2' = \operatorname{sgn} J |J|^s \mathfrak{g}_2.$$

We next put

$$\mathfrak{G} \stackrel{\text{def}}{=} \mathfrak{g}_1 \mathfrak{g}_2.$$

With this notation, we have

$$\mathfrak{G}' = \mathfrak{g}_1' \mathfrak{g}_2' = |J|^r \mathfrak{g}_1 \operatorname{sgn} J \cdot |J|^s \mathfrak{g}_2 = \operatorname{sgn} J \cdot |J|^{r+s} \mathfrak{g}_1 \mathfrak{g}_2 = \operatorname{sgn} J \cdot |J|^{r+s} \mathfrak{G},$$

which shows that $\mathfrak{G}$ is a G-density of weight $-(r+s)$.

Multiplication of different densities is illustrated diagrammatically in the table below

	W	G
W	W	G
G	G	W

Multiplication of densities by scalars or by W-scalars is given by the table

	σ	$\tilde{\sigma}$
W	W	G
G	G	W

Finally, the following table gives the multiplication of scalars

	σ	$\tilde{\sigma}$
σ	σ	$\tilde{\sigma}$
$\tilde{\sigma}$	$\tilde{\sigma}$	σ

All three tables are very similar and should therefore be easy to remember.

8. Densities of the same weight and of the same type can be added together, again yelding a density of the same kind and of the same weight.

Indeed, write

$$\mathfrak{g}_1' = J^r \mathfrak{g}_1, \quad \mathfrak{g}_2' = J^r \mathfrak{g}_2.$$

Then

$$\mathfrak{g}_1' + \mathfrak{g}_2' = J^r(\mathfrak{g}_1 + \mathfrak{g}_2).$$

It is thus seen that the set of all densities with the same weight constitute an algebraic field. Unlike densities (W, G) or like densities of different weights cannot be added together.

We now prove a theorem which is of considerable practical importance.

THEOREM. *If in a normal domain Ω*[1] *we are given a density field (of weight -1) with the property $\mathfrak{g} \neq 0$, then there exists a coordinate system (λ') in which the equality*

$$\mathfrak{g}' \overset{*}{=} \omega$$

holds, where $\omega(\xi)$ is an a priori continuous function of the point ξ of the domain Ω, with constant sign.

PROOF. Assume for definiteness that the domain Ω is normal with respect to the axis ξ^n, and consider the line

$$\xi^\nu = \overset{\nu}{c} \quad \text{for} \quad \nu = 1, 2, \ldots, n-1 \; (\overset{\nu}{c}-\text{const}),$$
$$\xi^n = \tau,$$

where τ is a parameter which varies along the line.

Denote the end values of the range of the parameter τ by $\psi(\overset{1}{c}, \overset{2}{c}, \ldots, \overset{n-1}{c})$ and $\chi(\overset{1}{c}, \overset{2}{c}, \ldots, \overset{n-1}{c})$; this means that the points

$$[\overset{1}{c}, \overset{2}{c}, \ldots, \overset{n-1}{c}, \psi(\overset{1}{c}, \overset{2}{c}, \ldots, \overset{n-1}{c})], \qquad [\overset{1}{c}, \overset{2}{c}, \ldots, \overset{n-1}{c}, \chi(\overset{1}{c}, \overset{2}{c}, \ldots, \overset{n-1}{c})]$$

lie at the boundary of the domain Ω.

Now consider the transformation $(\lambda) \to (\lambda')$ defined by the equations

$$\xi^{\nu'} = \delta^{\nu'}_{\nu} \xi^\nu \quad \text{for } \nu' = 1', 2', \ldots, (n-1)', \text{ where } \delta^{\nu'}_{\lambda} = \delta^{\nu}_{\lambda}$$

and

$$\xi^{n'} = \int\limits_{\psi}^{\xi^n} \frac{\omega(\xi^1, \xi^2, \ldots, \xi^{n-1}, \tau)}{\mathfrak{g}(\xi^1, \xi^2, \ldots, \xi^{n-1}, \tau)}\, d\tau, \tag{3.68}$$

where $\psi = \psi(\xi^1, \xi^2, \ldots, \xi^{n-1})$.

The Jacobian of this transformation is plainly given by

$$J = |A^{\lambda'}_{\lambda}| = A^{n'}_{n} = \frac{\partial \xi^{n'}}{\partial \xi^n} = \frac{\omega(\xi^1, \xi^2, \ldots, \xi^n)}{\mathfrak{g}(\xi^1, \xi^2, \ldots, \xi^n)}, \tag{3.69}$$

---

[1] A domain Ω is said to be *normal with respect to a coordinate system* (this concept is not invariant under transformations of group G_1) if there exists a λ_0, belonging to the sequence $1, 2, \ldots, n$, such that each parametric line $\xi^{\lambda_0} = \tau$, $\xi^\lambda = \text{const}$ for $\lambda \neq \lambda_0$ intersects the boundary of the domain Ω in at most two points or along a simply connected continuum.

and is therefore by assumption, of constant sign throughout the domain Ω. Furthermore it can easily be shown that the transformation (3.68) is globally invertible throughout the entire domain Ω. Since

$$\mathfrak{g}' = Jg,$$

then on substituting from (3.69) into this relation, we obtain

$$\mathfrak{g}' = \omega,$$

which is just what we had to prove.

In particular, we may require, and note that this requirement is compatible with the assumption that ω be a continuous and non-zero function, that the function ω be a constant C different from zero. In this event, we arrive in particular at the following

THEOREM. *For every non-trivial density* $\mathfrak{g}$ *there exists a coordinate system* (λ') *in which*

$$\mathfrak{g}' \equiv C,$$

where C *is a non-zero constant.*

33. Tensor Densities. We have seen above that tensors can be multiplied by scalars to yield once more a tensor of the same valence.

The question is whether tensors can be multiplied by densities, i.e. whether multiplication of a tensor by a density again produces a geometric object. A simple calculation shows that the answer is in the affirmative.

Suppose that we multiply the tensor $T^{\lambda_1\ldots\lambda_p}{}_{\mu_1\ldots\mu_q}$ by a G-density $\mathfrak{g}$ of weight $-r$. We now write

$$\mathfrak{T}^{\lambda_1\ldots\lambda_p}{}_{\mu_1\ldots\mu_q} \overset{\text{def}}{=} \mathfrak{g}T^{\lambda_1\ldots\lambda_p}{}_{\mu_1\ldots\mu_q}. \tag{3.70}$$

We have

$$\begin{aligned}\mathfrak{T}^{\lambda'_1\ldots\lambda'_p}{}_{\mu'_1\ldots\mu'_q} &= \mathfrak{g}' T'^{\lambda'_1\ldots\lambda'_p}{}_{\mu'_1\ldots\mu'_q} = \operatorname{sgn} J\cdot |J|^r A^{\lambda'_1\ldots\lambda'_p\mu_1\ldots\mu_q}_{\lambda_1\ldots\lambda_p\mu'_1\ldots\mu'_q}\,\mathfrak{g}T^{\lambda_1\ldots\lambda_p}{}_{\mu_1\ldots\mu_q}\\ &= \operatorname{sgn} J\cdot |J|^r A^{\lambda'_1\ldots\lambda'_p\mu_1\ldots\mu_q}_{\lambda_1\ldots\lambda_p\mu'_1\ldots\mu'_q}\,\mathfrak{T}^{\lambda_1\ldots\lambda_p}{}_{\mu_1\ldots\mu_q},\end{aligned} \tag{3.71}$$

and it can easily be verified that the transformation law (3.71) has the group property. Since the right-hand side of equation (3.71) depends only on the coordinates $\mathfrak{T}^{\lambda\ldots}_{\mu\ldots}$ in the coordinate system (λ) and on $A^{\lambda'}_{\lambda}$ since the Jacobian J is also a function of the variables $A^{\lambda'}_{\lambda}$, the transformation law (3.71) defines a geometric object of class one.

Objects of type (3.70) have been called *tensor densities*, although this name is not really appropriate. A more suitable name for them would be *density tensors*. The transformation law (3.71) emerged from the multiplication of a tensor by a density. On the other hand it can be shown conversely that if an object obeys the transformation law (3.71), that object can be obtained by multiplying a tensor T by an appropriate density of weight $-r$, and indeed, this can be done in an infinite number of ways. It suffices to put

$$T^{\lambda_1\dots\lambda_p}{}_{\mu_1\dots\mu_q} \stackrel{*}{=} \frac{\mathfrak{T}^{\lambda_1\dots\lambda_p}{}_{\mu_1\dots\mu_q}}{\varrho},$$

where ϱ denotes an arbitrary non-zero scalar, and then to put

$$\mathfrak{g} \stackrel{*}{=} \varrho.$$

We then have

$$\mathfrak{T}^{\lambda_1\dots\lambda_p}{}_{\mu_1\dots\mu_q} = \mathfrak{g}T^{\lambda_1\dots\lambda_p}{}_{\mu_1\dots\mu_q}.$$

Operations similar to those performed on tensors can be defined on tensor densities, except that addition is possible only when the valences (p, q) and the weights of the densities being added are the same. When tensor densities are multiplied, the weights are added. For example, the product of a density of weight $-r$ with a tensor density of weight $+r$ is a tensor.

If in particular we take a density n-vector $\mathfrak{g}^{\lambda_1\dots\lambda_n}$, i.e. an object with the property

$$\mathfrak{g}^{\lambda_1\dots\lambda_n} = \mathfrak{g}^{[\lambda_1\dots\lambda_n]} \tag{3.72}$$

and obeying the transformation law

$$\mathfrak{g}^{\lambda'_1\dots\lambda'_n} = \varepsilon|J|^r A^{\lambda'_1\dots\lambda'_n}_{\lambda_1\dots\lambda_n}\mathfrak{g}^{\lambda_1\dots\lambda_n}, \tag{3.73}$$

such an object has essentially only one coordinate

$$\mathfrak{g}^{1\dots n}. \tag{3.74}$$

All the other coordinates will either be zero, or—in the case when $\lambda_1, \lambda_2, \dots$ $\dots, \lambda_n$ is a permutation of the sequence $1, 2, \dots, n$—equal in absolute value to the coordinate (3.74), with the same sign for an even permutation and the opposite sign for an odd permutation. Reasoning just as on p. 83, we deduce that

$$\mathfrak{g}^{1'\dots n'} = \varepsilon|J|^{r+1}\mathfrak{g}^{1\dots n}, \tag{3.75}$$

i.e. that the essential coordinate (3.74) is a density of weight $-r-1$.

In particular, when object (3.72) is a vector density of weight $+1$, i.e. when $r = -1$, the coordinate (3.74) is a scalar.

Conversely, if ϱ is an arbitrary scalar and if we put

$$\mathfrak{g}^{\lambda_1\ldots\lambda_n} = \varrho\varepsilon^{\lambda_1\ldots\lambda_n}\ \text{[1]}, \tag{3.76}$$

the object so defined will be a vector density of weight $+1$.

To prove this, consider an n-vector constructed from the basis vectors $n!\, \underset{1}{e}^{[\lambda_1} \ldots \underset{n}{e}^{\lambda_n]}$ of the coordinate system (λ) and the density $\mathfrak{g}$ defined by equation

$$\mathfrak{g} \overset{*}{=} \varrho$$

together with the transformation law

$$\mathfrak{g}' = J^{-1}\mathfrak{g}.$$

Now set

$$\bar{\mathfrak{g}}^{\lambda_1\ldots\lambda_n} \overset{\text{def}}{=} \mathfrak{g}n!\underset{1}{e}^{[\lambda_1} \ldots \underset{n}{e}^{\lambda_n]},$$

which will be an n-vector density of weight $+1$.

Since

$$\underset{1}{e}^{[\lambda_1} \ldots \underset{n}{e}^{\lambda_n]} \overset{*}{=} \varepsilon^{\lambda_1\ldots\lambda_n}\ \text{[2]}$$

we have, by virtue of (3.76),

$$\bar{\mathfrak{g}}^{\lambda_1\ldots\lambda_n} \overset{*}{=} \varrho\varepsilon^{\lambda_1\ldots\lambda_n}.$$

We have, further,

$$\bar{\mathfrak{g}}^{1'\ldots n'} = J^{-1}A^{1'}_{\lambda_1} \ldots A^{n'}_{\lambda_n}\bar{\mathfrak{g}}^{\lambda_1\ldots\lambda_n} = J^{-1}A^{1'}_{\lambda_1} \ldots A^{n'}_{\lambda_n}\varrho\varepsilon^{\lambda_1\ldots\lambda_n} = J^{-1}\,\varrho J = \varrho.$$

It follows from this, by virtue of the above remark, that

$$\bar{\mathfrak{g}}^{\lambda'_1\ldots\lambda'_n} = \bar{\mathfrak{g}}^{1'\ldots n'}\varepsilon^{\lambda'_1\ldots\lambda'_n} = \varrho\varepsilon^{\lambda'_1\ldots\lambda'_n}.$$

Accordingly

$$\mathfrak{g}^{\lambda_1\ldots\lambda_n} = \bar{\mathfrak{g}}^{\lambda_1\ldots\lambda_n}$$

and the theorem is thus proved. In particular it follows that *the Ricci symbol* $\varepsilon^{\lambda_1\ldots\lambda_n}$ *is a contravariant density* ($\varrho = 1$) *of weight* $+1$.

In a similar way the Ricci symbol $\varepsilon_{\lambda_1\ldots\lambda_n}$ or, more generally, the object

$$\mathfrak{g}_{\lambda_1\ldots\lambda_n} \overset{\text{def}}{=} \varrho\varepsilon_{\lambda_1\ldots\lambda_n},$$

---

[1] For the definition of the symbol $\varepsilon^{\lambda_1\ldots\lambda_n}$ see p. 83. Since $\varepsilon^{\lambda_1\ldots\lambda_n}$ represents a density n-vector, we should for reasons of consistency use the gothic letter $\mathfrak{n}^{\lambda_1\ldots\lambda_n}$. However, in this case we retain the symbol ε, as this should remind the reader that $\varepsilon^2 = 1$ or $\varepsilon^2 = 0$.

[2] In another coordinate system (λ') this equality will no longer hold in general!

where ϱ is an arbitrary scalar, is shown to be a covariant density of weight -1.

This example shows that the same "system" of numbers $(0, +1, -1)$, represented by the symbols $\varepsilon^{\lambda_1 \dots \lambda_n}$ and $\varepsilon_{\lambda_1 \dots \lambda_n}$ may stand for different geometric objects depending on the configuration (it is difficult to speak of "order" when one has a large number of indices) in which this system is written.

Suppose we are given a two-index tensor density (contravariant, covariant, or mixed) of weight $-r$

$$\mathfrak{g}^{\lambda\mu}, \mathfrak{g}_{\lambda\mu}, \mathfrak{g}^{\lambda}_{\mu}.$$

For each of these densities we construct the determinant

$$\overset{0}{W} = |\mathfrak{g}^{\mu\lambda}|, \qquad \underset{0}{W} = |\mathfrak{g}_{\lambda\mu}|, \qquad W = |\mathfrak{g}^{\lambda}_{\mu}| \tag{3.77}$$

and we ask whether these determinants are geometric objects and if so, of what type.

To answer this question, we first recall Cauchy's theorem on multiplication from determinant theory and a generalization of this theorem. Cauchy's theorem states that if we are given two determinants

$$a = |a^{\nu}_{\lambda}|, \qquad b = |b^{\nu}_{\lambda}| \qquad (\lambda, \nu = 1, 2, \dots, n)$$

and if we set

$$c^{\nu}_{\lambda} \overset{\text{def}}{=} a^{\nu}_{\mu} b^{\mu}_{\lambda}$$

and

$$c = |c^{\nu}_{\lambda}|,$$

then the relation

$$c = ab$$

is valid. This theorem can easily be generalized, using mathematical induction, as follows:

Suppose we are given m determinants

$$\underset{i}{a} = |\underset{i}{a}{}^{\nu}_{\lambda}| \qquad (\lambda, \nu = 1, 2, \dots, n;\ i = 1, 2, \dots, m)$$

and let

$$b^{\nu}_{\lambda} \overset{\text{def}}{=} a^{\nu}_{\lambda_1} \underset{2}{a}{}^{\lambda_1}_{\lambda_2} \dots \underset{m-1}{a}{}^{\lambda_{m-2}}_{\lambda_{m-1}} \underset{m}{a}{}^{\lambda_{m-1}}_{\lambda}$$

and

$$b = |b^{\nu}_{\lambda}|.$$

Then we have the relation

$$b = \underset{1}{a}\underset{2}{a} \dots \underset{m}{a}.$$

We now return to the question posed above concerning the transformation properties of the determinants (3.77). By definition

$$\begin{aligned} \mathfrak{g}^{\lambda'\mu'} &= \varepsilon|J|^r A^{\lambda'\mu'}_{\lambda\ \mu}\,\mathfrak{g}^{\lambda\mu}, \\ \mathfrak{g}_{\lambda'\mu'} &= \varepsilon|J|^r A^{\lambda\ \mu}_{\lambda'\mu'}\,\mathfrak{g}_{\lambda\mu}, \\ \mathfrak{g}^{\lambda'}_{\mu'} &= \varepsilon|J|^r A^{\lambda'\mu}_{\lambda\ \mu'}\,\mathfrak{g}^{\lambda}_{\mu}. \end{aligned} \tag{3.78}$$

Using the theorem concerning the extraction of a common factor from a determinant together with the aforementioned generalization of Cauchy's theorem on the multiplication of determinants, we obtain

$$\begin{aligned} \overset{0}{W}{}' &= |\mathfrak{g}^{\lambda'\mu'}| = \varepsilon^n|J|^{rn}|A^{\lambda'}_{\lambda}|\cdot|A^{\mu'}_{\mu}|\cdot|\mathfrak{g}^{\lambda\mu}| = \varepsilon^n|J|^{rn+2}\overset{0}{W}, \\ \underset{0}{W}{}' &= |\mathfrak{g}_{\lambda'\mu'}| = \varepsilon^n|J|^{rn}|A^{\lambda}_{\lambda'}|\cdot|A^{\mu}_{\mu'}|\cdot|\mathfrak{g}_{\lambda\mu}| = \varepsilon^n|J|^{rn-2}\underset{0}{W}, \\ W' &= |\mathfrak{g}^{\lambda'}_{\mu'}| = \varepsilon^n|J|^{rn}|A^{\lambda'}_{\lambda}|\cdot|A^{\mu}_{\mu'}|\cdot|\mathfrak{g}^{\lambda}_{\mu}| = \varepsilon^n|J|^{rn}W. \end{aligned} \tag{3.79}$$

The formulae above imply in particular that if the quantities (3.78) are tensors ($r = 0$), the determinant $\overset{0}{W}$ will be a density of weight -2, the determinant $\underset{0}{W}$ a density of weight $+2$, and finally that the determinant W will be a scalar, namely a W-density for an even n, and a density of the same type as $\mathfrak{g}$ for an odd n.

We now recall the result from algebra which asserts that the rank of a matrix, is an invariant of any (non-singular [1]) affine transformation of the elements of the matrix. It follows from this result (since $J \neq 0$) that the orders of all three determinants W, $\overset{0}{W}$ and $\underset{0}{W}$ are geometric quantities associated with a given tensor density (or a given two-index tensor), and are independent of the coordinate system.

In the special case $n = 2$ consider the Ricci symbol $\varepsilon^{\lambda\mu}$. As we have shown above, this is a tensor density of weight $+1$, i.e. $r = -1$. The first formula of (3.79) implies that $\overset{0}{W} = |\mathfrak{g}^{\lambda\mu}|$ is a scalar ($rn+2 = -2+2 = 0$), which can also be verified by direct calculation

$$\overset{0}{W} = \begin{vmatrix} 0 & 1 \\ -1 & 0 \end{vmatrix} = 1 = \overset{0}{W}{}'.$$

(1) Non-singular means that the determinant of the coefficients of the transformation is non-zero.

The geometric objects defined so far, i.e. scalars, densities, vectors of both kinds, tensors of any valence and finally tensor densities, are given the collective name of *quantities* (geometric) [1].

34. Geometric Objects other than Quantities. We are now familiar with a large number of objects of various types. The question is whether still others exist, and it turns out that the answer is in the affirmative. Our next question therefore, is how to define all geometric objects. There is, of course, an infinite number of them: in fact, there is an infinite number of vectors alone. However, all contravariant vectors obey a common transformation law. One can thus ask, when two objects with the same transformation law are classified as being of the same type, how many different types of geometric objects are there? The answer to this question is again: an infinite number. Two tensors, one of valence (p, q) and the other of valence (r, s), are plainly of different types provided that the pairs (p, q) and (r, s) differ from one another. The concept of the similarity (equivalence) of two objects of different types [114] was introduced in order to simplify the classification of geometric objects. Using this idea it turns out that for a fixed dimension n of the space, a fixed number m of coordinates, and a fixed class s of objects, the number of different types up to similarity is, in general, finite.

As an example, consider an object a^λ with n coordinates: an object of class one with the transformation law

$$a^{\lambda'} = \sum_{\lambda} (-1)^{\lambda+\lambda'} a^\lambda A^{\lambda'}_{\lambda}. \tag{3.80}$$

We check that this law possesses the group property. Indeed, if we rearrange the right-hand side of the formula

$$a^{\lambda''} = \sum_{\lambda'} (-1)^{\lambda'+\lambda''} a^{\lambda'} A^{\lambda''}_{\lambda'},$$

we obtain

$$\sum_{\lambda'} (-1)^{\lambda'+\lambda''} a^{\lambda'} A^{\lambda''}_{\lambda'} = \sum_{\lambda'} (-1)^{\lambda'+\lambda''} A^{\lambda''}_{\lambda'} \sum_{\lambda} (-1)^{\lambda+\lambda'} a^\lambda A^{\lambda'}_{\lambda}$$

$$= \sum_{\lambda,\lambda'} (-1)^{\lambda'+\lambda''} (-1)^{\lambda+\lambda'} a^\lambda A^{\lambda''}_{\lambda'} A^{\lambda'}_{\lambda}$$

---

(1) This name is not the most fortunate, but its use is justified by the tradition associated with this name in physics.

$$= \sum_{\lambda,\lambda'} (-1)^{2\lambda'+\lambda+\lambda''} a^{\lambda} A^{\lambda''}_{\lambda'} A^{\lambda'}_{\lambda}$$

$$= \sum_{\lambda,\lambda'} (-1)^{\lambda''+\lambda} a^{\lambda} A^{\lambda''}_{\lambda'} A^{\lambda'}_{\lambda} = \sum_{\lambda} (-1)^{\lambda''+\lambda} a^{\lambda} \sum_{\lambda'} A^{\lambda''}_{\lambda'} A^{\lambda'}_{\lambda}$$

$$= \sum_{\lambda} (-1)^{\lambda''+\lambda} a^{\lambda} A^{\lambda''}_{\lambda'}.$$

We have thus established the group property for the law (3.80).

Objects with the law (3.80) were discovered by Jakubowicz [128]. It can be shown that these objects are equivalent to vectors, although they have a different transformation law.

Other class-one objects with a number of coordinates equal to the dimension of the space, which Jakubowicz called definors [128], were also discovered rather later. These are objects with n coordinates and with the transformation law

$$b^{\lambda'} = b^{\lambda} \Delta^{\lambda'}_{\lambda}, \tag{3.81}$$

where $\Delta^{\lambda'}_{\lambda}$ denotes the algebraic minor belonging to the element $A^{\lambda'}_{\lambda}$ in the determinant J. Somewhat later Zajtz [90] discovered these objects independently of Jakubowicz.

35. Conditions Sufficient for an Object to be a Tensor. We have seen above that tensor (and tensor densities) can be multiplied together, with the product having an appropriate new valence, which can be read off at once from the position of the indices. For instance, the multiplication of two covariant vectors a_λ and b_λ in this order gives rise to a two-index tensor

$$a_\lambda b_\mu = c_{\lambda\mu}.$$

We now ask conversely whether, when we are given a vector a_λ and a tensor $c_{\lambda\mu}$, we can find a covariant vector x_μ such that

$$a_\lambda x_\mu = c_{\lambda\mu}. \tag{3.82}$$

That such an operation (inverse to multiplication) does not exist in general can be seen from the simplest example of two-dimensional space (the case of one-dimensional space is trivial). In general, however, in n-dimensional space, it is plain that the set of equations (3.82) will not usually have any solutions, since the number of equations is n^2 (λ, μ vary independently over the values from 1 to n), while there are n unknowns; this means that for $n \geqslant 2$ there are more equations to be satisfied than there are unknowns, and such a system is in general not soluble.

Another question may, however, be asked. Suppose that at a particular point p of the space X_n we are given a geometric object with n coordinates a_λ and assume that for *each* vector b_μ the product

$$a_\lambda b_\mu$$

is a tensor. Can one then conclude that the object a_λ is a vector? The answer to this question is yes.

We write

$$c_{\lambda\mu} \stackrel{\text{def}}{=} a_\lambda b_\mu$$

and calculate $c_{\lambda'\mu'}$. Thus, on the one hand

$$c_{\lambda'\mu'} = a_{\lambda'} b_{\mu'} = a_{\lambda'} A^{\mu}_{\mu'} b_\mu,$$

while on the other hand

$$c_{\lambda'\mu'} = A^{\lambda\mu}_{\lambda'\mu'} c_{\lambda\mu} = A^{\lambda}_{\lambda'} A^{\mu}_{\mu'} a_\lambda b_\mu = A^{\lambda}_{\lambda'} a_\lambda A^{\mu}_{\mu'} b_\mu;$$

by comparison we obtain

$$a_{\lambda'} A^{\mu}_{\mu'} b_\mu = A^{\lambda}_{\lambda'} a_\lambda A^{\mu}_{\mu'} b_\mu.$$

We now multiply this by $A^{\mu'}_{\nu}$ and sum over the index μ'. We then have

$$a_{\lambda'} A^{\mu\mu'}_{\mu'\nu} b_\mu = A^{\lambda}_{\lambda'} a_\lambda A^{\mu\,\mu'}_{\mu'\nu} b_\mu,$$

or

$$a_{\lambda'} b_\nu = A^{\lambda}_{\lambda'} a_\lambda b_\nu \qquad (\lambda' = 1', 2', \ldots, n'; \ \nu = 1, 2, \ldots, n).$$

This identity is to hold for every vector b, i.e. for all b_ν's and this is possible only when the coefficients appearing in front of b on either side are equal; hence, the equality

$$a_{\lambda'} = A^{\lambda}_{\lambda'} a_\lambda,$$

must hold and this shows that a_λ is a covariant vector, which is what we set out to prove.

Similarly, if we are given a geometric object a_λ with n coordinates and if we knew that contraction with respect to λ in the product of the object a_λ with any contravariant vector b^λ always yields a scalar, i.e. that $a_\lambda b^\lambda$ is a scalar for every vector b^λ, then a_λ must be a covariant vector.

Indeed, put

$$a_\lambda b^\lambda = \sigma$$

and assume that σ is a scalar, i.e. that

$$a_{\lambda'} b^{\lambda'} = a_\lambda b^\lambda = \sigma.$$

Now b^λ is by assumption a contravariant vector, and so

$$b^{\lambda'} = A^{\lambda'}_{\lambda} b^\lambda,$$

or

$$a_{\lambda'} A^{\lambda'}_{\lambda} b^\lambda = a_\mu b^\mu$$

or in other words, changing the dummy index and putting both terms on one side of this equation, we have

$$(a_{\lambda'} A^{\lambda'}_{\lambda} - a_\lambda) b^\lambda = 0.$$

Since the equality above must hold for all b^λ, all the coefficients must be zero, i.e.

$$a_{\lambda'} A^{\lambda'}_{\lambda} = a_\lambda,$$

which proves that a_λ is a covariant vector.

The examples above can be generalized and the following theorem formulated:

THEOREM. *If we are given an object*

$$S(\lambda_1, \ldots, \lambda_p, \varrho_1, \ldots, \varrho_r, \mu_1, \ldots, \mu_q, \sigma_1, \ldots, \sigma_s)$$

which always yields a tensor of valence $(r+t, s+u)$ *when "transvected" onto any tensor*

$$T^{\mu_1 \ldots \mu_q \mu_{q+1} \ldots \mu_{q+t}}{}_{\lambda_1 \ldots \lambda_p \lambda_{p+1} \ldots \lambda_{p+u}},$$

where contraction is effected with respect to $\lambda_1, \lambda_2, \ldots, \lambda_p$ *and* $\mu_1, \mu_2, \ldots, \mu_q$, *then that object is a tensor, and more particularly it is a tensor of the type*

$$S^{\lambda_1 \ldots \lambda_p \varrho_1 \ldots \varrho_r}{}_{\mu_1 \ldots \mu_q \sigma_1 \ldots \sigma_s}.$$

Analogous theorems can be formulated on replacing the word tensor with tensor density.

36. The Basis (Fundamental) Tensor. Suppose that at some point p we are given a two-index tensor

$$g_{\lambda\mu} = g_{\mu\lambda}. \tag{3.83}$$

As we know, the rank of the coordinate matrix of this tensor

$$r = \text{rank}\,[g_{\lambda\mu}] \tag{3.84}$$

is an invariant under transformations of the coordinate system. Denote by $\mathfrak{g}$ the value of the determinant

$$\mathfrak{g} = |g_{\lambda\mu}|. \tag{3.85}$$

The reason for using a gothic letter is that, as we already know, the value of this determinant is not a scalar but is a density of weight 2, since $\mathfrak{g}$ transforms according to the law

$$\mathfrak{g}' = J^{-2}\mathfrak{g}.$$

Moreover, $\mathfrak{g}$ is a W-density since this law can be rewritten as

$$\mathfrak{g}' = |J|^{-2}\mathfrak{g}, \tag{3.86}$$

from which it can be seen that $\mathfrak{g}$ is of constant sign in every coordinate system. If

$$r = n, \tag{3.87}$$

then we always have

$$\mathfrak{g} \neq 0. \tag{3.88}$$

If $r < n$, then $\mathfrak{g}$ is a trivial zero density. We continue to assume that equation (3.87) holds.

Now consider the system of equations

$$g_{\lambda\mu} x^{\mu\nu} = A_\lambda^\nu \qquad (\lambda, \nu = 1, 2, \ldots, n),$$

where A_λ^ν is the unit tensor. This is a system of n^2 linear equations in n^2 unknowns which breaks down into n systems each consisting of n equations in n unknowns. If we fix the index ν_0 and write

$$y^\lambda \overset{\text{def}}{=} x^{\lambda\nu_0},$$

the set of equations can be rewritten as

$$g_{\lambda\mu} y^\mu = A_\lambda^{\nu_0} \qquad (\lambda = 1, 2, \ldots, n).$$

This is a set of n equations with determinant equal to $\mathfrak{g}$. If we assume the inequality (3.88), then the set is uniquely soluble for the unknowns y^λ, and from the theory of determinants and the solution of systems of linear equations it follows that

$$y^\lambda = \frac{1}{\mathfrak{g}} \begin{vmatrix} g_{11} & \cdots & g_{1,\lambda-1} & 0 & g_{1,\lambda+1} & \cdots & g_{1n} \\ \cdot & \cdot & \cdot & \cdot & \cdot & \cdot & \cdot \\ \cdot & \cdot & \cdot & 1 & \cdot & \cdot & \cdot \\ \cdot & \cdot & \cdot & \cdot & \cdot & \cdot & \cdot \\ g_{n1} & \cdots & g_{n,\lambda-1} & 0 & g_{n,\lambda+1} & \cdots & g_{nn} \end{vmatrix} \leftrightarrow \nu_0\text{-th row.}$$

Expanding the determinant by the λ-th column, we obtain

$$y^\lambda = \frac{1}{\mathfrak{g}} \operatorname{minor} g_{\nu_0\lambda} \qquad (\lambda = 1, 2, \ldots, n),$$

i.e. allowing the fixed index ν_0 to revert to the running index ν,

$$x^{\lambda\nu} = \frac{1}{\mathfrak{g}} \operatorname{minor} g_{\nu\lambda}.$$

We now introduce the notation

$$g^{\lambda\nu} \stackrel{\text{def}}{=} \frac{1}{\mathfrak{g}} \operatorname{minor} g_{\nu\lambda}. \tag{3.89}$$

It can be shown that $g^{\lambda\nu}$ is a tensor, which is only to be expected since transvecting onto the tensor $g_{\lambda\mu}$ yields the tensor

$$g_{\lambda\mu} g^{\mu\nu} = A^{\nu}_{\lambda}. \tag{3.90}$$

(REMARK. This fact does not, however, follow immediately from the theorem on p. 101).

Since

$$g_{\nu\lambda} = g_{\lambda\nu},$$

we also have

$$\operatorname{minor} g_{\nu\lambda} = \operatorname{minor} g_{\lambda\nu}, \text{ whence } g^{\lambda\nu} = g^{\nu\lambda}.$$

The tensor $g^{\lambda\nu}$ is called the *inverse of the tensor* $g_{\lambda\mu}$. Note that when $r < n$, i.e. when $\mathfrak{g} = 0$, the tensor $g_{\lambda\mu}$ does not have an inverse. Conversely, it can be shown that the relationships

$$g_{\lambda\mu} = \frac{\operatorname{minor} g^{\mu\lambda}}{\bar{\mathfrak{g}}} \tag{3.91}$$

hold, where $\bar{\mathfrak{g}}$ denotes the value of the determinant

$$\bar{\mathfrak{g}} \stackrel{\text{def}}{=} |g^{\lambda\nu}|.$$

Applying Cauchy's theorem concerning the multiplication of determinants to equation (3.90), we write

$$|g_{\lambda\mu}| \cdot |g^{\mu\nu}| = |A^{\nu}_{\lambda}|,$$

i.e.

$$\mathfrak{g}\bar{\mathfrak{g}} = 1,$$

so that equation (3.91) can be rewritten as

$$g_{\lambda\mu} = \mathfrak{g} \operatorname{minor} g^{\mu\lambda}. \tag{3.92}$$

The tensor $g_{\lambda\mu}$ is said equally to be the inverse of $g^{\lambda\mu}$.

REMARK. The operation of inversion can also be carried out on tensors which are not necessarily symmetric. However, a necessary condition for this operation to be possible is that the determinant constructed from the coordinates of the given tensor should not vanish. A mixed tensor T^{λ}_{ν} can also be inverted, provided that $|T^{\lambda}_{\nu}| \neq 0$. The inverse tensor is then also a mixed tensor, and therefore does not change valence with respect to the tensor being inverted.

We shall see below that when we are given a tensor (3.83) with property (3.87) we can define certain additional operations, and can in particular introduce a local length and angular metric. The tensor (3.83) has been called the basic (or fundamental) tensor, or the metric tensor.

37. The Operation of Raising and Lowering Indices. If we have a basic tensor $g_{\lambda\mu}$ with the property that

$$\mathfrak{g} = |g_{\lambda\mu}| \neq 0,$$

or, as we shall say, a non-singular tensor, we can use it to introduce the operation of raising or lowering any index in any tensor or tensor density.

DEFINITION. By *lowering* (*raising*) the contravariant index λ_i (covariant index μ_j) of a tensor density

$$\mathfrak{T}^{\lambda_1 \ldots \lambda_i \ldots \lambda_p}{}_{\mu_1 \ldots \mu_j \ldots \mu_q}$$

we shall mean transvecting the tensor $g_{\lambda_i \mu_{q+1}}$ ($g^{\mu_j \lambda_{p+1}}$) onto it.

This operation yields a density of the same weight but of different valence, viz. if the original density had a valence of (p, q) the new one has a valence of $(p-1, q+1)$ $[(p+1, q-1)]$. A quantity obtained by lowering (raising) the appropriate index will be indicated by the addition of a bar: $\overline{\mathfrak{T}}$. To signify in the new quantity $\overline{\mathfrak{T}}$ which index of the original quantity $\mathfrak{T}$ was raised or lowered the vertical position of a shifted, index is left unaltered while a dot is used to replace any index removed. Under this rule, contra- and co-variant indices in mixed tensors cannot henceforth be placed in the same vertical position. The only allowable departure from this rule occurs in the case of the unit tensor A^{ν}_{λ} which we shall continue to designate as before, rather than by $A^{\nu}_{\cdot\lambda}$. This exception does not lead to any ambiguity or confusion since lowering the index ν yields

$$A^{\nu}_{\lambda} g_{\nu\mu} = \overline{A}_{\mu\lambda}^{\cdot} = g_{\mu\lambda},$$

i.e. the results in the fundamental tensor. Similarly, raising the index λ yields

$$A^{\nu}_{\lambda} g^{\lambda\mu} = \overline{A}^{\nu\mu}_{\cdot} = g^{\nu\mu},$$

i.e., the inverse of the fundamental tensor, and both of these tensors are symmetric.

Conversely, raising one of the indices in the fundamental tensor itself gives the unit tensor — as follows from relation (3.90), whereas lowering one of the indices in the tensor $g^{\mu\lambda}$ also gives the unit tensor. The question now is what happens when both indices in the fundamental tensor $g_{\lambda\mu}$ are raised, or both indices in the inverse tensor $g^{\lambda\mu}$ are lowered. We raise both indices in turn

$$g_{\lambda\mu} g^{\lambda\nu} = \bar{g}^{\nu}_{\cdot\mu}, \qquad \bar{g}^{\nu}_{\cdot\mu} g^{\mu\sigma} = \bar{g}^{\nu\sigma}_{\cdot\cdot}.$$

But

$$\bar{g}^{\nu\sigma}_{\cdot\cdot} = g_{\lambda\mu} g^{\lambda\nu} g^{\mu\sigma}_{\cdot\cdot} = A^{\nu}_{\mu} g^{\mu\sigma} = g^{\nu\sigma},$$

so that by raising both indices in the fundamental tensor we get the inverse tensor.

Similarly, by lowering both indices in the inverse tensor, we arrive at the original fundamental tensor and consequently in this particular case it is superfluous to add a bar over the kernel letter. On the other hand, when applying the operation of raising or lowering indices to other quantities we should always add the bar over the kernel letter.

For example, by raising the two indices of a given tensor $a_{\lambda\mu}$ we obtain a new tensor

$$\bar{a}^{\lambda\mu}_{\cdot\cdot} = a_{\varrho\sigma} g^{\varrho\lambda} g^{\sigma\mu}.$$

It is now customary for all authors to omit the bar and even to call $\bar{a}^{\lambda\mu}$ the (contravariant) coordinates of the same tensor; from a strictly logical point of view, however, this is incorrect since the species and individuality of a tensor are determined not only by the values of the coordinates but also by the law governing their transformation under transition to another coordinate system. In view of the traditional, generally accepted method of notation, we shall also omit this bar over the kernel letter, especially as the very position (level) of the index next to the kernel letter indicates the transformation law.

It should be noted that if a particular index is raised in a given quantity and that index is then lowered in the resulting quantity the original quantity is recovered. (The very simple proof that this requires is omitted.)

We thus find that the operation of moving (i.e. raising or lowering) the same index is involutory [(1)].

---

(1) If the correspondence $y = f(x)$, which maps the set S into itself, where x and y are elements of that set, has the property that $f[f(x)] \equiv x$, we say that f is involutory or that it is an involution.

38. The Fundamental Tensor as a Metric Tensor. The adoption of a non-singular fundamental tensor at a point p enables us, as we shall see, to introduce a local metric, i.e. to measure the lengths of contravariant vectors attached at p and to measure the angles between vectors attached at p. We shall follows the reasoning of p. 17 and 72, where a metric was introduced for an orthogonal subgroup.

Suppose that we are given a vector v^λ at the point p. By multiplication we can construct the tensor $v^\lambda v^\mu$ from the vector v^λ. We now transvect the fundamental tensor onto this tensor, contracting both indices. In this way we obtain a scalar

$$g_{\lambda\mu} v^\lambda v^\mu.$$

Denoting the length of the vector v^λ by $|v|$, we set

$$|v| \overset{\text{def}}{=} |g_{\lambda\mu} v^\lambda v^\mu|^{1/2}. \tag{3.93}$$

The vector length $|v|$ is thus a scalar and, moreover, the property, already well-known for vectors in the space R_n, that when a vector is multiplied by a scalar factor ϱ the length of the vector is multiplied by $|\varrho|$, also holds here.

Indeed, if we set $w^\lambda = \varrho v^\lambda$, then by equation (3.93) we have

$$\begin{aligned} |w| &= |g_{\lambda\mu} w^\lambda w^\mu|^{1/2} = |g_{\lambda\mu} \varrho v^\lambda \varrho v^\mu|^{1/2} = |\varrho^2 g_{\lambda\mu} v^\lambda v^\mu|^{1/2} \\ &= |\varrho| \cdot |g_{\lambda\mu} v^\lambda v^\mu|^{1/2} = |\varrho| \cdot |v|. \end{aligned}$$

In definition (3.93), the absolute value of the scalar $g_{\lambda\mu} v^\lambda v^\mu$ was taken under the square root sign to make the length of the vector real. In the case when the quadratic form

$$g_{\lambda\mu} x^\lambda x^\mu \tag{3.94}$$

is positive definite i.e. when the value of expression (3.94) is greater than or equal to zero for every set of real values for x^λ, and zero only when all $x^\lambda = 0$, the absolute value sign can be omitted under the square root sign in equation (3.93). If the quadratic form (3.94) is not positive definite, the length of the vector may be zero even when the vector v is not itself a null vector. Such vectors are said to be *singular* or *isotropic* (with respect to the fundamental tensor naturally).

Suppose that an Euclidean space R_n with origin at the point p and Cartesian coordinates x^λ is "attached" at (i.e. assigned to) the point p. The equation

$$g_{\lambda\mu} x^\lambda x^\mu = 0 \tag{3.95}$$

then defines a cone in this space and vectors lying on the generatrices of that cone are singular vectors. This cone degenerates to the point p alone ($x^\lambda = 0$) when the form (3.94) is definite. If the quadratic form (3.94) is not definite, then for $n = 2$ equation (3.95) takes the form

$$g_{11}(x^1)^2 + 2g_{12}x^1x^2 + g_{22}(x^2)^2 = 0.$$

Under a suitable affine transformation of the variables x^i, this equation assumes the canonical form (p. 122)

$$h_{11}(y^1)^2 + h_{22}(y^2)^2 = 0, \tag{3.96}$$

where

$$h_{11} < 0, \qquad h_{22} > 0.$$

Equation (3.96) represents a system of two straight lines which intersect at the point $y^1 = y^2 = 0$ and which divide the plane into four domains. The left-hand side of equation (3.96), i.e. the value of the quadratic form, is of one sign in two opposite quadrants, and of the opposite sign in the other two quadrants. No one pair of quadrants can be singled out from the topological point of view, so that it is not possible to distinguish here between the interior and the exterior of cone (3.95). It is a different matter for $n > 2$. For example when $n = 3$, with the quadratic form (3.94) indefinite, cone (3.95) divides the space into three domains and in this case it is possible to distinguish certain regions, which we call the *interior* of the cone, from the rest of the region, called the *exterior* of the cone. We once again cast equation (3.95) into the canonical form

$$\Phi(y^1, y^2, y^3) = h_{11}(y^1)^2 + h_{22}(y^2)^2 + h_{33}(y^3)^2 = 0.$$

The indefiniteness of the form (3.95) implies that if the rank of the coordinate matrix of the tensor $g_{\lambda\mu}$ is 3, we always have either

$$h_{11} < 0, \qquad h_{22} > 0, \qquad h_{33} > 0,$$

or

$$h_{11} < 0, \qquad h_{22} < 0, \qquad h_{33} > 0.$$

In the first case, points (y^1, y^2, y^3) for which $\Phi(y^1, y^2, y^3) < 0$ are called *interior points* whereas points for which $\Phi(y^1, y^2, y^3) > 0$ are *exterior points*. In the second case, the converse is true. Both cases, however, are covered by the following general rule, viz.: the *interior of the cone* is the set of points for which

$$\Phi(y^1, y^2, y^3)\,\mathrm{sgn}(h_{11})\,\mathrm{sgn}(h_{22})\,\mathrm{sgn}(h_{33}) > 0,$$

while the *exterior* is the set of points for which

$$\Phi(y^1, y^2, y^3)\operatorname{sgn}(h_{11})\operatorname{sgn}(h_{22})\operatorname{sgn}(h_{33}) < 0.$$

For $n > 3$, the matter is more complicated and we shall not discuss it here in general. This concept is important for $n = 4$ in the special theory of relativity.

The reader should be warned against erroneously surmising at this point that even though it is definite at every point p of our space, the metric tensor makes possible an invariant determination of the distance between two points since it allows the length of the contravariant vector to be defined: this is not the case. Consider two points p, q with coordinates ξ^λ and η^λ, respectively, in the coordinate system (λ). Now, if we were to define the distance $\varrho(p, q)$ by setting

$$\varrho(p, q) = |(g_{\lambda\mu})_p(\eta^\lambda - \xi^\lambda)(\eta^\mu - \xi^\mu)|^{1/2},$$

which is suggested by the relevant formula from analytic geometry in Euclidean space (the symbol $(g_{\lambda\mu})_p$ denotes the coordinates of the fundamental tensor at the point p), we would see that such a definition is not invariant for the general group G_1.

It will be seen further on that for a given field of metric tensors, i.e. when $g_{\lambda\mu}$ is given at every point in space, the distance between any two points cannot be determined invariantly. However, this is possible for two infinitely close points, and this makes it possible to define the length of arcs situated in such spaces.

39. The Scalar Product. The fundamental tensor $g_{\lambda\mu}$ enables us further to define the concept of scalar product for two vectors attached at a point p. Suppose that two contravariant vectors $\underset{1}{v}^\lambda$ and $\underset{2}{v}^\lambda$ are given at the point p. We now form the scalar

$$\sigma[\underset{1}{v}, \underset{2}{v}] \stackrel{\text{def}}{=} g_{\lambda\mu}\underset{1}{v}^\lambda \underset{2}{v}^\mu. \tag{3.97}$$

This scalar has the property that

$$\sigma[\underset{2}{v}, \underset{1}{v}] = \sigma[\underset{1}{v}, \underset{2}{v}],$$

since

$$g_{\lambda\mu}\underset{2}{v}^\lambda \underset{1}{v}^\mu = g_{\mu\lambda}\underset{2}{v}^\mu \underset{1}{v}^\lambda = g_{\lambda\mu}\underset{2}{v}^\mu \underset{1}{v}^\lambda = g_{\lambda\mu}\underset{1}{v}^\lambda \underset{2}{v}^\mu,$$

which follows from a change of dummy index, the symmetry of the tensor $g_{\lambda\mu}$, and the fact that the ordinary product is independent of the order of the factors. The scalar σ will be called the *scalar product* of the vectors $\underset{1}{v}$ and $\underset{2}{v}$.

In the special case when

$$g_{\lambda\mu} \overset{*}{=} \delta_{\lambda\mu}{}^{(1)} \qquad (\lambda, \mu = 1, 2, \ldots, n),$$

then

$$|v| = \Big[\sum_{\lambda} (v^{\lambda})^2\Big]^{1/2}$$

and

$$\sigma[\underset{1}{v}, \underset{2}{v}] = \sum_{\lambda} \underset{1}{v}^{\lambda}\underset{2}{v}^{\lambda}$$

and in this case the general formulae for the length of a vector and for the scalar product of two vectors coincide with the formulae familiar from Cartesian Euclidean geometry.

The scalar $\sigma[\underset{1}{v}, \underset{2}{v}]$ defined by formula (3.97) has the property

$$\sigma[\alpha\underset{1}{v}, \beta\underset{2}{v}] = \alpha\beta\sigma[\underset{1}{v}, \underset{2}{v}], \tag{3.98}$$

since

$$\sigma[\alpha\underset{1}{v}, \beta\underset{2}{v}] = g_{\lambda\mu}\alpha\underset{1}{v}^{\lambda}\beta\underset{2}{v}^{\mu} = \alpha\beta g_{\lambda\mu}\underset{1}{v}^{\lambda}\underset{2}{v}^{\mu}.$$

We assert that the *scalar product defined by formula* (3.97) *has the property*

$$\sigma[\underset{1}{v}, \underset{2}{v}+\underset{3}{v}] = \sigma[\underset{1}{v}, \underset{2}{v}]+\sigma[\underset{1}{v}, \underset{3}{v}],$$

i.e., is distributive with respect to addition of vectors. This leads to the more general formula

$$\sigma\Big[\sum_{i=1}^{m} \underset{i}{v}, \sum_{j=1}^{l} \underset{j}{w}\Big] = \sum_{i=1}^{m}\sum_{j=1}^{l} \sigma[\underset{i}{v}, \underset{j}{w}]. \tag{3.99}$$

Indeed,

$$\sigma[\underset{1}{v}, \underset{2}{v}+\underset{3}{v}] = g_{\lambda\mu}\underset{1}{v}^{\lambda}(\underset{2}{v}^{\mu}+\underset{3}{v}^{\mu}) = g_{\lambda\mu}\underset{1}{v}^{\lambda}\underset{2}{v}^{\mu}+g_{\lambda\mu}\underset{1}{v}^{\lambda}\underset{3}{v}^{\mu} = \sigma[\underset{1}{v}, \underset{2}{v}]+\sigma[\underset{1}{v}, \underset{3}{v}].$$

The invariant

$$\sigma[\underset{1}{v}, \underset{2}{v}] = g_{\lambda\mu}\underset{1}{v}^{\lambda}\underset{2}{v}^{\mu}$$

can be rewritten in yet another, simpler form if the operation of raising

---

(1) The equality $g_{\lambda\mu} = \delta_{\lambda\mu}$ is not invariant for the general group G_1, but is invariant for the orthogonal subgroup G_m.

or lowering the indices is carried out. Since

$$g_{\lambda\mu}\underset{1}{v}^{\lambda} = \underset{1}{\bar{v}}_{\mu} \quad \text{and} \quad g_{\lambda\mu}\underset{2}{v}^{\mu} = \underset{2}{\bar{v}}_{\lambda},$$

it follows that

$$\sigma[\underset{1}{v}, \underset{2}{v}] = \underset{1}{\bar{v}}_{\mu}\underset{2}{v}^{\mu} = \underset{1}{v}^{\lambda}\underset{2}{\bar{v}}_{\lambda}$$

and it therefore turns out that the invariant σ was obtained by the transvection of two vectors, one covariant and one contravariant.

We now show that *when form* (3.94) *is positive definite, the real angle φ can be invariantly defined for every pair of non-zero vectors $\underset{1}{v}$ and $\underset{2}{v}$ by means of the formula*

$$\cos\varphi = \frac{\sigma[\underset{1}{v}, \underset{2}{v}]}{|\underset{1}{v}| \cdot |\underset{2}{v}|}. \tag{3.100}$$

In order to show this, we first note that since form (3.94) is positive definite and the vectors $\underset{i}{v}$ $(i = 1, 2)$ are non-zero, we have

$$|\underset{1}{v}| > 0, \quad |\underset{2}{v}| > 0,$$

so that the right-hand side of equation (3.100) has a finite numerical value. We must now show that the inequality

$$\frac{|\sigma[\underset{1}{v}, \underset{2}{v}]|}{|\underset{1}{v}| \cdot |\underset{2}{v}|} \leqslant 1 \tag{3.101}$$

holds, for only then will we obtain a real value for φ. Now consider the expression

$$F = \sigma[\alpha\underset{1}{v} + \beta\underset{2}{v}, \alpha\underset{1}{v} + \beta\underset{2}{v}],$$

where α, β denote arbitrary independent real variables. The inequality

$$\sigma[u, u] = g_{\lambda\mu}u^{\lambda}u^{\mu} \geqslant 0,$$

holds for all u^{λ} and hence for all numbers α, β we have

$$F = \sigma[\alpha\underset{1}{v} + \beta\underset{2}{v}, \alpha\underset{1}{v} + \beta\underset{2}{v}] \geqslant 0.$$

It further follows from formula (3.99) that

$$\begin{aligned} F &= \sigma[\alpha\underset{1}{v}, \alpha\underset{1}{v}] + \sigma[\alpha\underset{1}{v}, \beta\underset{2}{v}] + \sigma[\beta\underset{2}{v}, \alpha\underset{1}{v}] + \sigma[\beta\underset{2}{v}, \beta\underset{2}{v}] \\ &= \sigma[\alpha\underset{1}{v}, \alpha\underset{1}{v}] + 2\sigma[\alpha\underset{1}{v}, \beta\underset{2}{v}] + \sigma[\beta\underset{2}{v}, \beta\underset{2}{v}]. \end{aligned}$$

Moreover, by equation (3.98) we have

$$F = \alpha^2\sigma[\underset{1}{v}, \underset{1}{v}] + 2\alpha\beta\sigma[\underset{1}{v}, \underset{2}{v}] + \beta^2\sigma[\underset{2}{v}, \underset{2}{v}],$$

and since

$$\sigma[u, u] = |u|^2,$$

therefore

$$F = \alpha^2|\underset{1}{v}|^2 + 2\alpha\beta\sigma[\underset{1}{v}, \underset{2}{v}] + \beta^2|\underset{2}{v}|^2.$$

The form F, regarded as a quadratic form in the variables α, β, has a non-negative value for all α, β, and is therefore positive definite. Consequently, it follows from the elementary theory of quadratic forms that the determinant of this form must be non-positive, i.e. that

$$\sigma^2[\underset{1}{v}, \underset{2}{v}] - |\underset{1}{v}|^2|\underset{2}{v}|^2 \leqslant 0,$$

whence follows inequality (3.101), which we set out to prove.

We have thus shown that for any two non-zero vectors $\underset{1}{v}$ and $\underset{2}{v}$ an invariant (since both the numerator and denominator of the right-hand side of equation (3.100) are invariants) can be defined in terms of the fundamental tensor which can be taken as a measure of the angle between these vectors. This angle has the property that if each of the vectors is multiplied by any positive factor, the angle is unaltered. Indeed,

$$|\alpha\underset{1}{v}| = |\alpha| \cdot |\underset{1}{v}|, \qquad |\beta\underset{2}{v}| = |\beta| \cdot |\underset{2}{v}|,$$

$$\sigma[\alpha\underset{1}{v}, \beta\underset{2}{v}] = \alpha\beta\sigma[\underset{1}{v}, \underset{2}{v}],$$

and hence for $\alpha > 0$, $\beta > 0$ we have

$$\cos\varphi = \frac{\sigma[\alpha\underset{1}{v}, \beta\underset{2}{v}]}{|\alpha\underset{1}{v}| \cdot |\beta\underset{2}{v}|} = \frac{\alpha\beta\sigma[\underset{1}{v}, \underset{2}{v}]}{|\alpha| \cdot |\beta| \cdot |\underset{1}{v}| \cdot |\underset{2}{v}|} = \frac{\sigma[\underset{1}{v}, \underset{2}{v}]}{|\underset{1}{v}| \cdot |\underset{2}{v}|} = \cos\varphi.$$

When form (3.94) is not definite, formula (3.100) does not, of course, give a real angular measure for every pair of non-zero vectors.

A formula for $\sin^2\varphi$ can be obtained from equation (3.100) by making use of the Lagrange identity. The formula for $\sin^2\varphi$ can also be written as a biquadratic form: if we assume that the vectors $\underset{1}{v}^\nu$ and $\underset{2}{v}^\nu$ are already normalized, i.e. have length equal to 1, we have

$$\sin^2\varphi = 2g_{\lambda[\mu}g_{\omega]\nu}\underset{1}{v}^\lambda \underset{2}{v}^\omega \underset{1}{v}^\mu \underset{2}{v}^\nu. \tag{3.102}$$

The angular measure defined by formula (3.100) is additive in a sense. Consider the simple non-zero bivector

$$b^{\lambda\mu} = \underset{1}{v}^{[\lambda}\underset{2}{v}^{\mu]}$$

at the point p under consideration. At this point this bivector specifies a two-dimensional direction, i.e. the set of all bivectors linearly dependent on $b^{\lambda\mu}$ with a positive proportionality factor (in the special case of the space E_n, the bivector determines an oriented plane passing through the point p).

Now take three vectors $\underset{1}{w}^{\lambda}, \underset{2}{w}^{\mu}, \underset{3}{w}^{\lambda}$ belonging to this two-dimensional direction, i.e. such that the trivectors

$$\underset{1}{v}^{[\lambda}\underset{2}{v}^{\mu}\underset{i}{w}^{\nu]} = 0 \qquad (i = 1, 2, 3).$$

The property that the angular measure, as defined by formula (3.100) is additive, means that for each of the three vectors $\underset{i}{w}$ $(i = 1, 2, 3)$ belonging to the same two-dimensional direction we have

$$\varphi[\underset{1}{w}, \underset{2}{w}]+\varphi[\underset{2}{w}, \underset{3}{w}] = \varphi[\underset{1}{w}, \underset{3}{w}]$$

or

$$\varphi[\underset{1}{w}, \underset{2}{w}]+\varphi[\underset{2}{w}, \underset{3}{w}] = 2\pi-\varphi[\underset{1}{w}, \underset{3}{w}],$$

or else the relation

$$\varphi[\underset{1}{w}, \underset{2}{w}]-\varphi[\underset{2}{w}, \underset{3}{w}] = \varphi[\underset{1}{w}, \underset{3}{w}].$$

A detailed treatment of the principle of additivity (of Chasles) for angular measure can be found in a paper by Gołąb [113]. The book by Duschek–Mayer [23], gives a rigorous proof of the additivity of angular measures in Riemannian spaces, but makes use of complex numbers to do this. A proof not employing complex numbers, but utilizing the integral calculus, and therefore not purely algebraic, has been outlined in another paper by Gołąb [105]. In a paper written jointly with Bielecki [91] we have proved that among all Finsler spaces, Riemannian spaces are the most general in which angular measure is additive.

40. The Perpendicularity of Vectors. If a non-singular fundamental tensor $g_{\lambda\mu}$ is defined at a point p, the concept of the perpendicularity of two vectors $\underset{1}{v}$ and $\underset{2}{v}$ at that point can be defined by means of that tensor.

DEFINITION. Two non-zero vectors $\underset{1}{v}$ and $\underset{2}{v}$ are said to be *perpendicular* to each other if their scalar product is zero, i.e. if

$$g_{\lambda\mu}\underset{1}{v}^{\lambda}\underset{2}{v}^{\mu} = 0. \tag{3.103}$$

NOTE. The symmetry of the fundamental tensor implies that the perpendicularity relation possesses symmetry. If the form (3.94) is definite, a vector cannot be perpendicular to itself. If the form (3.94) is not definite, only singular vectors (lying on the cone (3.95)) are perpendicular to themselves.

41. The Unit Vector of a Vector. If a vector v^{λ} is non-singular, i.e. if

$$|v| > 0,$$

we can uniquely define a vector $\overset{*}{v}$, the *unit vector* of v, by means of the equality

$$\overset{*}{v} \overset{\text{def}}{=} \frac{v}{|v|}, \tag{3.104}$$

i.e., writing it out explicitly, the equality

$$\overset{*}{v}{}^{\lambda} = \frac{v^{\lambda}}{|g_{\nu\mu}v^{\nu}v^{\mu}|^{1/2}}.$$

The unit vector (3.104) is thus linearly dependent on the vector v and has the same sense, since the factor $\frac{1}{|v|}$ is positive, and, moreover, has length equal to 1; hence the name unit vector since

$$|\overset{*}{v}| = \left|\frac{1}{|v|}\right| \cdot |v| = \frac{|v|}{|v|} = 1.$$

The concept of a unit vector cannot be defined for null or singular vectors; this only arises in the case when the form (3.94) is not definite.

42. The Geometric Interpretation of the Tensor $g_{\lambda\mu}$. Now assume that our space is based on the group G_c and that in this space we are given the fundamental tensor $g_{\lambda\mu}$. The equation

$$g_{\lambda\mu}\xi^{\lambda}\xi^{\mu} = 1 \tag{3.105}$$

then defines a figure which is invariant under transformations belonging to the centred affine group (Fig. 3).

For $n = 2$, equation (3.105) represents a conic section; for $n = 3$, a quadric; for any arbitrary n, a hypersurface of the second degree, i.e. a hyperquadric. When the form (3.94) is definite if we want the hyperquadric to be real, we must consider the equation

$$g_{\lambda\mu}\xi^{\lambda}\xi^{\mu} = \operatorname{sgn} g_{11}, \tag{3.106}$$

rather than equation (3.105). For the non-definite form, equation (3.105) always defines a non-empty real figure. A hyperquadric with equation (3.106) or (3.105) can be regarded as the geometric image of the tensor $g_{\lambda\mu}$, since this tensor defines it uniquely.

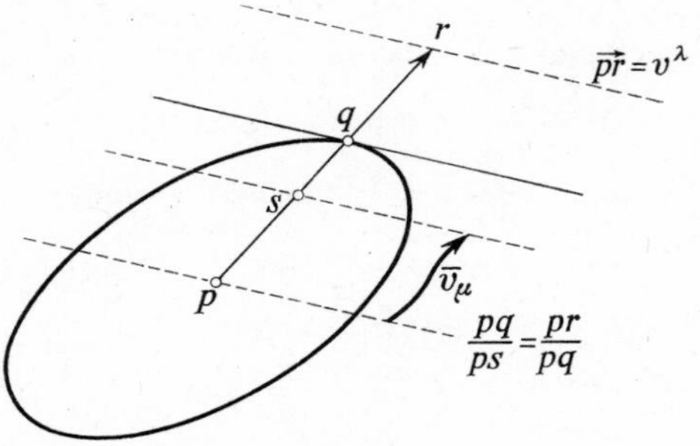

Fig. 3.

The geometric interpretation of raising and lowering indices will be as follows. Suppose that we are given a vector v^{λ} which we attach at the origin of the system. In this case, the hyperplane of the covariant vector

$$\overline{v}_{\mu} = v^{\lambda} g_{\lambda\mu}$$

has the direction of a hyperplane tangent to the hyperquadric at the point where the radius drawn through the vector v intersects the hyperquadric. Conversely, if we take a covariant vector w_{μ}, then to find the direction of the vector

$$w_{\lambda} g^{\lambda\mu}, \tag{3.107}$$

we must draw a hyperplane π tangent to hyperquadric (3.106), so that the contravariant vector of the first hyperplane passing through the origin, and the second hyperplane π are linearly dependent on w with a positive proportionality factor. The point of tangency will then determine the direction and sense of vector (3.107).

Just as the usual interpretation of contravariant and covariant vectors in the space E_n fails in a space based on the group G_1, so the interpretation above is not available in spaces more general than E_n.

One can only imagine that a local space E_n is "attached" at each point of the space X_n, as if it was tangent to X_n if X_n were embedded in a space of higher dimension; each allowable coordinate system in X_n will then induce a unique rectilinear coordinate system in all the local spaces E_n and $g_{\lambda\mu}$ can be interpreted geometrically in these local spaces E_n if a field of tensors $g_{\lambda\mu}$ is given in X_n. The space X_n is then called a *Riemannian space* based on the fundamental tensor $g_{\lambda\mu}$ and we denote it by the symbol V_n.

As we have seen, the fundamental tensor $g_{\lambda\mu}$, whose coordinate matrix is of rank n, determines a non-trivial density $\mathfrak{g} = |g_{\lambda\mu}|$ of weight $+2$

$$\mathfrak{g}' = J^{-2}\mathfrak{g}.$$

This is a W-density, and consequently the space cannot be endowed with (by means of $\mathfrak{g}$ at any rate) a local orientation determined by the ordinary density

$$\mathfrak{v}' = J\mathfrak{v}.$$

With the aid of the density $\mathfrak{g}$, however, it is possible to define volume measure by means of the integral

$$I \overset{\text{def}}{=} \int_{D(\lambda)} |\mathfrak{g}|^{1/2} d\xi^1 \ldots d\xi^n.$$

The metric tensor $g_{\lambda\mu}$ (non-singular) also makes possible the introduction of an invariant p-dimensional measure, where $1 \leqslant p \leqslant n-1$. For the time being, we prove this for the space E_n. Consider a p-dimensional hyperparallelepiped spanned by the vectors $\underset{1}{v}, \underset{2}{v}, \ldots, \underset{p}{v}$, and set

$$'g_{ab} = \underset{a}{v^\lambda}\underset{b}{v^\mu} g_{\lambda\mu} \qquad (a, b = 1, 2, \ldots, p). \tag{3.108}$$

Now write

$$'\mathfrak{g} \overset{\text{def}}{=} |'g_{ab}|. \tag{3.109}$$

The p-dimensional volume of this hyperparallelepiped is defined

$$V \overset{\text{def}}{=} |'\mathfrak{g}|^{1/2}. \tag{3.110}$$

Consider now the special case of $p = 2$ when the vectors $\underset{1}{v^\lambda}, \underset{2}{v^\lambda}$ determine a parallelogram. By the formula above, the surface measure of this parallelogram is

$$[(\underset{1}{v^\lambda}\underset{1}{v^\mu} g_{\lambda\mu})(\underset{2}{v^\lambda}\underset{2}{v^\mu} g_{\lambda\mu}) - (\underset{1}{v^\lambda}\underset{2}{v^\mu} g_{\lambda\mu})^2]^{1/2} = [|\underset{1}{v}|^2 |\underset{2}{v}|^2 - \sigma^2[\underset{1}{v}, \underset{2}{v}]]^{1/2}. \tag{3.111}$$

On the other hand, let us calculate $\sin\varphi$, where φ denotes the angle between the vectors $\underset{1}{v}$ and $\underset{2}{v}$. We have

$$\sin\varphi = [1-\cos^2\varphi]^{1/2} = \left[1 - \frac{\sigma^2[\underset{1}{v},\underset{2}{v}]}{|\underset{1}{v}|^2\,|\underset{2}{v}|^2}\right]^{1/2} = \frac{1}{|\underset{1}{v}|\cdot|\underset{2}{v}|}\left[|\underset{1}{v}|^2\,|\underset{2}{v}|^2 - \sigma^2[\underset{1}{v},\underset{2}{v}]\right]^{1/2}.$$

Accordingly

$$|\underset{1}{v}|\cdot|\underset{2}{v}|\sin\varphi = \left[|\underset{1}{v}|^2\,|\underset{2}{v}|^2 - \sigma^2[\underset{1}{v},\underset{2}{v}]\right]^{1/2},$$

so that, for $p = 2$ we have agreement with the familiar formula for the area of a parallelogram in Euclidean geometry.

43. Ricci *n*-vectors. A contravariant n-vector

$$\underset{(\lambda)}{E}^{\lambda_1\ldots\lambda_n} \overset{\text{def}}{=} n!\underset{1}{e}^{\lambda_1}\ldots\underset{n}{e}^{\lambda_n} \tag{3.112}$$

and a covariant n-vector

$$\overset{(\lambda)}{E}_{\lambda_1\ldots\lambda_n} \overset{\text{def}}{=} n!\overset{1}{e}_{[\lambda_1}\ldots\overset{n}{e}_{\lambda_n]} \tag{3.113}$$

are associated with every coordinate system (λ) at each point. These vectors, under the general group G_1, are not invariant, however, but change from system to system.

With the introduction of the fundamental tensor $g_{\lambda\mu}$, however, we are in a position to define invariantly two n-vectors, one contravariant and the other covariant (the so-called Ricci n-vectors); we do this by writing

$$e^{\lambda_1\ldots\lambda_n} \overset{\text{def}}{=} \frac{1}{\sqrt{|\mathfrak{g}|}}\varepsilon^{\lambda_1\ldots\lambda_n}, \qquad e_{\lambda_1\ldots\lambda_n} \overset{\text{def}}{=} \sqrt{|\mathfrak{g}|}\,\varepsilon_{\lambda_1\ldots\lambda_n}, \tag{3.114}$$

where ε is the familiar Ricci symbol (density) (cf. p. 83) and $\mathfrak{g}$ is given by (3.85). Strictly speaking, $e^{\lambda_1\ldots\lambda_n}$ and $e_{\lambda_1\ldots\lambda_n}$ are G-tensors; $\varepsilon^{\lambda_1\ldots\lambda_n}$ ($\varepsilon_{\lambda_1\ldots\lambda_n}$) is a density of weight $+1$ (-1), whereas $\sqrt{|\mathfrak{g}|}$ is a W-density of weight $+1$; accordingly, the transformation law is

$$e^{\lambda'_1\ldots\lambda'_n} = \frac{1}{\sqrt{|\mathfrak{g}'|}}\varepsilon^{\lambda'_1\ldots\lambda'_n} = \frac{1}{\sqrt{|\mathfrak{g}|}\,|J|^{-1}}J^{-1}\varepsilon^{\lambda_1\ldots\lambda_n}A^{\lambda'_1\ldots\lambda'_n}_{\lambda_1\ldots\lambda_n}$$

$$= \operatorname{sgn}J\cdot A^{\lambda'_1\ldots\lambda'_n}_{\lambda_1\ldots\lambda_n}e^{\lambda_1\ldots\lambda_n} \tag{3.115}$$

and, similarly,

$$e_{\lambda'_1\ldots\lambda'_n} = \operatorname{sgn}J\cdot A^{\lambda_1\ldots\lambda_n}_{\lambda'_1\ldots\lambda'_n}e_{\lambda_1\ldots\lambda_n}. \tag{3.116}$$

For the Ricci n-vectors (3.114), the following identity can be established

$$e^{\varrho_1 \ldots \varrho_m \lambda_{m+1} \ldots \lambda_n} e_{\varrho_1 \ldots \varrho_m \mu_{m+1} \ldots \mu_n} = m!(n-m)! A^{[\lambda_{m+1} \ldots \lambda_n]}_{[\mu_{m+1} \ldots \mu_n]}. \qquad (3.117)$$

Similar identities can be obtained with the same right-hand sides, when e is replaced by ε, or by E, on the left-hand side of formula (3.117). It may also be shown that

$$e^{\lambda_1 \ldots \lambda_n} = (\operatorname{sgn} \mathfrak{g}) g^{\lambda_1 \mu_1} \ldots g^{\lambda_n \mu_n} e_{\mu_1 \ldots \mu_n}. \qquad (3.118)$$

The n-vectors (3.114) can be used to set up certain correspondences between contravariant p-vectors and covariant $(n-p)$-vectors at the same point.

Suppose we are given a contravariant p-vector

$$v^{\lambda_1 \ldots \lambda_p} = v^{[\lambda_1 \ldots \lambda_p]},$$

where $1 \leqslant p \leqslant n-1$. We now set

$$\overset{*}{v}_{\mu_{p+1} \ldots \mu_n} \overset{\text{def}}{=} v^{\lambda_1 \ldots \lambda_p} e_{\lambda_1 \ldots \lambda_p \mu_{p+1} \ldots \mu_n}. \qquad (3.119)$$

Thus $\overset{*}{v}$ is a covariant $(n-p)$-vector uniquely associated with the p-vector v and it is for this reason that we have used the same kernel letter and merely added an asterisk.

Similarly, a contravariant $(n-p)$-vector can be assigned uniquely to every covariant p-vector

$$w_{\lambda_1 \ldots \lambda_p} = w_{[\lambda_1 \ldots \lambda_p]}$$

by means of the formula

$$\overset{*}{w}^{\mu_{p+1} \ldots \mu_n} \overset{\text{def}}{=} w_{\lambda_1 \ldots \lambda_p} e^{\lambda_1 \ldots \lambda_p \mu_{p+1} \ldots \mu_n}. \qquad (3.120)$$

44. Vector Products. In particular, under this correspondence, a vector always corresponds to an $(n-1)$-vector.

We now consider the sequence of vectors

$$\underset{i}{v}^{\lambda} \quad (i = 1, 2, \ldots, n-1) \qquad (3.121)$$

and use it to construct the $(n-1)$-vector

$$v^{\lambda_1 \ldots \lambda_{n-1}} \overset{\text{def}}{=} \underset{1}{v}^{[\lambda_1} \ldots \underset{n-1}{v}^{\lambda_{n-1}]}. \qquad (3.122)$$

Consider the vector $\overset{*}{v}_{\mu}$ which is associated with it by formula (3.119)

$$\overset{*}{v}_{\mu} = v^{\lambda_1 \ldots \lambda_{n-1}} e_{\lambda_1 \ldots \lambda_{n-1} \mu}.$$

By raising the index, the convariant vector v can in turn be uniquely assigned the contravariant vector

$$\bar{\overset{*}{v}}{}^{\lambda} = g^{\lambda\mu}\overset{*}{v}_{\mu}.$$

We thus have a contravariant vector uniquely associated with the $(n-1)$-vector (3.122), viz.

$$\bar{\overset{*}{v}}{}^{\lambda} = g^{\lambda\mu} v^{\lambda_1 \dots \lambda_{n-1}} e_{\lambda_1 \dots \lambda_{n-1}\mu}. \tag{3.123}$$

This vector will be referred to as the *vector product* of the sequence of vectors (3.121).

The vector product is also sometimes denoted by the symbol

$$[\underset{1}{v}, \dots, \underset{n-1}{v}]. \tag{3.124}$$

Note that the vector product defined above has the following properties:

1. If the vectors $\underset{i}{v}{}^{\lambda}$ are linearly dependent the vector product is a null vector. This is because the $(n-1)$-vector (3.122) is then a null vector.

2. If any one of the vectors in sequence (3.121) is multiplied by a scalar factor α, the vector product is also multiplied by α.

3. If in sequence (3.121) we permute the vectors $\underset{i}{v}$, the vector product will, or will not, change sign according as the permutation is even or odd.

4. If the space is Euclidean, then length of the vector product represents the volume of a hyperparallelepiped spanned by the vectors $\underset{1}{v}, \dots, \underset{n-1}{v}$.

We prove this for $n = 3$. The volume V of this parallelepiped is by definition

$$V = |\det(\underset{a}{v}{}^{\lambda} \underset{b}{v}{}^{\mu} g_{\lambda\mu})|^{1/2} \qquad (a, b = 1, 2),$$

which, by formula (3.111), is equal to

$$(|\underset{1}{v}|^2 \, |\underset{2}{v}|^2 - \sigma^2 [\underset{1}{v}, \underset{2}{v}])^{1/2}.$$

On the other hand,

$$\bar{\overset{*}{v}}{}^{\lambda} = g^{\lambda\mu} \underset{1}{v}{}^{[\varrho} \underset{2}{v}{}^{\sigma]} e_{\varrho\sigma\mu} = \tfrac{1}{2} g^{\lambda\mu} e_{\varrho\sigma\mu} (\underset{1}{v}{}^{\varrho} \underset{2}{v}{}^{\sigma} - \underset{1}{v}{}^{\sigma} \underset{2}{v}{}^{\varrho}),$$

whence

$$\begin{aligned} |\bar{\overset{*}{v}}{}^{\lambda}|^2 &= g_{\lambda\mu} \bar{\overset{*}{v}}{}^{\lambda} \bar{\overset{*}{v}}{}^{\mu} \\ &= \tfrac{1}{4} g_{\lambda\mu} g^{\lambda\omega} e_{\varrho\sigma\omega} (\underset{1}{v}{}^{\varrho} \underset{2}{v}{}^{\sigma} - \underset{1}{v}{}^{\sigma} \underset{2}{v}{}^{\varrho}) g^{\mu\tau} e_{\alpha\beta\tau} (\underset{1}{v}{}^{\alpha} \underset{2}{v}{}^{\beta} - \underset{1}{v}{}^{\beta} \underset{2}{v}{}^{\alpha}) \end{aligned}$$

$$= \tfrac{1}{4} g^{\omega\tau} e_{\varrho\sigma\omega} e_{\alpha\beta\tau} (\underset{1}{v}^{\varrho} \underset{2}{v}^{\sigma} - \underset{1}{v}^{\sigma} \underset{2}{v}^{\varrho}) (\underset{1}{v}^{\alpha} \underset{2}{v}^{\beta} - \underset{1}{v}^{\beta} \underset{2}{v}^{\alpha})$$

$$= \tfrac{1}{4} g^{\omega\tau} e_{\varrho\sigma\omega} e^{\gamma\delta\eta} g_{\gamma\alpha} g_{\delta\beta} g_{\eta\tau} (\underset{1}{v}^{\varrho} \underset{2}{v}^{\sigma} - \underset{1}{v}^{\sigma} \underset{2}{v}^{\varrho}) (\underset{1}{v}^{\alpha} \underset{2}{v}^{\beta} - \underset{1}{v}^{\beta} \underset{2}{v}^{\alpha})$$

$$= \tfrac{1}{4} A_{\eta}^{\omega} e_{\varrho\sigma\omega} e^{\gamma\delta\eta} (\underset{1}{v}^{\varrho} \underset{2}{v}^{\sigma} - \underset{1}{v}^{\sigma} \underset{2}{v}^{\varrho}) (\underset{1}{v}_{\gamma} \underset{2}{v}_{\delta} - \underset{1}{v}_{\delta} \underset{2}{v}_{\gamma})$$

$$= \tfrac{1}{2} A_{[\varrho\sigma]}^{[\gamma\delta]} (\underset{1}{v}^{\varrho} \underset{2}{v}^{\sigma} - \underset{1}{v}^{\sigma} \underset{2}{v}^{\varrho}) (\underset{1}{v}_{\gamma} \underset{2}{v}_{\delta} - \underset{1}{v}_{\delta} \underset{2}{v}_{\gamma})$$

$$= \tfrac{1}{2} A_{\varrho\sigma}^{\gamma\delta} (\underset{1}{v}^{\varrho} \underset{2}{v}^{\sigma} - \underset{1}{v}^{\sigma} \underset{2}{v}^{\varrho}) (\underset{1}{v}_{\gamma} \underset{2}{v}_{\delta} - \underset{1}{v}_{\delta} \underset{2}{v}_{\gamma})$$

$$= \tfrac{1}{2} (\underset{1}{v}^{\gamma} \underset{2}{v}^{\delta} - \underset{1}{v}^{\delta} \underset{2}{v}^{\gamma}) (\underset{1}{v}^{\alpha} \underset{2}{v}^{\beta} - \underset{1}{v}^{\beta} \underset{2}{v}^{\alpha}) g_{\alpha\gamma} g_{\beta\delta}$$

$$= \tfrac{1}{2} [g_{\alpha\gamma} \underset{1}{v}^{\gamma} \underset{1}{v}^{\alpha} g_{\beta\delta} \underset{2}{v}^{\beta} \underset{2}{v}^{\delta} + g_{\beta\delta} \underset{1}{v}^{\delta} \underset{1}{v}^{\beta} g_{\alpha\gamma} \underset{2}{v}^{\gamma} \underset{2}{v}^{\alpha} - g_{\alpha\gamma} \underset{2}{v}^{\alpha} \underset{1}{v}^{\gamma} g_{\beta\delta} \underset{2}{v}^{\delta} \underset{1}{v}^{\beta} -$$
$$- g_{\alpha\gamma} (\underset{1}{v}^{\alpha} \underset{2}{v}^{\gamma} g_{\beta\delta} \underset{2}{v}^{\beta} \underset{1}{v}^{\delta})]$$

$$= \tfrac{1}{2} (|\underset{1}{v}|^2 |\underset{2}{v}|^2 + |\underset{1}{v}|^2 |\underset{2}{v}|^2 - \sigma^2 [\underset{1}{v}, \underset{2}{v}] - \sigma^2 [\underset{1}{v}, \underset{2}{v}])$$

$$= |\underset{1}{v}|^2 |\underset{2}{v}|^2 - \sigma^2 [\underset{1}{v}, \underset{2}{v}],$$

which is what we had to prove.

5. The sense of the vector product $[\underset{1}{v}, \ldots, \underset{n-1}{v}]$ is such that the orientation of the space, assuming now that $\underset{1}{v}, \ldots, \underset{n-1}{v}$ are linearly independent, by means of the vector sequence $\underset{1}{v}^{\lambda}, \underset{2}{v}^{\lambda}, \ldots, \underset{n-1}{v}^{\lambda}, [\underset{1}{v}, \ldots, \underset{n-1}{v}]$ agrees with the orientation determined by the coordinate system (λ).

To show this, it suffices to prove that the determinant

$$\mathfrak{w} = |\underset{1}{v}^{\lambda}, \underset{2}{v}^{\lambda}, \ldots, \underset{n-1}{v}^{\lambda}, \bar{\overset{*}{v}}^{\lambda}|$$

has a positive value.

The calculation may be carried out directly, or one may alternatively reason as follows. Suppose we have chosen the coordinate system so that

$$\underset{i}{v}^{\lambda} \overset{*}{=} \delta_i^{\lambda} \qquad (i = 1, 2, \ldots, n-1;\ \lambda = 1, 2, \ldots, n).$$

The value of the determinant $\mathfrak{w}$ is then most easily obtained by expanding with respect to the elements of the last row, in which the n-th coordinates of all the individual vectors appear. We then have

$$\mathfrak{w} = \bar{\overset{*}{v}}^{n} = g^{n\lambda} \underset{1}{v}^{[\lambda_1} \ldots \underset{n-1}{v}^{\lambda_{n-1}]} e_{\lambda_1 \ldots \lambda_{n-1} \lambda}.$$

By virtue of the antisymmetry of the tensor e, we can when summing omit the brackets in the products $\underset{1}{v}^{[\lambda_1} \dots \underset{n-1}{v}^{\lambda_{n-1}]}$ to get

$$\mathfrak{w} = g^{n\lambda}\underset{1}{v}^{\lambda_1} \dots \underset{n-1}{v}^{\lambda_{n-1}} e_{\lambda_1\dots\lambda_{n-1}\lambda} \overset{*}{=} g^{n\lambda}\delta^{\lambda_1\dots\lambda_{n-1}}_{1\dots n-1} e_{\lambda_1\dots\lambda_{n-1}\lambda} = g^{n\lambda}_{\cdot} e_{1\dots n-1,\lambda}$$

$$= g^{nn}\sqrt{|\mathfrak{g}|}\,.$$

Now

$$g^{nn} = \frac{1}{\mathfrak{g}}\operatorname{minor} g_{nn},$$

and so

$$\mathfrak{w} = \frac{\operatorname{minor} g_{nn}}{\sqrt{|\mathfrak{g}|}}\operatorname{sgn}\mathfrak{g}$$

and since $\mathfrak{g} > 0$ and the minor $g_{nn} > 0$, which follows from the fact that the form with coefficients $g_{\lambda\mu}$ is real, positive, and definite, we have $\mathfrak{w} > 0$. In coordinate systems (λ') with the same orientation as (λ) (i.e. when $J > 0$), the determinant $\mathfrak{w}' = J\mathfrak{w} > 0$, and our assertion is proved.

6. The vector product $[\underset{1}{v}, \dots, \underset{n-1}{v}]$ is perpendicular to each vector $\underset{i}{v}$ of the sequence (3.121).

Indeed, if we construct the appropriate scalar product we have

$$\sigma[\overset{*}{v}, \underset{i}{v}] = g_{\varrho\sigma}\overset{*}{v}^{\varrho}\underset{i}{v}^{\sigma} = g_{\varrho\sigma}g^{\varrho\mu}e_{\lambda_1\dots\lambda_{n-1}\mu}\underset{1}{v}^{[\lambda_1} \dots \underset{n-1}{v}^{\lambda_{n-1}]}\underset{i}{v}^{\sigma}.$$

$$= A^{\mu}_{\sigma}e_{\lambda_1\dots\lambda_{n-1}\mu}\underset{1}{v}^{[\lambda_1} \dots \underset{n-1}{v}^{\lambda_{n-1}]}\underset{i}{v}^{\sigma}$$

$$= e_{\lambda_1\dots\lambda_{n-1}\sigma}\underset{1}{v}^{[\lambda_1} \dots \underset{n-1}{v}^{\lambda_{n-1}]}\underset{i}{v}^{\sigma} = e_{\lambda_1\dots\lambda_n}\underset{1}{v}^{[\lambda_1} \dots \underset{n-1}{v}^{\lambda_{n-1}]}\underset{i}{v}^{\lambda_n}$$

$$= e_{\lambda_1\dots\lambda_n}\underset{1}{v}^{[[\lambda_1} \dots \underset{n-1}{v}^{\lambda_{n-1}]}\underset{i}{v}^{\lambda_n]} = e_{\lambda_1\dots\lambda_n}\underset{1}{v}^{[\lambda_1} \dots \underset{n-1}{v}^{\lambda_{n-1}}\underset{i}{v}^{\lambda_n]}.$$

Now the n-vector $\underset{1}{v}^{[\lambda_1} \dots \underset{n-1}{v}^{\lambda_{n-1}}\underset{i}{v}^{\lambda_n]}$ is a null vector if it is constructed from the vectors $\underset{1}{v}, \dots, \underset{i}{v}, \dots, \underset{n-1}{v}, \underset{i}{v}$ which are linearly dependent.

7. In the special case, when $n = 3$, when the space is Euclidean (with the group G_m), and when (λ) is a rectangular Cartesian system of coordinates, the definition of the vector product (3.123) coincides with the classical definition.

Indeed, the assumption that (λ) is a system of rectangular Cartesian coordinates means that

$$g_{\lambda\mu} \overset{*}{=} \delta_{\lambda\mu}, \quad g^{\lambda\mu} \overset{*}{=} \delta^{\lambda\mu}.$$

Hence

$$\mathfrak{g} \overset{*}{=} 1$$

and

$$e_{\lambda\mu\nu} \overset{*}{=} \{0, \text{ or } 1, \text{ or } -1\},$$

depending on the order in which the indices λ, μ, ν appear.

Consequently,

$$\bar{\overset{*}{v}} \overset{*}{=} \delta^{\lambda\mu} e_{\varrho\sigma\mu} \underset{1}{v}^{[\varrho} \underset{2}{v}^{\sigma]} \overset{*}{=} e_{\varrho\sigma\lambda} \underset{1}{v}^{[\varrho} \underset{2}{v}^{\sigma]} \overset{*}{=} \tfrac{1}{2} e_{\varrho\sigma\lambda} (\underset{1}{v}^{\varrho} \underset{2}{v}^{\sigma} - \underset{1}{v}^{\sigma} \underset{2}{v}^{\varrho}).$$

Thus, in turn, we have

$$\begin{aligned}
\bar{\overset{*}{v}}^1 &\overset{*}{=} \tfrac{1}{2} e_{\varrho\sigma 1} \{\underset{1}{v}^{\varrho} \underset{2}{v}^{\sigma} - \underset{1}{v}^{\sigma} \underset{2}{v}^{\varrho}\} \\
&= \tfrac{1}{2} e_{231} \{\underset{1}{v}^2 \underset{2}{v}^3 - \underset{1}{v}^3 \underset{2}{v}^2\} + \tfrac{1}{2} e_{321} \{\underset{1}{v}^3 \underset{2}{v}^2 - \underset{1}{v}^2 \underset{2}{v}^3\} \\
&= \tfrac{1}{2} \{\underset{1}{v}^2 \underset{2}{v}^3 - \underset{1}{v}^3 \underset{2}{v}^2\} - \tfrac{1}{2} \{\underset{1}{v}^3 \underset{2}{v}^2 - \underset{1}{v}^2 \underset{2}{v}^3\} \\
&= \underset{1}{v}^2 \underset{2}{v}^3 - \underset{1}{v}^3 \underset{2}{v}^2 = \begin{vmatrix} \underset{1}{v}^2 & \underset{1}{v}^3 \\ \underset{2}{v}^2 & \underset{2}{v}^3 \end{vmatrix}.
\end{aligned}$$

Similarly, we obtain

$$\bar{\overset{*}{v}}^2 = \begin{vmatrix} \underset{1}{v}^3 & \underset{1}{v}^1 \\ \underset{2}{v}^3 & \underset{2}{v}^2 \end{vmatrix}$$

and, finally

$$\bar{\overset{*}{v}}^3 = \begin{vmatrix} \underset{1}{v}^1 & \underset{1}{v}^2 \\ \underset{2}{v}^1 & \underset{2}{v}^2 \end{vmatrix},$$

which is in agreement with the classical formulae for the coordinates of the vector product.

In n-dimensional Euclidean space we have, in Cartesian coordinates, the equations

$$\bar{\overset{*}{v}}^{\lambda} \overset{*}{=} \pm M^{\lambda}$$

(M^{λ} denotes the determinant constructed from the matrix $[\underset{i}{v}^{\lambda}]$ by deleting the λ-th row) with the appropriate sign $\pm$.

We have seen that the fundamental tensor enables each contravariant (covariant) p-vector to be assigned a particular covariant (contravariant) $(n-p)$-vector in an invariant manner.

In a space based on the group G_1 somewhat less can be achieved using the Ricci symbols which, as is now familiar, are

$$\begin{aligned} \overset{*}{\mathfrak{v}}_{\mu_{p+1}\ldots\mu_n} &\overset{\text{def}}{=} v^{\lambda_1\ldots\lambda_p}\varepsilon_{\lambda_1\ldots\lambda_p\mu_{p+1}\ldots\mu_n}, \\ \overset{*}{\mathfrak{v}}^{\mu_{p+1}\ldots\mu_n} &\overset{\text{def}}{=} v_{\lambda_1\ldots\lambda_p}\varepsilon^{\lambda_1\ldots\lambda_p\mu_{p+1}\ldots\mu_n}. \end{aligned} \tag{3.125}$$

We associate invariantly with each contravariant p-vector a covariant $(n-p)$-vector density (of weight -1) and with each covariant p-vector, we associate a contravariant $(n-p)$-vector density (of density $+1$).

The n-vectors E defined on p. 116 can, it is true, be used to associate $(n-p)$-vectors with p-vectors by means of the equations

$$\begin{aligned} \overset{*}{v}_{\mu_{p+1}\ldots\mu_n} &= \frac{1}{(n-p)!}\overset{(\lambda)}{E}_{\lambda_1\ldots\lambda_p\mu_{p+1}\ldots\mu_n} v^{\lambda_1\ldots\lambda_p}, \\ \overset{*}{v}^{\mu_{p+1}\ldots\mu_n} &= \frac{1}{p!}\underset{(\lambda)}{E}^{\lambda_1\ldots\lambda_p\mu_{p+1}\ldots\mu_n} v_{\lambda_1\ldots\lambda_p} \end{aligned} \tag{3.126}$$

but this association is not invariant since the n-vectors $\overset{(\lambda)}{E}$, $\underset{(\lambda)}{E}$ depend on the choice of coordinate system, as the symbols themselves indicate.

45. The Canonical Form of the Fundamental Tensor. The theory of quadratic forms tells us that a real quadratic form

$$g_{\lambda\mu}x^\lambda x^\mu \tag{3.127}$$

can be put, by means of the non-singular affine transformation

$$x^{\lambda'} = \alpha_\lambda^{\lambda'} x^\lambda, \qquad |\alpha_\lambda^{\lambda'}| \neq 0$$

into the *canonical form*

$$g_{\lambda'\mu'}x^{\lambda'}x^{\mu'}$$

This canonical form has the properties

$$\begin{aligned} g_{\lambda'\mu'} &= 0 \quad \text{for} \quad \lambda' \neq \mu', \\ g_{\lambda'\lambda'} &= \pm 1 \quad \text{or} \quad 0, \end{aligned}$$

where, if the rank r of the matrix $[g_{\lambda\mu}]$ has the highest possible value, i.e. if it is equal to n, then all the $g_{\lambda'\lambda'}$'s are equal to ± 1. By suitably numbering the variables $x^{\lambda'}$, replacing the variables $x^{\lambda'}$ with y^λ and correspondingly altering the coefficients $g_{\lambda'\lambda'}$ to $h_{\lambda\mu}$, we can always arrange it so that

$$h_{\lambda\lambda} = \begin{cases} -1 & \text{for } \lambda = 1, \ldots, s, \\ 1 & \text{for } \lambda = s+1, \ldots, r, \\ 0 & \text{for } \lambda = r+1, \ldots, n. \end{cases} \tag{3.128}$$

The matrix $[h_{\lambda\mu}]$ is thus of the form

$$\begin{matrix} & \overleftarrow{\;s\;}\!\!\!\!\!\rightarrow & & \overleftarrow{n-r}\!\!\!\!\!\rightarrow \\ \begin{matrix} \\ \\ \uparrow \\ r-s \\ \downarrow \\ \\ \\ \end{matrix} & \left[\begin{matrix} -1 \ldots & 1 & 0 \ldots & 0\ 0 \ldots 0 \\ 0 \ldots & -1 & 0 \ldots & 0\ 0 \ldots 0 \\ 0 \ldots & & +1 \ldots & 0\ 0 \ldots 0 \\ 0 \ldots & & & +1\ 0 \ldots 0 \\ 0 \ldots & & & 0 \ldots 0 \\ 0 \ldots & & & 0 \end{matrix}\right] \end{matrix} \tag{3.129}$$

The sequence of signs $-, -, \ldots -, + \ldots +$, where there are s minus signs and $r-s$ plus signs, is called the *signature* of the quadratic form and the number s is known as the index of the form.

Given a quadratic form with coefficients $g_{\lambda\mu}$, we ask how it can be put into canonical form, i.e. how to find a tensor $P^{\lambda}{}_{\mu}$ with the property that the $h_{\lambda\mu}$ defined by

$$h_{\lambda\mu} \overset{\text{def}}{=} P^{\varrho}{}_{\lambda} P^{\sigma}{}_{\mu} g_{\varrho\sigma}$$

have a matrix of the form (3.129). This problem leads to the solution of the *characteristic equation* (of degree n)[1]

$$|g_{\lambda\mu} - \delta_{\lambda\mu}\zeta| = 0; \tag{3.130}$$

the solution to this problem is provided by the theory of elementary divisors originated by K. Weierstrass.

If the form (3.127) is definite, then $r = n$ and the canonical form is

$$\pm h_{\lambda\lambda}\, y^{\lambda} y^{\lambda},$$

where the sign (+ or −) depends on what the signature is. In the canonical system, the formula for the length of a vector v with coordinates v^{λ} reduces to

$$|v| = |h_{\lambda\mu} v^{\lambda} v^{\mu}|^{1/2} \overset{*}{=} |h_{\lambda\lambda} v^{\lambda} v^{\lambda}|^{1/2} \overset{*}{=} \Big| -\sum_{1}^{s} (v^{\lambda})^2 + \sum_{s+1}^{n-r} (v^{\mu})^2 \Big|^{1/2}.$$

---

[1] This equation is sometimes called the *secular equation*.

In the case of a definite form (3.127), however, this formula takes on the simplest possible form:

$$|v| \stackrel{*}{=} \Big[\sum_{1}^{n} (v^{\lambda})^2\Big]^{1/2}, \tag{3.131}$$

which is familiar form Euclidean geometry. If we take the fundamental vectors $\underset{\lambda}{e}^{\nu}$, we find that their lengths are

$$|\underset{\lambda}{e}| = |g_{\varrho\sigma}\underset{\lambda}{e}^{\varrho}\underset{\lambda}{e}^{\sigma}|^{1/2} = |g_{\lambda\lambda}|^{1/2}.$$

In the case of a definite form, we have for the fundamental vectors of the canonical system the relations

$$|\underset{\lambda}{e}| \stackrel{*}{=} 1 \qquad (\lambda = 1, 2, \ldots, n), \tag{3.132}$$

so that, in this case the fundamental vectors are also unit vectors (i.e. of length 1).

We now take any two arbitrary, but different, fundamental vectors $\underset{\lambda}{e}, \underset{\mu}{e}$ $(\lambda \neq \mu)$ of the canonical system with a definite form of the metric tensor, and calculate their scalar product. We have

$$\sigma[\underset{\lambda}{e}, \underset{\mu}{e}] = g_{\varrho\sigma}\underset{\lambda}{e}^{\varrho}\underset{\mu}{e}^{\sigma} \stackrel{*}{=} g_{\varrho\sigma}\delta^{\varrho}_{\lambda}\delta^{\sigma}_{\mu} = g_{\lambda\mu} \stackrel{*}{=} \delta_{\lambda\mu} = 0,$$

and so these vectors are perpendicular. We therefore have the following result: for a definite form of the metric tensor the set of fundamental vectors $\underset{\lambda}{e}^{\nu}$ $(\lambda = 1, 2, \ldots, n)$ belonging to the canonical system constitutes a set of unit vectors which are mutually orthogonal or, as we shall say in future, an *orthonormal* set.

46. The Eigenvalues of a Tensor. Given a mixed tensor

$$T^{\nu}_{\lambda}, \tag{3.133}$$

we study the equation

$$T^{\nu}_{\lambda}x^{\lambda} = \varrho x^{\nu} \tag{3.134}$$

with an unknown scalar ϱ and an unknown vector x^{ν}. If there exists a scalar ϱ_0 and a vector $\underset{0}{x}^{\nu}$ such that relation (3.134) is satisfied, ϱ_0 is called an *eigenvalue of tensor* (3.133) and $\underset{0}{x}^{\nu}$ an *eigenvector of tensor* (3.133). In order to find all the eigenvalues we have only to solve the equation

$$|T^{\nu}_{\lambda} - \varrho A^{\nu}_{\lambda}| = 0, \tag{3.135}$$

which is an n-th degree equation in the unknown ϱ, and therefore has at most n roots.

It the quadratic form belonging to the metric tensor is definite and if the tensor

$$\overline{T}_{\mu\lambda} = g_{\mu\lambda} T^{\nu}{}_{\lambda}$$

is symmetric, (3.135) is equivalent to the equation

$$|\overline{T}_{\mu\lambda} - \varrho g_{\mu\lambda}| = 0, \tag{3.136}$$

which has real roots only ([53], Part I, p. 260).

We now show that eigenvectors associated with different eigenvalues are perpendicular to each other.

Let ϱ_1 and ϱ_2 be different roots of equation (3.136). Let $\underset{1}{v}^{\lambda}$ and $\underset{2}{v}^{\lambda}$ denote eigenvectors belonging to the eigenvalues ϱ_1 and ϱ_2. We then have

$$T^{\nu}_{\lambda} \underset{1}{v}^{\lambda} = \varrho_1 \underset{1}{v}^{\nu}, \qquad T^{\nu}_{\lambda} \underset{2}{v}^{\lambda} = \varrho_2 \underset{2}{v}^{\nu}.$$

We also have

$$\begin{aligned}\overline{T}_{\mu\lambda} \underset{1}{v}^{\mu} \underset{2}{v}^{\lambda} &= g_{\mu\nu} T^{\nu}_{\lambda} \underset{1}{v}^{\mu} \underset{2}{v}^{\lambda} = g_{\mu\nu} \underset{1}{v}^{\mu} T^{\nu}_{\lambda} \underset{2}{v}^{\lambda} \\ &= g_{\mu\nu} \underset{1}{v}^{\mu} \varrho_2 \underset{2}{v}^{\nu} = \varrho_2 g_{\mu\nu} \underset{1}{v}^{\mu} \underset{2}{v}^{\nu}.\end{aligned}$$

On the other hand, using the assumption that the tensor $\overline{T}$ is symmetric we find

$$\begin{aligned}\overline{T}_{\mu\lambda} \underset{1}{v}^{\mu} \underset{2}{v}^{\lambda} &= \overline{T}_{\lambda\mu} \underset{1}{v}^{\mu} \underset{2}{v}^{\lambda} = g_{\lambda\nu} T^{\nu}_{\mu} \underset{1}{v}^{\mu} \underset{2}{v}^{\lambda} \\ &= g_{\lambda\nu} \underset{2}{v}^{\lambda} T^{\nu}_{\mu} \underset{1}{v}^{\mu} = g_{\lambda\nu} \underset{2}{v}^{\lambda} \varrho_1 \underset{1}{v}^{\nu} \\ &= \varrho_1 g_{\lambda\nu} \underset{1}{v}^{\nu} \underset{2}{v}^{\lambda} = \varrho_1 g_{\nu\lambda} \underset{1}{v}^{\nu} \underset{2}{v}^{\lambda} \\ &= \varrho_1 g_{\mu\nu} \underset{1}{v}^{\mu} \underset{2}{v}^{\nu}.\end{aligned}$$

Comparing the two results we see that

$$\varrho_2 g_{\mu\nu} \underset{1}{v}^{\mu} \underset{2}{v}^{\nu} = \varrho_1 g_{\mu\nu} \underset{1}{v}^{\mu} \underset{2}{v}^{\nu},$$

or, equivalently,

$$(\varrho_2 - \varrho_1) g_{\mu\nu} \underset{1}{v}^{\mu} \underset{2}{v}^{\nu} = 0,$$

whence it follows, since $\varrho_1 \neq \varrho_2$, that

$$g_{\mu\nu} \underset{1}{v}^{\mu} \underset{2}{v}^{\nu} = 0,$$

which shows that the eigenvectors $\underset{1}{v}$ and $\underset{2}{v}$ are perpendicular to each other.

Moreover, we have the relation

$$\overline{T}_{\dot{\lambda}\mu}\underset{1}{v}^{\lambda}\underset{2}{v}^{\mu} = 0.$$

In fact, even when the characteristic equation (3.136) has multiple roots, it can still be shown that there always exists an orthogonal system of vectors $\underset{i}{v}$ $(i = 1, 2, \ldots, n)$ with the property that

$$\overline{T}_{\dot{\lambda}\mu}\underset{i}{v}^{\lambda}\underset{j}{v}^{\mu} = 0 \qquad (i, j = 1, 2, \ldots, n;\ i \neq j);$$

taking the unit vectors

$$\underset{k}{i} = \frac{1}{|\underset{k}{v}|}\underset{k}{v} \qquad (k = 1, 2, \ldots, n)$$

of this system, we obtain an orthogonal system with the property that

$$\overline{T}_{\dot{i}j} \overset{*}{=} \overline{T}_{\dot{\lambda}\mu}\underset{i}{i}^{\lambda}\underset{j}{i}^{\mu} = 0 \qquad (i, j = 1, 2, \ldots, n;\ i \neq j), \tag{3.137}$$

and thus for every tensor T there exists an orthogonal system, for which the fundamental vectors are mutually perpendicular unit vectors, in which all the coordinates of the tensor with different indices are equal to zero.

The assumption that the tensor $\overline{T}$ is symmetric is essential in the above theorem. This follows from the fact that symmetry is known to be an invariant property, and equations (3.137) assert in perticular that the tensor $\overline{T}$ is symmetric in the canonical system.

CHAPTER IV

A SUPPLEMENT TO TENSOR ALGEBRA

47. Restriction of Groups and Type of Quantity. As stated previously, raising or lowering an index leads to tensors of another valence and although the operation is involutory this does not justify the identification of the new tensors with the original ones.

If we follow the practice of other authors and omit the bar over the kernel letter of a tensor (or tensor density), we must still bear in mind that we are dealing with a different quantity. The exception, when we are allowed to omit this bar and to identify the resultant quantity with the original, occurs when we are based on the orthogonal group G_m of the Euclidean space R_n in which, moreover, the allowable coordinate systems adopted are rectangular Cartesian systems, i.e. systems in which the coordinates of the fundamental tensor are given by the simple relations

$$g_{\lambda\mu} \stackrel{*}{=} \delta_{\lambda\mu}. \tag{4.1}$$

Following Schouten, we agree to denote the rectangular Cartesian systems of Euclidean space (canonical systems) by lower-case Latin letters. Thus, for instance,

$$T^i{}_j$$

denote the coordinates of a mixed tensor in any of the canonical systems.

In this case, the operation of lowering the index i yields

$$T_{ij} = g_{ik} T^k{}_j \stackrel{*}{=} \delta_{ik} T^k{}_j = T^i{}_j,$$

so that, as we see here, the coordinates (as a matrix) are unaltered. Recall that a geometric object has been defined as a function which at a given point in space assigns a sequence of numbers uniquely to every allowable system of coordinates (in the special case under consideration, rectangular Cartesian coordinate systems are the allowable ones). Consequently, we must in this case state that (with a certain agreed order) the two sequences $T^i{}_j$, T_{ij} are the same and hence are one and the same function, or in other words that the two tensors are identical while merely being denoted in different way. In this case, it would even be misleading to put a bar over T.

The difference between covariant and contravariant coordinates thus disappears for tensors in Euclidean space and in rectangular coordinate systems. The type of tensor is determined by only by the number and not by the type of the indices.

In particular, the difference between the unit tensor and the fundamental tensor vanishes. So does the difference between contravariant vectors and covariant vectors. Only two different geometric interpretations of analytic quantities defined identically remain. A vector can be interpreted geometrically either as an ordered pair of points, or as an ordered pair of parallel hyperplanes, the distance between these planes being the inverse distance of these points.

The difference between tensor densities (W-densities) and tensors of corresponding valence also vanishes since with the group G_m we always have

$$J^2 = 1. \tag{4.2}$$

The difference between G-densities and tensors remains, however, except that all G-densities reduce to G-tensors, i.e. objects with the transformation law

$$\tilde{T}^{\lambda'_1 \ldots \lambda'_p}{}_{\mu'_1 \ldots \mu'_q} = \operatorname{sgn} J \cdot A^{\lambda'_1 \ldots \lambda'_p \mu_1 \ldots \mu_q}_{\lambda_1 \ldots \lambda_p \mu'_1 \ldots \mu'_q} \tilde{T}^{\lambda_1 \ldots \lambda_p}{}_{\mu_1 \ldots \mu_q}. \tag{4.3}$$

In a geometry based on the group G_m it is also possible, albeit bending the logical-formal point of view somewhat, to bring about the "identification" of objects with various numbers of indices.

Consider, for instance, the contravariant $(n-1)$-vector

$$v^{\lambda_1 \ldots \lambda_{n-1}} = \underset{1}{v}^{[\lambda_1} \ldots \underset{n-1}{v}^{\lambda_{n-1}]}, \tag{4.4}$$

which formally has n^{n-1} coordinates although, as we already know, the number of different non-zero coordinates, in absolute value at least, is at most $\binom{n}{n-1} = n$, or as many as the vector has coordinates. We also know that if a fundamental non-singular tensor is given, the $(n-1)$-vector (4.4) can be uniquely assigned the vector

$$\overset{*}{v}^{\lambda} = g^{\lambda\mu} v^{\lambda_1 \ldots \lambda_{n-1}} e_{\lambda_1 \ldots \lambda_{n-1} \mu}.$$

For rectangular Cartesian coordinates we have

$$\overset{*}{v}^i \overset{*}{=} \delta^{ij} v^{m_1 \ldots m_{n-1}} e_{m_1 \ldots m_{n-1} j} = e_{m_1 \ldots m_{n-1} i} v^{m_1 \ldots m_{n-1}}.$$

But in the systems (i)

$$e_{m_1 \ldots m_n} \overset{*}{=} \varepsilon_{m_1 \ldots m_n}$$

and consequently $\overset{*}{v}^i$ is equal to the minor formed by removing the i-th row from the matrix

$$[\underset{a}{v}^k] \quad (a = 1, 2, \ldots, n-1; k = 1, 2, \ldots, n),$$

where k corresponds to rows and a to columns.

We thus have formal identity between the coordinates of the vector $\overset{*}{v}^i$ and certain essential coordinates (i.e. non-zero) of the $(n-1)$-vector in every system (i). Consequently, vectors can be identified with bivectors in Cartesian coordinate systems in R_3 and hence, also, vector multiplication of vectors can be regarded as an "inner" operation, even though it was not originally such an operation in the sense that it is an association of contravariant (covariant) bivectors with covariant (contravariant) vectors.

48. Another Definition of Tensors. A tensor can also be defined in a different manner, viz. referring to the original definition of tensor which first came into being in connection with the concept of stress tensor in the theory of elasticity [85].

This is more general than the method used above in that it enables tensors of valence 2 to be defined in terms of vectors (contra- and co-variant), provided the latter have been introduced axiomatically and not analytically (as geometric objects). A drawback of this method is that tensors of higher valence must be introduced inductively, i.e. tensors with three indices by means of two-index tensors and vectors, and so on.

Suppose that we are given an operator F which with each (contravariant) vector x of a space of fixed dimension n uniquely associates a particular (contravariant) vector y of the same space

$$y = F(x). \tag{4.5}$$

Furthermore, assume that this operator has the following properties:

(i) $F(\alpha x) = \alpha F(x)$ for every α and x,

(ii) $F(\underset{1}{x} + \underset{2}{x}) = F(\underset{1}{x}) + F(\underset{2}{x})$ for all vectors $\underset{1}{x}, \underset{2}{x}$.

An operator of this kind is called a *linear operator*. It is easily shown that if the null vector (neutral element of addition) is denoted by θ, then we must have

$$F(\theta) = \theta.$$

Such an operator will be called a *tensor*. If the operator F has the additional property:

(iii) There exists a set of n linearly independent vectors

$$\overset{0}{\underset{1}{x}}, \overset{0}{\underset{2}{x}}, \dots, \overset{0}{\underset{n}{x}} \tag{4.6}$$

such that the set

$$F(\overset{0}{\underset{1}{x}}), F(\overset{0}{\underset{2}{x}}), \dots, F(\overset{0}{\underset{n}{x}}) \tag{4.7}$$

constitutes a set of linearly independent vectors; the tensor is then said to have (highest) *rank* equal to n.

Now consider set (4.6) which we shall call the basis. If such a set exists, any vector x can be "expanded" uniquely in terms of the basis vectors, i.e. for every x there exists a unique set of numbers $\varrho^1, \dots, \varrho^n$ such that $x = \varrho^\lambda \overset{0}{\underset{\lambda}{x}}$. By virtue of properties (i) and (ii), we have

$$F(x) = F(\varrho^\lambda \overset{0}{\underset{\lambda}{x}}) = \sum_\lambda F(\varrho^\lambda \overset{0}{\underset{\lambda}{x}}) = \sum_\lambda \varrho^\lambda F(\overset{0}{\underset{\lambda}{x}}) = \varrho^\lambda \overset{0}{\underset{\lambda}{y}},$$

where

$$\overset{0}{\underset{\lambda}{y}} \overset{\text{def}}{=} F(\overset{0}{\underset{\lambda}{x}}) \qquad (\lambda = 1, 2, \dots, n).$$

On the other hand, any vector $\overset{0}{\underset{\lambda}{y}}$ can be expressed in terms of the basis vectors (4.6); hence, there exists a set of numbers

$$P^\lambda_\mu \qquad (\lambda, \mu = 1, 2, \dots, n),$$

with the property that

$$\overset{0}{\underset{\lambda}{y}} = P^\mu_\lambda \overset{0}{\underset{\mu}{x}}.$$

Accordingly, we can write

$$F(x) = \varrho^\lambda P^\mu_\lambda \overset{0}{\underset{\mu}{x}} = \sigma^\mu \overset{0}{\underset{\mu}{x}},$$

where

$$\sigma^\mu \overset{\text{def}}{=} \varrho^\lambda P^\mu_\lambda. \tag{4.8}$$

When the basis (4.6) is taken as a "coordinate system" (for vectors only), the ϱ^λ may be regarded as the coordinates of the variable vector x, and the σ^λ as the coordinates of the transformed vector $F(x)$. The rule for calculating the coordinates σ^λ when the coordinates ϱ^λ are given is of the form (4.8) which, as we see. is just like the familiar rule for transvection of the vector ϱ onto the tensor P. This rule is defined once the matrix P is given. But this matrix is "random" in the sense that we have started with one particular basis (4.6), although we could equally well start with another. We therefore see that the functional F can be defined analytically in terms of a quadratic matrix, i.e. a two-index object. In this case we are dealing with a mixed two-index tensor.

By considering linear operations which associate covariant vectors with contravariant vectors, we arrive at the concept of the covariant tensor, and when we consider operations which set up a correspondence between contravariant vectors and covariant vectors, we arrive at the concept of the contravariant tensor.

Linear operations assigning covariant vectors to covariant vectors lead only to mixed tensors, and not to a new species of tensor.

Furthermore we can consider operations which associate vectors with two-index tensors or conversely; such operations in turn provide a basis for introducing the concept of three-index tensor (of various valences). In this way, using linear operations with various domains and counter-domains which are the spaces of tensors already defined, we are able gradually to define tensors of higher valence.

If we wish to "calculate" with these objects, however, we must "stick" to one basis and introduce coordinates, and this analyzation leads us back to the original definition and the former point of view.

49. Dyadics (Dual Sums). In 1901 J. W. Gibbs introduced the concept of so-called *dyadics* (Hoborski [35] refers to this concept as *dual sum*) and expounded their algebra; dyadics have found extensive application in physics and in differential geometry (C. E. Weatherburn, P. Delens). It is known that the Euclidean analytic geometry of linear objects is simplified considerably by the introduction of the vector method. The conventional vector method does not, however, embrace objects of the second degree. However, such objects can be studied to advantage with the aid of dyadic calculus. The definition of a dyadic is as follows.

In a space where the scalar product of vectors is defined, i.e. in a space where the distance metric and angular metric of vectors are defined, suppose that we are given a sequence of $2n$ fixed contravariant vectors

$$\underset{1}{u}, \ldots, \underset{n}{u}, \underset{1}{v}, \ldots, \underset{n}{v}. \tag{4.9}$$

Using the sequence (4.9), we construct the operator

$$y = F(x)$$

which, if sequence (4.9) is regarded as fixed, assigns the vector y uniquely to each variable vector x according to the formula

$$y = \sum_{i=1}^{n} \underset{i}{u}(\underset{i}{v}x), \tag{4.10}$$

where the symbol $(\underset{i}{v}x)$ denotes scalar multiplication.

It is evident that if α is a number, then

$$F(\alpha x) = \alpha F(x)$$

and similarly, that

$$F(\underset{1}{x}+\underset{2}{x}) = \sum_{i=1}^{n} \underset{i}{u}[\underset{i}{v}(\underset{1}{x}+\underset{2}{x})] = \sum_{i=1}^{n} \underset{i}{u}[(\underset{i}{v}\underset{1}{x})+(\underset{i}{v}\underset{2}{x})]$$
$$= \sum_{i=1}^{n} \underset{i}{u}(\underset{i}{v}\underset{1}{x})+ \sum_{i=1}^{n} \underset{i}{u}(\underset{i}{v}\underset{2}{x}) = F(\underset{1}{x})+F(\underset{2}{x}).$$

This operator is therefore linear, i.e. it is a special case of the two-index tensor. If we write in the coordinates of the vectors by adding the contravariant index, we have

$$y^{\lambda} = \sum_{i=1}^{n} \underset{i}{u}^{\lambda}(g_{\alpha\beta}\underset{i}{v}^{\alpha}x^{\beta}). \tag{4.11}$$

In the special case, when a sequence of basis vectors $\underset{\mu}{e}$, belonging to the coordinate system (λ) is taken for the sequence $\underset{i}{u}$, we obtain

$$y^{\lambda} = \sum_{\mu=1}^{n} \underset{\mu}{\varrho}\underset{\mu}{e}^{\lambda},$$

where, for brevity, we have put

$$\underset{\mu}{\varrho} = g_{\alpha\beta}\underset{\mu}{v}^{\alpha}x^{\beta}.$$

For Euclidean spaces and Cartesian coordinate systems, we have

$$g_{ik} \overset{*}{=} \delta_{ik}, \qquad \underset{i}{e}^{k} \overset{*}{=} \delta_{i}^{k}$$

and then

$$\underset{i}{\varrho} = \delta_{jk}\underset{i}{v}^j x^k = \sum_j \underset{i}{v}^j x^j$$

and formula (4.11) becomes

$$y^k = \sum_{i=1}^{n} \underset{i}{\varrho}\underset{i}{e}^k = \underset{k}{\varrho} = \sum_{j}^{n} \underset{k}{v}^j x^j .$$

Sequence (4.9), which we use to define the vector-vector correspondence by means of formula (4.11) or (for Cartesian systems) by the simpler formula

$$y^\lambda = \sum_{i=1}^{n} \sum_{\beta=1}^{m} \underset{i}{u}^\lambda \underset{i}{v}^\beta x^\beta, \qquad \text{where } m \text{ is the dimension of the space,}$$

is called a *dyadic* and is denoted briefly by the symbol $(\underset{i}{u}, \underset{i}{v})$.

The reader who wishes to acquaint himself with dyadic theory is referred to the book by Hoborski where applications are also dealt with. The reader who has become accustomed to our symbols is warned that he will have to make a switch in order to master the obsolete symbols used in the book mentioned above.

50. Anholonomic Systems. We have seen that with the most general group of transformations for a given coordinate system (λ), the set of basis vectors

$$\underset{\lambda}{e} \qquad (\lambda = 1, 2, \ldots, n) \tag{4.12}$$

which satisfy the relations

$$\underset{\lambda}{e}^\nu \overset{*}{=} \delta^\nu_\lambda \qquad (\lambda, \nu = 1, 2, \ldots, n) \tag{4.13}$$

is defined uniquely at every point p of a space, or of a region which is part of the space. The basis vectors constitute a set of linearly independent vectors since the order of the matrix $[\underset{\lambda}{e}^\nu]$ is n. Hence, any vector v attached at the point p can be expressed linearly in a unique manner in terms of the basis vectors, i.e. we have

$$v = \sum_\lambda \overset{\lambda}{v}\underset{\lambda}{e} = \overset{\lambda}{v}\underset{\lambda}{e}, \tag{4.14}$$

which when written in terms of coordinates, yields

$$v^\nu = \overset{\lambda}{v}\underset{\lambda}{e}^\nu,$$

whence, in view of equation (4.13)

$$v^\nu \overset{*}{=} \overset{\nu}{v},$$

which means that the scalar coefficients $\overset{\lambda}{v}$ in (4.14) are simply the coordinates of the vector v in the coordinate system (λ). If we use the terminology introduced in Section 48, systems of basis vectors can be described as local bases. Of course, if the group G_1 reduces to a subgroup G_a, it suffices to have a single basis (attached at any point p) for every vector attached at any other point to admit an expansion in terms of the vectors of another basis.

The following problem now arises. Suppose that at every point p we have selected a basis consisting of n independent vectors

$$\underset{K}{u} \quad (K = \mathrm{I}, \mathrm{II}, \ldots, N). \tag{4.15}$$

Analytically, this means that we are given n^2 functions

$$\underset{K}{u}^\nu \quad (\nu = 1, 2, \ldots, n;\ K = \mathrm{I}, \mathrm{II}, \ldots, N), \tag{4.16}$$

in the independent variables ξ^ν, with the property that the matrix

$$[\underset{K}{u}^\nu] \tag{4.17}$$

is of rank n at every point p. Functions (4.16) are, of course, assumed to be regular (of class at least C^1).

The question is whether there exists a coordinate system (K) in which the basis vectors of (K) are at each point the same as the vectors of the bases adopted *a priori*.

To answer this question, we first establish necessary conditions. If a coordinate system (K) such that (4.15) is the system of basis vectors for (K) does indeed exist, then

$$\underset{K}{u}^I \overset{*}{=} \underset{K}{e}^I = \delta^K_I.$$

But

$$\underset{K}{u}^I = A^I_\lambda \underset{K}{u}^\lambda,$$

and hence the system of equations

$$\underset{K}{u}^\lambda A^I_\lambda = \delta^I_K, \tag{4.18}$$

must be satisfied. When the coordinate system (λ) is used, the $\underset{K}{u}^\lambda$ are given functions of the variables ξ^ν, whereas we are seeking functions

$$\psi^I(\xi^1, \xi^2, \ldots, \xi^n) \qquad (I = \mathrm{I}, \mathrm{II}, \ldots, N)$$

(i.e. we are looking for a transformation from the original coordinate system ξ^λ to a new system $\xi^I = \psi^I(\xi^\lambda)$) such that

$$A^I_\lambda = \frac{\partial \psi^I}{\partial \xi^\lambda},$$

i.e., such that the system

$$\underset{K}{u}^\lambda \frac{\partial \psi^I}{\partial \xi^\lambda} = \delta^I_K \qquad (I, K = \mathrm{I}, \mathrm{II}, \ldots, N)$$

is satisfied.

This is a system of partial differential equations of the first order in the functions ψ which we are looking for. The integrability conditions for this system can easily be written down.

To this end, we introduce a set of covariant vectors $\overset{K}{\bar{u}}$ inverse to (4.15), i.e. satisfying the relations

$$\overset{K}{\bar{u}}_\lambda \underset{I}{u}^\lambda = \delta^K_I \qquad (I, K = \mathrm{I}, \mathrm{II}, \ldots, N)$$

or, the equivalent relations

$$\overset{K}{\bar{u}}^\lambda \underset{K}{u}^\mu = \delta^\mu_\lambda \qquad (\lambda, \mu = 1, 2, \ldots, n). \tag{4.19}$$

Equations (4.18) can now be replaced by an equivalent set in normal form (solved in the partial derivatives of the functions sought) by multiplying both sides by $\overset{K}{\bar{u}}_\mu$ and summing with respect to K

$$\overset{K}{\bar{u}}_\mu \underset{K}{u}^\lambda \frac{\partial \psi^I}{\partial \xi^\lambda} = \delta^I_K \overset{K}{\bar{u}}_\mu = \overset{K}{\bar{u}}_\mu,$$

i.e. after simplifying the left-hand side by using equation (4.19), we have

$$\frac{\partial \psi^I}{\partial \xi^\lambda} = \overset{I}{\bar{u}}_\lambda \qquad (\lambda = 1, 2, \ldots, n;\ I = \mathrm{I}, \mathrm{II}, \ldots, N). \tag{4.20}$$

We differentiate (4.20) with respect to ξ^μ and introduce the convenient symbol $\partial_\mu = \partial/\partial\xi^\mu$. Noting that $\partial^2\psi^I/\partial\xi^\lambda\partial\xi^\mu = \partial^2\psi^I/\partial\xi^\mu\partial\xi^\lambda$, we obtain

$$\partial_\lambda \overset{I}{\bar{u}}_\mu = \partial_\mu \overset{I}{\bar{u}}_\lambda \qquad (\lambda, \mu = 1, 2, \ldots, n;\ I = \mathrm{I}, \mathrm{II}, \ldots, N),$$

i.e.,

$$\partial_{[\lambda}\overset{I}{\bar{u}}_{\mu]} = 0. \tag{4.21}$$

Conversely, it turns out that the integrability conditions (4.21) are sufficient for solutions of (4.20) to exist and hence for the existence of solutions of the equivalent system of equations (4.18).

Since conditions (4.21) are in general not satisfied for arbitrary given functions (4.16), the set of vectors (4.15) is not in general a system of basis vectors for a coordinate system, or, as we shall put it, is not a holonomic system. If conditions (4.21) are not satisfied the system is said to be *non-holonomic*. In general, system (4.15) is called an *anholonomic system* (or reference frame). When conditions (4.21) are satisfied, the anholonomic system becomes holonomic; otherwise, it is non-holonomic.

Equations (4.21) are in general not satisfied for arbitrary $\underset{K}{u}$'s, and so the system obtained is in general non-holonomic. A set of n families of curves, viz. the K-th family, belongs to system (4.16); this is a congruence of curves for which the vectors $\underset{K}{u}$ are tangents. The question of whether for the system $\underset{K}{u}$ there exists a coordinate system such that the parametric lines of this coordinate system coincide with the lines of the aforementioned congruences is a different matter. This is a weaker requirement than the holonomicity condition, since the vectors $\underset{K}{u}$ need not be basis vectors for the coordinate system sought, but have only to be linearly dependent on $\underset{K}{e}$. Coordinate systems with this property, always exist for $n = 2$, for instance, whereas for $n \geqslant 3$ they in general do not exist. This important problem is associated with the integration of Pfaff systems, but it will not be considered here.

Anholonomic systems will be indicated by upper-case Roman letters. Remember that no coordinate system belongs to a non-holonomic system. This does not prevent us from representing each tensor or density in anholonomic reference frames if in this anholonomic system (K) we define the coordinates of the quantity under consideration suitably.

If in an anholonomic system (λ) a quantity $\mathfrak{T}$ has coordinates

$$\mathfrak{T}^{\lambda_1 \ldots \lambda_p}{}_{\mu_1 \ldots \mu_q}, \tag{4.22}$$

then by its coordinates in the anholonomic system (K) we mean the quantities given by

$$\mathfrak{T}^{K_1\dots K_p}{}_{M_1\dots M_q} \overset{\text{def}}{=} \varepsilon|H|^{-r}\mathfrak{T}^{\lambda_1\dots\lambda_p}{}_{\mu_1\dots\mu_q}\overset{K_1}{\bar{u}}_{\lambda_1}\cdots\overset{K_p}{\bar{u}}_{\lambda_p}\underset{M_1}{u}^{\mu_1}\cdots\underset{M_q}{u}^{\mu_q}, \tag{4.23}$$

where

$$H = |\overset{K}{\bar{u}}_\lambda|\ ^{(1)}, \qquad \varepsilon = 1 \text{ or } \varepsilon = \operatorname{sgn} H. \tag{4.24}$$

This definition is justified for the following reasons:

1. If the system (K) is holonomic, then

$$\overset{K}{\bar{u}}_\lambda = A^K_\lambda$$

and relations (4.23) reduce to the very familiar transformation law for the coordinates of a tensor density, and $H = J$.

2. In the special case, when $\mathfrak{T}$ is a vector v^λ, relation (4.23) becomes

$$v^K = v^\lambda \overset{K}{\bar{u}}_\lambda, \tag{4.25}$$

i.e. it represents the coordinates of the vector v when the system of vectors $\underset{K}{u}^\lambda$ is taken as the basis.

Indeed, since the rank of matrix (4.17) is n by assumption the vector v can be represented uniquely in terms of the basis vectors (4.15), i.e.

$$v = \overset{K}{\sigma}\underset{K}{u}.$$

We now calculate the scalar coefficients $\overset{K}{\sigma}$. We have

$$v^\lambda = \overset{K}{\sigma}\underset{K}{u}^\lambda.$$

If we multiply both sides of this relation by $\overset{I}{\bar{u}}_\lambda$ and sum over λ, we obtain

$$v^\lambda\overset{I}{\bar{u}}_\lambda = \overset{K}{\sigma}\underset{K}{u}^\lambda\overset{I}{\bar{u}}_\lambda = \overset{K}{\sigma}\delta^I_K = \overset{I}{\sigma}.$$

The coordinates of the vector v in the basis (4.15) are therefore

$$\overset{I}{\sigma} = v^\lambda\overset{I}{\bar{u}}_\lambda,$$

i.e., comparing with (4.25), they are equal to the quantities v^I.

3. The rule (4.23) possesses the group property with respect to any three anholonomic systems (K), (K'), (K''). The proof is left to the reader.

---

(1) We do not use the letter J because H is not in general a Jacobian for any transformation of coordinates.

To put equation (4.23) into simpler form, and at the same time to make the calculation more mechanical we introduce the following symbols. If the system (K) is non-holonomic (defined by the basis (4.15)), then

$$A^K_\lambda \overset{\text{def}}{=} \overset{K}{\bar{u}}_\lambda, \qquad A^\lambda_K \overset{\text{def}}{=} \underset{K}{u}^\lambda. \tag{4.26}$$

On the other hand, if (K) and (K') are two non-holonomic systems we adopt the convention

$$A^{K'}_K \overset{\text{def}}{=} A^{K'}_\lambda A^\lambda_K, \qquad A^K_{K'} \overset{\text{def}}{=} A^K_\lambda A^\lambda_{K'}. \tag{4.27}$$

In order to justify this convention, we must show that the left-hand sides do not depend on the coordinate system (λ). Indeed, we calculate

$$A^{K'}_{\mu'} A^{\mu'}_{K'} = \overset{K'}{\bar{u}}_{\mu'} \underset{K'}{u}^{\mu'}.$$

However, the definition implies that

$$\underset{K'}{u}^{\mu'} = \underset{K'}{u}^\lambda A^{\mu'}_\lambda .$$

Similarly,

$$\overset{K'}{\bar{u}}_{\mu'} = \overset{K'}{\bar{u}}_\lambda A^\lambda_{\mu'} = \overset{K'}{\bar{u}}_\mu A^\mu_{\mu'},$$

and so

$$\begin{aligned} \overset{K'}{\bar{u}}_{\mu'} \underset{K'}{u}^{\mu'} &= \overset{K'}{\bar{u}}_\mu A^\mu_{\mu'} \underset{K'}{u}^\lambda A^{\mu'}_\lambda = \overset{K'}{\bar{u}}_\mu \underset{K'}{u}^\lambda A^{\mu\mu}_{\mu'} \\ &= \overset{K'}{\bar{u}}_\mu \underset{K'}{u}^\lambda A^\mu_\lambda = \overset{K'}{\bar{u}}_\lambda \underset{K'}{u}^\lambda = A^{K'}_\lambda A^\lambda_K \end{aligned}$$

which is what was to be proved.

With the adoption of this convention generalizing the meaning of the symbol A, in the special case when the system (K) is holonomic, A^K_λ and A^λ_K assume their original meaning. Moreover, it can be shown that the mnemotechnical rule

$$A^K_{K'} A^{K'}_{K''} = A^K_{K''}, \tag{4.28}$$

is preserved, regardless of whether or not the systems (K), (K'), (K'') are holonomic or non-holonomic; furthermore it is always true that

$$A^K_L = \delta^K_L. \tag{4.29}$$

In this event, instead of the notation (4.24) we can return to the old notation

$$J = |A^K_\lambda|.$$

except that we must always remember that when (K) is a non-holonomic system, J is not the Jacobian of the transformation since the coordinates ξ^K do not exist. Relation (4.23) then becomes

$$\mathfrak{T}^{K_1 \dots K_p}{}_{M_1 \dots M_q} = \varepsilon |J| A^{K_1 \dots K_p \, \mu_1 \dots \mu_q}_{\lambda_1 \dots \lambda_p \, M_1 \dots M_q} \mathfrak{T}^{\lambda_1 \dots \lambda_p}{}_{\mu_1 \dots \mu_q}, \tag{4.30}$$

i.e. it takes a form of the kind that we already have for the transition from one coordinate system to another.

We therefore find that it is also possible to speak of the coordinates of tensors and of tensor densities in relation to non-holonomic systems.

Note that, as can easily be shown, the property of symmetry, or anti-symmetry, with respect to indices of one and the same kind also holds for non-holonomic systems.

It is not meaningful, however, to speak of non-holonomic coordinates, although some authors do it.

Moreover, in the case of multi-index tensors one can speak of "mixed" coordinates in the sense that different indices belong to different coordinate systems or frames of reference. Thus, for the tensor $T^\lambda{}_{\mu\nu}$, for instance, we define

$$T^K{}_{I'J''} \stackrel{\text{def}}{=} T^\lambda{}_{\mu\nu} A^{K\mu\nu}_{\lambda I' J''}. \tag{4.31}$$

This definition fully justifies the use of the symbol A instead of u and $\bar{u}$, since with the convention above A^K_λ and A^λ_K do indeed become mixed coordinates of the unit tensor A.

When working with anholonomic systems, we find it convenient to introduce yet another convention, viz.

$$\partial_K \stackrel{\text{def}}{=} A^\lambda_K \partial_\lambda. \tag{4.32}$$

It must only be borne in mind that in general

$$\partial_K \partial_L \neq \partial_L \partial_K, \tag{4.33}$$

whereas we always have

$$\partial_\lambda \partial_\mu = \partial_\mu \partial_\lambda.$$

Hence, we also have

$$\partial_{K'} = A^K_{K'} \partial_K,$$

for, on the one hand,

$$\partial_{K'} = A^\lambda_{K'} \partial_\lambda,$$

while on the other hand,

$$A^K_{K'} \partial_K = A^K_{K'} A^\lambda_K \partial_\lambda = A^\lambda_{K'} \partial_\lambda.$$

Note also that the Leibniz rule

$$\partial_K(\varphi\psi) = \varphi\partial_K\psi + (\partial_K\varphi)\psi,$$

applies to the symbol ∂_K; this follows at once from the definition.

Now assume that the matrix of coordinates of the fundamental tensor $g_{\lambda\mu}$ is of rank n. By putting this tensor into canonical form (p. 122), we can arrange matters so that

$$g_{ij} \overset{*}{=} \begin{cases} 0, & \text{when } i \neq j, \\ -1, & \text{when } i = j = 1, \ldots, s, \\ +1, & \text{when } i = j = s+1, \ldots, n. \end{cases}$$

When the tensor $g_{\lambda\mu}$ gives rise to a positive definite form, $s = 0$ and we then have

$$g_{ij} \overset{*}{=} \delta_{ij}.$$

In these special coordinate systems (i), when we denote the fundamental vectors by $\underset{i}{e}$ and the inverse system of covariant wectors by $\overset{j}{e}$, we have

$$g_{ij} \overset{*}{=} \sum_k \overset{k}{e}_i \overset{k}{e}_j .$$

The vectors e constitute an orthonormal system, as we have already seen on p. 124.

On the other hand, if we take an orthonormal system of vectors

$$\underset{K}{u}^{\lambda} \quad (\lambda = 1, \ldots, n;\ K = \mathrm{I}, \mathrm{II}, \ldots, N),$$

at every point in the space V_n, this system will not in general be holonomic.

Systems of this kind, as Einstein demonstrated, serve to define the parallel displacement of vectors and he based his theory of "teleparallelism" (or distant parallelism) in electrodynamics on such systems (cf. p. 252).

51. The Object of Anholonomicity. Schouten and van Dantzig [148] introduced the so-called *object of anholonomicity* $\Omega_{IJ}{}^K$, which they defined as follows

$$\Omega_{IJ}{}^K \overset{\text{def}}{=} A^{\lambda\mu}_{IJ}\partial_{[\lambda}A^K_{\mu]} \quad (I, J, K = \mathrm{I}, \mathrm{II}, \ldots, N). \tag{4.34}$$

We shall show that an object so defined is independent of the original coordinate system (λ).

To this end, let (λ') be any other arbitrary coordinate system and write

$$\Omega'_{IJ}{}^K = A^{\lambda'\mu'}_{I\,J}\partial_{[\lambda'}A^K_{\mu']}.$$

Accordingly,

$$\begin{aligned} A_{I\ J}^{\lambda'\mu'}\partial_{\lambda'}A_{\mu'}^{K} &= A_{I\ J}^{\lambda'\mu'}\partial_{\lambda'}(A_{\mu}^{K}A_{\mu'}^{\mu}) \\ &= A_{I\ J}^{\lambda'\mu'}A_{\mu}^{K}\partial_{\lambda'}A_{\mu'}^{\mu}+A_{I\ J}^{\lambda'\mu'}A_{\mu'}^{\mu}\partial_{\lambda'}A_{\mu}^{K} \\ &= A_{\mu}^{K}A_{I\ J'}^{\lambda'\mu}\partial_{\lambda'}A_{\mu'}^{\mu}+A_{I\ J\ \mu'}^{\lambda'\mu'\mu}A_{\lambda'}^{\lambda}\partial_{\lambda}A_{\mu}^{K} \\ &= A_{\mu}^{K}A_{I\ J}^{\lambda'\mu'}\partial_{\lambda'}A_{\mu'}^{\mu}+A_{IJ}^{\lambda\mu}\partial_{\lambda}A_{\mu}^{K}. \end{aligned}$$

Similarly,

$$A_{I\ J}^{\lambda'\mu'}\partial_{\mu'}A_{\lambda'}^{K} = A_{\mu}^{K}A_{I\ J}^{\lambda'\mu'}\partial_{\mu'}A_{\lambda'}^{\mu}+A_{IJ}^{\lambda\mu}\partial_{\mu}A_{\lambda}^{K}.$$

Therefore

$$\Omega'_{IJ}{}^{K} = A_{\mu}^{K}A_{I\ J}^{\lambda'\mu'}\partial_{[\lambda'}A_{\mu']}^{\mu}+A_{IJ}^{\lambda\mu}\partial_{[\lambda}A_{\mu]}^{K}.$$

But $\partial_{[\lambda'}A_{\mu']}^{\mu} = 0$, since (λ) and (λ') are coordinate systems; consequently,

$$\Omega'_{IJ}{}^{K} = A_{IJ}^{\lambda\mu}\partial_{[\lambda}A_{\mu]}^{K} = \Omega_{IJ}{}^{K}, \tag{4.35}$$

which was to be proved.

Note that since the rule (4.37) possesses the group property (this is proved later on), the proof of the above property which follows from the group property result as a special case could be omitted here. We now show that in (4.34) we have a geometric object of class two.

Indeed, let (λ) be a coordinate system and let (K), (K'), (K'') be three anholonomic systems. We have

$$\begin{aligned} \Omega_{IJ}{}^{K} &= A_{IJ}^{\lambda\mu}\partial_{[\lambda}A_{\mu]}^{K}, \\ \Omega_{I'J'}{}^{K'} &= A_{I'J'}^{\lambda\ \mu}\partial_{[\lambda}A_{\mu]}^{K'}, \\ \Omega_{I''J''}{}^{K''} &= A_{I''J''}^{\lambda\ \mu}\partial_{[\lambda}A_{\mu]}^{K''}. \end{aligned}$$

We first look for the transformation law. It is

$$\begin{aligned} \Omega_{I'J'}{}^{K'} &= A_{I'J'}^{\lambda\ \mu}\partial_{[\lambda}A_{\mu]}^{K'} = A_{I'J'K}^{\lambda\ \mu\ K'}\partial_{[\lambda}A_{\mu]}^{K}+A_{I'J'}^{\lambda\ \mu}A_{[\mu}^{K}\partial_{\lambda]}A_{K}^{K'} \\ &= A_{I'IJ'JK}^{I\ \lambda J\ \mu K'}\partial_{[\lambda}A_{\mu]}^{K}-A_{I'J'}^{\lambda\ \mu}A_{[\lambda}^{K}\partial_{\mu]}A_{K}^{K'} \\ &= A_{I'J'K}^{I\ J\ K'}\Omega_{IJ}{}^{K}-A_{I'J'}^{\lambda\ \mu}A_{[\lambda}^{K}\partial_{\mu]}A_{K}^{K'}. \end{aligned} \tag{4.36}$$

The object now is formally to eliminate the Greek indices from the second term on the right-hand side. Accordingly, we may write

$$A_{I'J'}^{\lambda\ \mu}A_{\lambda}^{K}\partial_{\mu}A_{K}^{K'} = A_{I'\lambda}^{\lambda\ K}A_{J'}^{\mu}\partial_{\mu}A_{K}^{K'} = A_{I'}^{K}\partial_{J'}A_{K}^{K'}.$$

Similarly,

$$A_{I'J'}^{\lambda\ \mu}A_{\mu}^{K}\partial_{\lambda}A_{K}^{K'} = A_{J'}^{K}\partial_{I'}A_{K}^{K'}.$$

Using this, we rewrite formula (4.36) as

$$\Omega_{I'J'}{}^{K'} = A^{I\,J\,K'}_{I'J'K}\,\Omega_{IJ}{}^{K} - A^{K}_{[I'}\partial_{J']}A^{K'}_{K}\,. \tag{4.37}$$

This is the transformation law for the coordinates of an object of anholonomicity under transition from system (K) to system (K').

The fact that this object is of class two follows from the fact that second-order partial derivatives of the functions transforming the coordinates appear (implicitly) in the transformation law.

We now show that the rule (4.37) possesses the group property. We therefore write another formula, for the transition from (K') to (K''), and eliminate $\Omega_{I'J'}{}^{K'}$

$$\Omega_{I''J''}{}^{K''} = A^{I'\,J'\,K''}_{I''J''K'}\,\Omega_{I'J'}{}^{K'} - A^{K'}_{[I''}\,\partial_{J'']}A^{K''}_{K'}\,. \tag{4.38}$$

On inserting the expression (4.37) into (4.38), we obtain

$$\begin{aligned}\Omega_{I''J''}{}^{K''} &= A^{I'\,J'\,K''}_{I''J''K'}\,[A^{I\,J\,K'}_{I'J'K}\,\Omega_{IJ}{}^{K} - A^{K}_{[I'}\partial_{J']}A^{K'}_{K}] - A^{K'}_{[I''}\partial_{J'']}A^{K''}_{K'}\\ &= A^{I'\,J'\,K''\,I\,J\,K'}_{I''J''K'I'J'K}\,\Omega_{IJ}{}^{K} - A^{I'\,J'\,K''K}_{I''J''K'[I'}\partial_{J']}A^{K'}_{K} - A^{K'}_{[I''}\partial_{J'']}A^{K''}_{K'}\\ &= A^{I\,J\,K''}_{I''J''K}\,\Omega_{IJ}{}^{K} - A^{I'\,J'\,K''K}_{I''J''K'[I'}\partial_{J']}A^{K'}_{K} - A^{K'}_{[I''}\partial_{J'']}A^{K''}_{K'}\,.\end{aligned} \tag{4.39}$$

When we rearrange the second term on the right

$$A^{I'\,J'\,K''K}_{I''J''K'I'}\partial_{J'}A^{K'}_{K} = A^{K\,I'\,K''}_{I''J''K'}\partial_{J'}A^{K'}_{K} = A^{K\,K''}_{I''K'}\partial_{J''}A^{K'}_{K}\,.$$

and we thus arrive at

$$A^{I'\,J'\,K''K}_{I''J''K'[I'}\partial_{J']}A^{K'}_{K} = A^{K''}_{K'}A^{K}_{[I''}\partial_{J'']}A^{K'}_{K}\,.$$

When we apply the Leibniz rule to the second and third terms of the formula (4.39), we have further

$$\begin{aligned}&A^{K''}_{K'}A^{K}_{I''}\partial_{J''}A^{K'}_{K} + A^{K'}_{I''}\partial_{J''}A^{K''}_{K'}\\ &\quad = A^{K''}_{K'}\{\partial_{J''}(A^{K'K}_{K\,I''}) - A^{K'}_{K}\partial_{J''}A^{K}_{I''}\} + \partial_{J''}(A^{K''K'}_{K'I''}) - A^{K''}_{K'}\partial_{J''}A^{K'}_{I''}\\ &\quad = A^{K''}_{K'}\partial_{J''}A^{K'}_{I''} - A^{K''}_{K}\partial_{J''}A^{K}_{I''} + \partial_{J''}A^{K''}_{I''} - A^{K''}_{K'}\partial_{J''}A^{K'}_{I''}\\ &\quad = -A^{K''}_{K}\partial_{J''}A^{K}_{I''} = A^{K}_{I''}\partial_{J''}A^{K''}_{K}\,,\end{aligned}$$

since $\partial_{J''}A^{K''}_{I''} = 0$. The sum of the second and third terms on the right-hand side of (4.39) thus is $-A^{K}_{[I''}\partial_{J'']}A^{K''}_{K}$ and finally

$$\Omega_{I''J''}{}^{K''} = A^{I\,J\,K''}_{I''J''K}\,\Omega_{IJ}{}^{K} - A^{K}_{[I''}\partial_{J'']}A^{K''}_{K}\,,$$

which establishes the group property for the transformation law (4.37).

Note that the object Ω is antisymmetric with respect to the lower indices.

Indeed,

$$\Omega_{JI}{}^{K} = A^{\lambda\mu}_{JI}\partial_{[\lambda}A^{K}_{\mu]} = A^{\mu\lambda}_{IJ}\partial_{[\mu}A^{K}_{\lambda]} = -A^{\mu\lambda}_{JI}\partial_{[\lambda}A^{K}_{\mu]} = -A^{\lambda\mu}_{IJ}\partial_{[\lambda}A^{K}_{\mu]} = -\Omega_{IJ}{}^{K};$$

this property is, moreover, invariant as is immediately evident from the transformation law (4.37).

If the system (K) is holonomic, then

$$A^{K}_{\lambda} = \frac{\partial\xi^{K}}{\partial\xi^{\lambda}} = \partial_{\lambda}\xi^{K}$$

and

$$\partial_{[\lambda}A^{K}_{\mu]} = 0,$$

which consequently yields

$$\Omega_{IJ}{}^{K} \overset{h}{=} 0. \tag{4.40}$$

On the other hand, equation (4.40) does not hold for non-holonomic systems (K). An object of anholonomicity thus has all its coordinates zero only in holonomic systems, but it does not have a property possessed by tensors or tensor densities, viz. that if all their coordinates are zero in one system they are also zero in every other system, holonomic or non-holonomic.

We now calculate the coordinates of a contracted object of anholonomicity

$$\Omega_{J} \overset{\text{def}}{=} \Omega_{IJ}{}^{I} = A^{\lambda\mu}_{IJ}\partial_{[\lambda}A^{J}_{\mu]}. \tag{4.41}$$

Reasoning as on p. 28, we proceed from the relation

$$A^{\lambda}_{K} = \frac{\partial\ln|J|}{\partial A^{K}_{\lambda}},$$

where (λ) is an arbitrary coordinate system, and (K) is an arbitrary anholonomic system. When we multiply both sides of this relation by $\partial_{\mu}A^{K}_{\lambda}$, we obtain

$$A^{\lambda}_{K}\partial_{\mu}A^{K}_{\lambda} = \frac{\partial\ln|J|}{\partial A^{K}_{\lambda}}\partial_{\mu}A^{K}_{\lambda} = \partial_{\mu}\ln|J|. \tag{4.42}$$

We now rewrite formula (4.34) in the form

$$2\Omega_{IJ}{}^{K} = A^{\lambda\mu}_{IJ}\partial_{\lambda}A^{K}_{\mu} - A^{\lambda\mu}_{IJ}\partial_{\mu}A^{K}_{\lambda},$$

whence we have

$$A^{\lambda\mu}_{IJ}\partial_{\mu}A^{K}_{\lambda} = A^{\lambda\mu}_{IJ}\partial_{\lambda}A^{K}_{\mu} - 2\Omega_{IJ}{}^{K},$$

$$\partial_{\mu}A^{\lambda}_{K} = \partial_{\lambda}A^{K}_{\mu} - 2A^{IJ}_{\lambda\mu}\Omega_{IJ}{}^{K}.$$

Multiplying both sides by A^{λ}_{K}, and, of course, summing twice over λ and K, we obtain

$$A^{\lambda}_{K}\partial_{\mu}A^{K}_{\lambda} = \partial_{K}A^{K}_{\mu}-2A^{IJ\lambda}_{\lambda\mu K}\Omega_{IJ}{}^{K} = \partial_{K}A^{K}_{\mu}-2A^{J}_{\mu}\Omega_{KJ}{}^{K} = \partial_{K}A^{K}_{\mu}-2A^{J}_{\mu}\Omega_{J}.$$

Substituting from relation (4.42) into the left-hand side, we get

$$\partial_{\mu}\ln|J| = \partial_{K}A^{K}_{\mu}-2A^{J}_{\mu}\Omega_{J}.$$

From this Ω_J is easily found to be given by

$$2A^{J}_{\mu}\Omega_{J} = \partial_{K}A^{K}_{\mu}-\partial_{\mu}\ln|J|$$

or

$$2A^{J\mu}_{\mu I}\Omega_{J} = A^{\mu}_{I}\partial_{K}A^{K}_{\mu}-A^{\mu}_{I}\partial_{\mu}\ln|J|,$$

i.e.

$$2\Omega_{I} = -A^{K}_{\mu}\partial_{K}A^{\mu}_{I}-\partial_{I}\ln|J| = -\partial_{\mu}A^{\mu}_{I}-\partial_{I}\ln|J|,$$

whence, finally,

$$\Omega_{I} = -\tfrac{1}{2}\{\partial_{\mu}A^{\mu}_{I}+\partial_{I}\ln|J|\}. \tag{4.43}$$

The transformation law for the object Ω_I is obtained either from the law (4.37) when in it we contract with respect to the indices K' and I'

$$\Omega_{J'} = A^{I}_{K'}{}^{J}_{J'}{}^{K'}_{K}\Omega_{IJ}{}^{K}-A^{K}_{[K'}\partial_{J']}A^{K'}_{K} = A^{J}_{J'}\Omega_{J}-A^{K}_{[K'}\partial_{J']}A^{K'}_{K}$$

and then rearrange the second term on the right-hand side, or from formula (4.43) by writing

$$\Omega_{J'} = -\tfrac{1}{2}\{\partial_{\mu}A^{\mu}_{J'}+\partial_{J'}\ln|J_1|\},$$

where $J_1 = |A^{K'}_{\mu}|$ (whereas $J = |A^{K}_{\mu}|$).

Writing $J_2 = |A^{K'}_{K}|$ and bearing in mind that $J_1 = J_2 J$, we have

$$\begin{aligned}\Omega_{J'} &= -\tfrac{1}{2}\{\partial_{\mu}(A^{\mu}_{J}A^{J}_{J'})+A^{J}_{J'}\partial_{J}\ln|J_1|\}\\ &= -\tfrac{1}{2}\{A^{J}_{J'}\partial_{\mu}A^{\mu}_{J}+A^{\mu}_{J}\partial_{\mu}A^{J}_{J'}+A^{J}_{J'}\partial_{J}\ln|J_1|\}\\ &= -\tfrac{1}{2}\{A^{J}_{J'}\partial_{\mu}A^{\mu}_{J}+\partial_{J}A^{J}_{J'}+A^{J}_{J'}\partial_{J}\ln|J|+A^{J}_{J'}\partial_{J}\ln|J_2|\}\\ &= A^{J}_{J'}\Omega_{J}-\tfrac{1}{2}\{\partial_{J}A^{J}_{J'}+\partial_{J'}\ln|J_2|\}.\end{aligned}$$

Finally, changing notation, if we put $J = |A^{K'}_{K}|$ the transformation law for the object Ω_J assumes the form

$$\Omega_{J'} = A^{J}_{J'}\Omega_{J}-\tfrac{1}{2}\{\partial_{J}A^{J}_{J'}+\partial_{J'}\ln|J|\}, \qquad J = |A^{J'}_{J}|. \tag{4.44}$$

Further on we shall come to know another very important geometric object, also with three indices, viz. the object of *parallel displacement*. Its transformation law will be similar to that of the object of anholonomicity of Schouten and van Dantzig.

One might ask whether, if we are given the functions

$$\Omega^{K}_{IJ} = \Omega^{K}_{IJ}(\xi)$$

as functions of the variables ξ^λ with the property that they are antisymmetric with respect to the lower indices, there exists in space an anholonomic system (K) for which the given system of functions represents an object of anholonomicity. It turns out [107] that for $n = 2$ the answer is always in the affirmative (the system contains two functions), for $n \geqslant 4$ the answer is in general negative, while for $n = 3$, the answer is in general affirmative, though exceptional cases do exist.

52. The Hessian of a Scalar Field. We now give an example of yet another geometric object which is not a special object but which is a concomitant [(1)] of another object and which becomes a special object when the group G_1 is restricted to the subgroup G_a.

Suppose we are given a scalar field $\sigma = \sigma(\xi)$, which we assume is of class C^2. It is known that $s_\lambda = \partial_\lambda \sigma$ then represents a field of covariant vectors. We now set

$$\sigma_{\lambda\mu} \stackrel{\text{def}}{=} \partial_\lambda \partial_\mu \sigma \tag{4.45}$$

and we denote by $\mathfrak{h}$ the determinant constructed from the elements $\sigma_{\lambda\mu}$:

$$\mathfrak{h} \stackrel{\text{def}}{=} |\sigma_{\lambda\mu}| \tag{4.46}$$

which is called the *Hessian* of the field σ.

We ask how the determinant $\mathfrak{h}$ transforms under the transition from the coordinate system (λ) to (λ'). With this in mind, we write

$$\overset{(\lambda')}{\mathfrak{h}} = |\sigma_{\lambda'\mu'}|,$$

so that we have

$$\sigma_{\mu'} = A^\mu_{\mu'} \sigma_\mu,$$

and consequently

$$\sigma_{\lambda'\mu'} = \partial_{\lambda'}(A^\mu_{\mu'} \sigma_\mu) = A^\mu_{\mu'} \partial_{\lambda'} \sigma_\mu + \sigma_\mu \partial_{\lambda'} A^\mu_{\mu'} = A^{\mu\lambda}_{\mu'\lambda'} \partial_\lambda \sigma_\mu + \sigma_\mu \partial_{\lambda'} A^\mu_{\mu'}$$
$$= A^{\mu\lambda}_{\mu'\lambda'} \sigma_{\lambda\mu} + \sigma_\mu \partial_{\lambda'} A^\mu_{\mu'}.$$

With the general group G_1 we see that the $\sigma_{\mu\lambda}$'s do not transform according

---

(1) The concept of concomitant, known for quite some time in the theory of invariants, was made precise in the theory of geometric objects in a paper published by Gołąb [116]. An object with coordinates Ω_j $(j = 1, 2, \ldots, q)$ is said to be the *concomitant* of an object with coordinates ω_a $(a = 1, 2, \ldots, p)$, if the coordinates Ω_j are single-valued functions of the coordinates ω_a. If we are concerned with fields of objects, the coordinates Ω_j may then depend not only on the coordinates ω_a themselves, but also on the derivatives of these coordinates with respect to the coordinates of the point of space; in this case we say that the object Ω is a *differential concomitant* of the object ω.

to the tensor law; the object $\sigma_{\lambda\mu}$ is a comitant of the vector field $\underset{(\lambda)}{\sigma}$. Similarly, the value of the determinant $\overset{(\lambda')}{\mathfrak{h}}$ will depend not only on $\overset{(\lambda)}{\mathfrak{h}}$ and on $A_\lambda^{\lambda'}$, $A_{\lambda'}^{\lambda}$, $\partial_{\lambda'} A_{\mu'}^{\mu}$, but also on the values σ_μ, $\sigma_{\lambda\mu}$, i.e. the Hessian of the field σ is in general a comitant of the objects σ_μ and $\sigma_{\lambda\mu}$ (as already stated, the object $\sigma_{\lambda\mu}$ is itself a comitant of σ_λ) and is not a special geometric object.

However, if we restrict the group G_1 to the affine subgroup G_a, then $A_\lambda^{\lambda'}$ will be a constant and, furthermore,

$$\partial_{\lambda'} A_{\mu'}^{\mu} = 0$$

so that $\sigma_{\lambda\mu}$ will then be a tensor, and the Hessian consequently becomes a density of weight $+2$.

53. Pentsov Objects. Among all geometric objects of the first class with a single coordinate we have come to know scalars, biscalars, G-densities, and W-densities. It is easily seen that the invertible functions of density are also geometric objects of the first class.

One may ask if it is the case that there are no other geometric objects of the first class with one coordinate. It turns out that for $n \geqslant 3$ there are indeed no more [111]. However, for $n = 2$ others do exist [1]. Consider, for instance, an arbitrary contravariant vector v^λ ($\lambda = 1, 2$) and write

$$\overset{(\lambda)}{\omega} \overset{\text{def}}{=} \frac{v^2}{v^1}. \tag{4.47}$$

Next, we calculate $\overset{(\lambda')}{\omega}$, discover the transformation law and verify that we do in fact have a geometric object of the first class.

We have

$$\omega' = \overset{(\lambda')}{\omega} = \frac{v^{2'}}{v^{1'}} = \frac{A_1^{2'} v^1 + A_2^{2'} v^2}{A_1^{1'} v^1 + A_2^{1'} v^2} = \frac{A_1^{2'} + A_2^{2'} \omega}{A_1^{2'} + A_2^{1'} \omega}. \tag{4.48}$$

We find that ω' does indeed depend only on the value of ω and on the values of $A_\lambda^{\lambda'}$ defined by the transformation $(\lambda) \to (\lambda')$. It must now be shown that the law above possesses the group property.

To do this, we start with the formula

$$\omega'' = \frac{A_{1'}^{2''} + A_{1'}^{2''} \omega'}{A_{1'}^{1''} + A_{2'}^{1''} \omega'}$$

---

[1] Objects of this type were first found by Pentsov [141].

and eliminate ω'. In this way we arrive at

$$\omega'' = \frac{A_{1'}^{2''}+A_{2'}^{2''}\dfrac{A_1^{2'}+A_2^{2'}\omega}{A_1^{1'}+A_2^{1'}\omega}}{A_{1'}^{1''}+A_{2'}^{1''}\dfrac{A_1^{2'}+A_2^{2'}\omega}{A_1^{1'}+A_1^{2'}\omega}} = \frac{A_{1'}^{2''}A_1^{1'}+A_{2'}^{2''}A_1^{2'}+(A_{1'}^{2''}A_2^{1'}+A_{2'}^{2''}A_2^{2'})\omega}{A_{1'}^{1''}A_1^{1'}+A_{2'}^{1''}A_2^{1'}+(A_{1'}^{1''}A_2^{1'}+A_{2'}^{1''}A_2^{2'})\omega}.$$

Using the formula

$$A_{\lambda'}^{\lambda''}A_{\lambda}^{\lambda'} = A_{\lambda}^{\lambda''},$$

we find that

$$\omega'' = \frac{A_1^{2''}+A_2^{2''}\omega}{A_1^{1''}+A_2^{1''}\omega}$$

in keeping with the law (4.48).

With the rotation group G_0 in the Euclidean plane the object ω represents the tangent of the angle α that the vector v makes with the axis ξ^1. If the coordinate system (λ) is turned through an angle φ, we have

$$A_1^{1'} = \cos\varphi, \quad A_2^{1'} = \sin\varphi, \quad A_1^{2'} = -\sin\varphi, \quad A_2^{2'} = \cos\varphi$$

and formula (4.48) becomes

$$\tan\alpha' = \frac{-\sin\varphi+\cos\varphi\tan\alpha}{\cos\varphi+\sin\varphi\tan\alpha} = \frac{\sin\alpha\cos\varphi-\cos\alpha\sin\varphi}{\sin\alpha\sin\varphi+\cos\alpha\cos\varphi}$$

$$= \frac{\sin(\alpha-\varphi)}{\cos(\alpha-\varphi)} = \tan(\alpha-\varphi),$$

from which we deduce that the vector v will make an angle of $\alpha-\varphi$ with the axis $\xi^{1'}$ turned through an angle φ.

Pentsov objects can also be formed for $n \geqslant 2$.

If a contravariant vector v is given in X_n, we set in general

$$\omega^\lambda = \frac{v^\lambda}{v^1} \quad (\lambda = 2, \ldots, n) \tag{4.49}$$

and we ask about the transformation law for ω^λ. We obtain

$$\omega^{\lambda'} = \frac{A_1^{\lambda'}+A_\mu^{\lambda'}\omega^\mu}{A_1^{1'}+A_\mu^{1'}\omega^\mu} \tag{4.50}$$

and it is easily verified that this law does possess the group property, so that ω^λ represents a geometric object (with non-linear transformation law) with $n-1$ coordinates. This object, just as in the special case of $n = 2$, can be interpreted geometrically as a direction without sense.

The question now is what object, i.e. what object with what transformation law, can be interpreted as a direction with sense. A certain result for $n = 2$ [1] gives reason to suppose that there do not exist geometric objects with $n-1$ coordinates which represent direction with sense. It turns out instead that there exist [121] geometric objects with n coordinates which represent direction with sense. We set

$$\omega^1 = \operatorname{sgn} v^1, \qquad \omega^\lambda = \frac{v^\lambda}{v^1} \quad \text{for} \quad \lambda = 2, \ldots, n. \tag{4.51}$$

and look for the transformation law for ω^1. We have

$$\omega^{1'} = \omega^1 \operatorname{sgn}(A_1^{1'} + A_\lambda^{1'} \omega^\lambda). \tag{4.52}$$

A relatively simple calculation shows that this law has the group property. It is true that ω^1 is not in itself a geometric object, but the set (ω^1, ω^λ for $\lambda = 2, \ldots, n$) does represent a geometric object and it can be interpreted as a direction with sense for when the original vector v^λ is replaced by $\tilde{v}^\lambda = \varrho v^\lambda$, with $\varrho > 0$, we obtain $\tilde{\omega}^\lambda = \omega^\lambda$, while for $\varrho < 0$ we have $\tilde{\omega}^\lambda \neq \omega^\lambda$.

54. The Density Gradient. We have seen that $s_\lambda = \partial_\lambda \sigma$ is a field of covariant vectors if σ is an arbitrary scalar field (of class C^1).

The question arises as to whether

$$\mathfrak{v}_\lambda \overset{\text{def}}{=} \partial_\lambda \mathfrak{v}, \tag{4.53}$$

is an object, and if so of what kind, if $\mathfrak{v}$ is the density field

$$\mathfrak{v}' = \varepsilon |J|^r \mathfrak{v}. \tag{4.54}$$

Let us now examine the transformation law for $\mathfrak{v}_\lambda$, making use of more general anholonomic systems. We have

$$\begin{aligned} \mathfrak{v}_{K'} = \partial_{K'} \mathfrak{v}' &= \partial_{K'}(\varepsilon |J|^r \mathfrak{v}) = \varepsilon(r|J|^{r-1} \operatorname{sgn} J \mathfrak{v} \partial_{K'} J + |J|^r \partial_{K'} \mathfrak{v}) \\ &= \varepsilon |J|^r (\mathfrak{v} r \partial_{K'} \ln |J| + \partial_{K'} \mathfrak{v}) \\ &= \varepsilon |J|^r A_{K'}^{K} (\mathfrak{v}_K + r \mathfrak{v} \partial_K \ln |J|). \end{aligned} \tag{4.55}$$

If $r \neq 0$ (i.e. if the density does not reduce to a scalar), if $\mathfrak{v}$ is not a trivial zero density, if the group does not reduce to G_a (for in this latter case the $A_\lambda^{\lambda'}$'s are constant and J is constant; the result then is that $\partial_K \ln |J| = 0$) and if the group does not reduce to the unimodular group G_u (for then $\ln J = 0$), then an extra term appears in the rule because of this extra term $\mathfrak{v}_K$ is not a tensor density but an object with a more complicated transformation law. The object $\mathfrak{v}_K$ becomes a density of the same weight,

$-r$, as $\mathfrak{v}$ and of the same type, G or W with the group G_a or with the group G_u. Nor is the object $\mathfrak{v}_K$ a geometric object, since the $\mathfrak{v}_{K'}$'s do not depend only on $\mathfrak{v}_K$ and the transformation law $(K) \rightarrow (K')$; however, $\mathfrak{v}_K$ is an object since the law (4.55) does possess the group property, as is easily verified: the simple proof relies only on the fact that the determinant of the combined transformation is equal to the product of the determinants of the compound transformations.

The inconvenient fact that ordinary differentiation of a density does not yield a tensor density will not be removed until we come to tensor analysis in Part Two of this book. The object (4.53) is called a *density gradient*.

55. Split Tensors. The concept of a tensor which we have hitherto learned always referred to a particular point in space, based on a certain pseudogroup of transformations. There is a need, however, to consider even more general concepts.

Suppose we have two spaces X_n and Y_m and a pseudogroup of transformations, not necessarily the same, in each. Let us denote the coordinates in the space X_n as before by

$$\xi^\lambda, \xi^{\lambda'}, \ldots \qquad (\lambda = 1, 2, \ldots, n),$$

and the coordinates in the space Y_m by

$$\eta^a, \eta^{a'}, \ldots \qquad (a = 1, 2, \ldots, m).$$

We now form the Cartesian product of the two spaces

$$Z_{n+m} = X_n \times Y_m$$

taking the pair (ξ^λ, η^a) as the coordinates of a point in the space Z_{n+m}, and we allow the coordinate transformations $(\lambda) \rightarrow (\lambda')$, $(a) \rightarrow (a')$ to be carried out independently of one another. We now introduce the notations

$$A_\lambda^{\lambda'} = \frac{\partial \xi^{\lambda'}}{\partial \xi^\lambda}, \qquad B_a^{a'} = \frac{\partial \eta^{a'}}{\partial \eta^a}, \tag{4.56}$$

i.e. the mixed coordinates of unit tensors in the two spaces X_n and Y_m, and we introduce the following definition:

DEFINITION. A *tensor density* is said to be defined at a particular point of the space Z if a set of numbers obeying the transformation law

$$\mathfrak{T}^{\lambda'_1\ldots\lambda'_p}{}_{\mu'_1\ldots\mu'_q}{}^{a'_1\ldots a'_r}{}_{b'_1\ldots b'_s} \tag{4.57}$$
$$= \varepsilon|J|^t\varepsilon_1|J_1|^w\mathfrak{T}^{\lambda_1\ldots\lambda_p}{}_{\mu_1\ldots\mu_q}{}^{a_1\ldots a_r}{}_{b_1\ldots b_s}A^{\lambda'_1\ldots\lambda'_p\mu_1\ldots\mu_q}_{\lambda_1\ldots\lambda_p\mu'_1\ldots\mu'_q}B^{a'_1\ldots a'_r b_1\ldots b_s}_{a_1\ldots a_r b'_1\ldots b'_s},$$

where $J = |A^{\lambda}_{\lambda'}|$, $J_1 = |B^{a}_{a'}|$, is given for every coordinate system. A quantity of this kind will be referred to as a *split quantity* because it "split" so that some indices are in the space X_n, and others in space Y_m [(1)].

The algebra of these quantities resembles that of ordinary tensors. They can be multiplied by scalars or by densities. Quantities of the same valence (p, q, r, s, t, w) can be added. Contraction can be carried out with respect to indices of the same kind (Latin or Greek) and symmetrization or anti-symmetrization can be performed only within a group of indices of the same level and species.

Another, similar situation obtains when the space Y_m is part of the space X_n. Then, instead of the space Z_{n+m} we take the surrounding space X_n, and split quantities are defined only for points of the subspace Y_m. The transformations of coordinates ξ^λ in the surrounding space and the coordinates η^a in the embedded space remain independent, of course, so that nothing changes in definition (4.57) of split quantities.

In this case, we introduce the intermediate tensor [(2)], following Schouten—merely changing the notation, writing

$$C^\lambda_a \stackrel{\text{def}}{=} \frac{\partial \xi^\lambda}{\partial \eta^a} = \frac{\partial \varphi^\lambda}{\partial \eta^a} \quad (\lambda = 1, 2, \ldots, n;\ a = 1, 2, \ldots, m), \tag{4.58}$$

if the equations

$$\xi^\lambda = \varphi(\eta^1, \eta^2, \ldots, \eta^m) \quad (\lambda = 1, 2, \ldots, n) \tag{4.59}$$

are the parametric equations of the space Y_m embedded in X_n. It is easily verified that C^λ_a is indeed a split tensor in accordance with the definition given above. By using the tensor C^λ_a we can now explain what it means to say that a certain quantity (defined at a point p of the embedded space Y_m) is a quantity of the space Y_m, even though it has been defined a priori as a quantity of the surrounding space X_n. Furthermore, certain quantities $\mathfrak{T}$ of the space X_n, which are defined at the point $p \in Y_m$, can now be uniquely assigned a quantity of the space Y_m which we can call the *projection* of the given quantity onto the space Y_m.

---

(1) Schouten calls such quantities *Verbindungsgrößen*, i.e. *binding quantities*.

(2) Our intermediate tensor C^λ_a is denoted B^λ_a by Schouten. Quantities B^ν_λ are not abbreviated at all by Schouten. The quantity C^λ_a of Schouten is denoted E^λ_a by us.

In the space X_n, take a vector v^ν which is defined at a point $p \in Y_m$ belonging to an m-vector tangent to the space Y_m

$$s^{\lambda_1 \dots \lambda_m} \overset{\text{def}}{=} C_1^{[\lambda_1} \dots C_m^{\lambda_m]}, \tag{4.60}$$

i.e., such that

$$s^{[\lambda_1 \dots \lambda_m} v^{\lambda_{m+1}]} = 0 . \tag{4.61}$$

Since each of the vectors

$$\underset{a}{t}{}^\nu \overset{\text{def}}{=} C_a^\nu$$

is tangent to the surface Y_m, the m-vector $s^{\lambda_1 \dots \lambda_m}$ is called a *tangent* to Y_m. The vector v^ν is then said to *lie* in the space Y_m.

In general, a contravariant quantity $\mathfrak{T}^{\lambda_1 \dots \lambda_p}$ at a point $p \in Y_m$ is said *to lie in the space* Y_m if

$$s^{[\lambda_1 \dots \lambda_m} \mathfrak{T}^{|\lambda_{m+1} \dots \lambda_{m+j-1}| \lambda_{m+j}] \lambda_{m+j+1} \dots \lambda_{m+p}} = 0 \quad \text{for} \quad j = 1, 2, \dots, p, \tag{4.62}$$

where, as usual, the indices between the vertical bars are not subject to antisymmetrization.

In order now to define the concept of the projection of a quantity $\mathfrak{T}$ of the space X_n onto the space Y_m, two cases must be distinguished. For purely covariant quantities the concept of projection can be defined without any further special assumptions. For quantities with contravariant indices, however, it is necessary to make a further assumption, viz. that the space Y_m has been 'bristled' (1).

To understand this difference, consider first the simplest case when the space Y_2 is embedded in the space E_3, and take some point $p \in Y_2$. At this point we attach an arbitrary vector w_λ, i.e. we take a pair of parallel planes, the first of which passes through the point p. This pair of planes will uniquely cut out of the plane tangent to Y_2 at p a pair of straight lines which can be regarded as a covariant vector of the plane tangent to the surface Y_2, that is, to the space Y_2. This vector can be called the *projection* of w_λ onto Y_2. On the other hand, if we take a contravariant vector v^ν attached at a point $p \in Y_2$, then in order to obtain its projection onto Y_2 we must know the projection of the terminal point of the vector v (the space is E_3, and thus the contravariant vector can therefore be thought of as a pair of points!). The projection will be uniquely defined if a direction of projection is known, and this in turn can be given most easily if a non-

(1) This term has been proposed by A. Hoborski whereas J. A. Schouten uses the term taut which is rather less suggestive.

zero contravariant vector n^ν which does not lie in the plane tangent to Y_2 is taken at the point p. The projection of the vector v^ν will then be a vector, with origin p and terminal point q, lying in Y_2 so that the vector $\overrightarrow{qr}$, where r is the terminal point of the vector v, is parallel to n^ν.

The aim now is to get an analytic expression for the projection so defined. If $\mathfrak{T}_{\lambda_1 \ldots \lambda_p}$ is a purely covariant quantity attached at a point $p \in Y_m$, its projection onto Y_m will be the quantity

$$'\mathfrak{T}_{a_1 \ldots a_p} \overset{\text{def}}{=} \mathfrak{T}_{\lambda_1 \ldots \lambda_p} C_{a_1}^{\lambda_1} \ldots C_{a_p}^{\lambda_p}. \tag{4.63}$$

On the other hand, to obtain the projection of the contravariant quantity $\mathfrak{T}^{\lambda_1 \ldots \lambda_p}$ attached at a point $p \in Y_m$, we define at the point p an auxiliary system of $(n-m)$ vectors

$$\underset{\mathfrak{a}}{b}^{\lambda} \quad (\mathfrak{a} = m+1, m+2, \ldots, n) \tag{4.64}$$

so that the set of vectors

$$\underset{\nu}{b}^{\lambda}, \tag{4.65}$$

where we make the additional assumption

$$\underset{a}{b}^{\lambda} \overset{*}{=} C_a^{\lambda}, \tag{4.66}$$

is a set of linearly independent vectors.

The set of vectors (4.64) is said to be the "bristling" of the space Y_m at the point p. The quantity $\mathfrak{T}^{\lambda_1 \ldots \lambda_p}$ will be projected onto Y_m parallel to the $(n-m)$-direction which is determined by the bristling $(n-m)$-vector

$$\underset{m+1}{b}^{[\lambda_{m+1}} \ldots \underset{n}{b}^{\lambda_n]}. \tag{4.67}$$

Denoting by

$$\overset{\nu}{b}_{\lambda} \tag{4.68}$$

a set of covariant vectors inverse to (4.65) and writing

$$C_{\lambda}^{a} \overset{\text{def}}{=} \overset{c}{b}_{\lambda} \underset{c}{b}^{a} \tag{4.69}$$

we can now define the projection $'\mathfrak{T}$ by means of the relation

$$'\mathfrak{T}^{a_1 \ldots a_p} \overset{\text{def}}{=} \mathfrak{T}^{\lambda_1 \ldots \lambda_p} C_{\lambda_1}^{a_1} \ldots C_{\lambda_p}^{a_p}. \tag{4.70}$$

The projections of both quantities (4.63) and (4.70) are given by similar formulae. Note, however, that the coefficients $C_a^{\lambda'}$ are determined automatically when a space $Y_m \subset X_n$ is given, whereas the quantities C_{λ}^{a} depend on the bristling of the space Y_m.

The definition of the projection of a mixed quantity can now be easily guessed, viz.

$$'\mathfrak{T}^{a_1\dots a_p}{}_{b_1\dots b_q} = \mathfrak{T}^{\lambda_1\dots\lambda_p}{}_{\mu_1\dots\mu_q} C^{a_1\dots a_p \mu_1\dots\mu_q}_{\lambda_1\dots\lambda_p b_1\dots b_q}. \tag{4.71}$$

It may be shown that the relations

$$C^a_\lambda C^\lambda_b = \delta^a_b \quad (a, b = 1, 2, \dots, m), \tag{4.72}$$

hold, whereas the similar relations

$$C^\lambda_a C^a_\mu = \delta^\lambda_\mu$$

in general do not.

It is true that relations (4.72) would, to some extent, justify replacing the letter C by the letter B (as Schouten does) for then, because $B^b_a = \delta^b_a$, where B is the unit tensor of the embedded space Y_m, the relations

$$B^a_\lambda B^\lambda_b = B^a_b$$

would be easy to remember. However, we prefer to retain both notations so that A remain a tensor of the space X_n, and B a tensor of the space Y_m, while C is a tensor intermediary to both spaces.

A criterion for the quantity $\mathfrak{T}^{\lambda_1\dots\lambda_p}{}_{\mu_1\dots\mu_q}$ to lie in the space Y_m and not to project beyond Y_m can now be written easily in terms of the split tensors C^λ_a and C^a_λ. This criterion is

$$\mathfrak{T}^{\lambda_1\dots\lambda_p}{}_{\mu_1\dots\mu_q} = \mathfrak{T}^{\varrho_1\dots\varrho_p}{}_{\sigma_1\dots\sigma_q} C^{\lambda_1\dots\lambda_p}_{\varrho_1\dots\varrho_p} C^{\sigma_1\dots\sigma_q}_{\mu_1\dots\mu_q}, \tag{4.73}$$

where the symbol C^ν_λ stands for

$$C^\nu_\lambda \overset{\text{def}}{=} C^\nu_a C^a_\lambda. \tag{4.74}$$

As an example let us calculate the projection of the $(n-m)$-vector (4.67) onto the space Y_m. By definition we have

$$\underset{m+1}{'b}{}^{[a_{m+1}} \dots \underset{n}{'b}{}^{a_n]} = \underset{m+1}{b}{}^{[\lambda_{m+1}} \dots \underset{n}{b}{}^{\lambda_n]} C^{a_{m+1}\dots a_n}_{\lambda_{m+1}\dots\lambda_n} \overset{*}{=} \underset{m+1}{b}{}^{[\lambda_{m+1}} \dots \underset{n}{b}{}^{\lambda_n]} \overset{a_{m+1}}{b}{}_{\lambda_{m+1}} \dots \overset{a_n}{b}{}_{\lambda_n}.$$

But

$$\underset{m+1}{b}{}^{\lambda_{m+1}} \dots \underset{n}{b}{}^{\lambda_n} \overset{a_{n+1}}{b}{}_{\lambda_{m+1}} \dots \overset{a_n}{b}{}_{\lambda_n} = \delta^{a_{m+1}\dots a_n}_{m+1\dots n}$$

and consequently

$$'b = \delta^{[a_{m+1}}_{m+1} \dots \delta^{a_n]}_n .$$

The right-hand side of the last equation, however, is zero, as can easily be shown by induction. Thus, $b = 0$ whence it follows in particular that the projection of every quantity lying in a subspace forming the bristling is zero.

56. Strong Quantities. Suppose that in the space X_n we are given a special subgroup of transformations of the form

$$\begin{aligned} \eta^{a'} &= \varphi^{a'}(\eta^1, \ldots, \eta^m) && (a = 1, 2, \ldots, m), \\ \zeta^{\mathfrak{a}'} &= \psi^{\mathfrak{a}'}(\zeta^{m+1}, \ldots, \zeta^{m+s}; \eta^1, \ldots, \eta^m) && (\mathfrak{a} = m+1, \ldots, m+s = n). \end{aligned} \tag{4.75}$$

Geometries based on such a special subgroup have been studied by many authors. A systematic investigation of these geometries has been carried out by two Japanese geometers, Kawaguchi [131] and Hokari [127].

Denoting the coordinates of a variable point of the space X_n, belonging to the systems $(a)+(\mathfrak{a})$, by

$$\xi^\nu \quad (\nu = 1, 2, \ldots, m, m+1, \ldots, n), \tag{4.76}$$

we consider a quantity $\mathfrak{T}$ which transforms according to the law

$$\mathfrak{T}^{\lambda'_1 \ldots \lambda'_p}{}_{\mu'_1 \ldots \mu'_q} = \varepsilon |J|^{-r} \varepsilon_1 |j|^{-t} \mathfrak{T}^{\lambda_1 \ldots \lambda_p}{}_{\mu_1 \ldots \mu_q} A^{\lambda'_1 \ldots \lambda'_p \mu_1 \ldots \mu_q}_{\lambda_1 \ldots \lambda_p \mu'_1 \ldots \mu'_q}, \tag{4.77}$$

where, by (4.56), we have

$$\begin{gathered} A^a_{a'} = \frac{\partial \eta^a}{\partial \eta^{a'}}, \quad A^a_{\mathfrak{a}'} = 0, \quad A^{\mathfrak{a}}_{a'} = \frac{\partial \zeta^{\mathfrak{a}}}{\partial \eta^{a'}}, \quad A^{\mathfrak{a}}_{\mathfrak{a}'} = \frac{\partial \zeta^{\mathfrak{a}}}{\partial \zeta^{\mathfrak{a}'}}, \\ A^{a'}_a = \frac{\partial \eta^{a'}}{\partial \eta^a}, \quad A^{a'}_{\mathfrak{a}} = 0, \quad A^{\mathfrak{a}'}_a = \frac{\partial \zeta^{\mathfrak{a}'}}{\partial \eta^a}, \quad A^{\mathfrak{a}'}_{\mathfrak{a}} = \frac{\partial \zeta^{\mathfrak{a}'}}{\partial \zeta^{\mathfrak{a}}}, \\ J = \left| \frac{\partial \zeta^{\mathfrak{a}'}}{\partial \zeta^{\mathfrak{a}}} \right|, \quad j = \left| \frac{\partial \eta^{a'}}{\partial \eta^a} \right|. \end{gathered} \tag{4.78}$$

If a quantity $\mathfrak{T}$ possesses the property that the relations

$$\mathfrak{T}^{\mathfrak{a}'_1 \ldots \mathfrak{a}'_p}{}_{\mathfrak{b}'_1 \ldots \mathfrak{b}'_q} = \varepsilon |J|^{-r} \mathfrak{T}^{\mathfrak{a}_1 \ldots \mathfrak{a}_p}{}_{\mathfrak{b}_1 \ldots \mathfrak{b}_q} A^{\mathfrak{a}'_1 \ldots \mathfrak{a}'_p \mathfrak{b}_1 \ldots \mathfrak{b}_q}_{\mathfrak{a}_1 \ldots \mathfrak{a}_p \mathfrak{b}'_1 \ldots \mathfrak{b}'_q}, \tag{4.79}$$

are valid, we call it a *strong quantity* of the space X_n.

Suppose we are given the quantity $\mathfrak{T}$ at a point in the space X_n. The equations

$$\eta^a = \underset{0}{\eta}{}^a \quad (a = 1, 2, \ldots, m) \tag{4.80}$$

define a surface Y_s in the space X_n. We choose the constants $\underset{0}{\eta}{}^a$ in such a way that this surface passes through the point at which the quantity $\mathfrak{T}$ is attached. Assume that $\underset{a}{b}{}^\lambda$ is some vector tangent to the parametric line, that $\eta^b = \text{const}$, and $b \neq a$, $\xi^{\mathfrak{a}} = \text{const}$, and bristle the surface Y_s using the set of vectors $\underset{a}{b}{}^\lambda$. By $'\mathfrak{T}$ we denote the projection of $\mathfrak{T}$ onto Y_s following the law given in the preceding section. Accordingly, if $\mathfrak{T}$ is a strong quantity, the coordinates of the projection $'\mathfrak{T}$ are precisely $\mathfrak{T}^{\mathfrak{a}_1 \cdots \mathfrak{a}_p}{}_{\mathfrak{b}_1 \cdots \mathfrak{b}_q}$.

The concept of strong quantities is important in mechanics (in this case $m = 1$); it was introduced into science by Wundheiler [158], one of the originators of the exact definition of geometric object.

57. Multipoint Tensor Fields. In addition to ordinary tensor fields, where a tensor with a certain valence is defined at every point p in a certain region of the space X_n, one can consider tensor fields where a tensor with a particular valence is associated with a set of r points $(p_1, \dots, p_r)$. This kind of field was first studied by Michal [138]. We now give a few simple examples of such fields.

EXAMPLE 1. Consider an n-dimensional Riemannian space V_n. It is known (cf. p. 299) that two points p, q sufficiently close together can be joined by a single geodesic [(1)]. The length of this geodesic is denoted by $\varrho(p, q)$ which is called the *geodesic distance* between the points p, q. The distance $\varrho(p, q)$ is an example of a two-point scalar field.

EXAMPLE 2. Set

$$\sigma \overset{\text{def}}{=} \tfrac{1}{2}\varrho^2(p, q) \tag{4.81}$$

and let ξ^ν denote the coordinates of a point p, while η^ν denotes the coordinates of another point q. We now consider

$$\partial\sigma/\partial\eta^\nu \tag{4.82}$$

and examine how they transform. We have

$$\frac{\partial\sigma}{\partial\eta^{\nu'}} = \frac{\partial\sigma}{\partial\eta^\nu} \cdot \frac{\partial\eta^\nu}{\partial\eta^{\nu'}} = \frac{\partial\sigma}{\partial\eta^\nu}(A^\nu_{\nu'})_q. \tag{4.83}$$

The quantities (4.82) constitute a two-point field: this field is scalar with respect to the first point p and covariant vectorial with respect to the second point q.

EXAMPLE 3. A *correlation tensor* field [52], which is a two-point field of contravariant tensors of valence (2, 0), is introduced in turbulence theory.

By $v(t, \xi^1, \xi^2, \xi^3) = v^i(t, p)$ we denote the coordinates of the velocity field of a flowing medium at some time t and at the point p. We denote

(1) This concept will be introduced in Part II, Section 103.

the metric tensor by $g_{ik}(p)$, and put

$$C^{ik}(p, q) = \frac{M[v^i(t, p) \cdot v^k(t, p)]}{[g_{rs}(p) M[v^r(t, p) v^s(t, p)] g_{jl}(q) M[v^j(t, q) v^l(t, q)]]^{1/2}}, \tag{4.84}$$

where $M[\varphi(t)]$ stands for the average value of the function φ in a certain fixed interval $\langle t_1, t_2 \rangle$. It may be shown that C^{ik} is a two-point (p, q) field of two-index contravariant tensors. This tensor is given the name of correlation tensor.

58. A Backward Glance. Before we go on to Part II, i.e. to tensor analysis, let us sum up the results established in Part I.

Some space consisting of element called *points* always forms the basis for our studies. In order to make our ideas amenable to calculation, we consider coordinate systems in these spaces, i.e. mappings of points of space, or of part of space, onto analytic entities or, in other words, sequences of numbers. All our discussions concerning coordinate systems were of a local character. The topological structure of the space plays no part in tensor algebra.

Since we are compelled to consider various coordinate systems and go over from one to another, it becomes necessary to specify the set of allowable coordinate systems; they are selected by the choice of transformation group — or more generally, pseudogroup, G — whose choice together with the choice of one system of protocoordinates completely defines the set of all allowable systems. Only in certain geometries can the choice of this protosystem be made more specific [108]; this possibility, however, is not particularly important.

In addition to coordinate systems, it becomes necessary to consider so-called non-holonomic systems which are defined by means of so-called local bases, i.e. systems built up from n (dimension of the space) linearly independent vectors (reference frames).

The choice of the (pseudo)group of transformations is the fundamental element throughout tensor calculus. The role this plays has been demonstrated with striking examples in which the species of the geometric object under consideration changed completely with a change of group.

The fundamental, general concept in tensor calculus is that of *geometric object*. This modern concept embodies in particular the concept of tensors and tensor densities which play the cardinal role.

In addition to objects with one coordinate—scalars, biscalars, W-scalars, and densities, we have objects with n coordinates, the most important being vectors (contra- and co-variant) and vector densities. Tensors have more than n coordinates in general. An algebra is constructed for these objects, i.e. certain operations involving these quantities are defined.

Tensors are certainly "constructs" which are closely associated with a particular point in space. In general, operations are defined for quantities defined at the same point in space. Only if an affine group G_a is employed is it possible in the case of tensors to "abstract" from the "point of attachment" and to define free quantities, just as we go over to the concept of free vector from the concept of attached vector in Euclidean or affine space.

This possibility has its origins in the following fact. Having defined (with the group G_a) some quantity $\mathfrak{T}$ at a point p of space, we can "transplant" that quantity to the whole of the space (or throughout a neighbourhood of p), expanding it to a constant function, i.e. one taking the same coordinates at all other points. Such an operation is invariant because the transformation law for the quantity $\mathfrak{T}$ contains the values of the partial derivatives of the transforming functions at the appropriate point of space; with the group G_a these values are constant and consequently this function retains its constant character in any other allowable coordinate system.

On the other hand, with a group more general than the affine, where the $A_{\lambda}^{\lambda'}$'s are not constant, the extension of the quantity $\mathfrak{T}$ to the entire field by taking this function to be constant with respect to the points of the field would no longer be invariant.

It would thus be meaningless to speak of such an extension.

The fact that two quantities $\mathfrak{T}$ of the same species, but attached at two different points in space, cannot be compared causes serious difficulties in tensor analysis, as we shall see in Part II of the book.

Restricting the transformation group may enrich the geometry under consideration in certain properties; on the other hand, restriction of the group impoverishes the space in regard to the variety of existing geometric objects. The group G_m, for instance, allows a metric (distance and angular) to be introduced, but at the same time the restriction to this subgroup obliterates the difference between tensor densities and tensors, and between contravariant and covariant quantities. Restriction of, say, the group G_m to the translation subgroup would make tensor geometry utterly banal, and the transformation laws for tensor coordinates would become illusory since these coordinates would not change at all under transition from one coordinate system to another.

If a non-trivial Weyl density is fixed in spaces based on the group G_a, and a field of such densities is fixed in spaces based on the group G_i ($i = 1, 2, \dots$), it becomes possible to introduce an n-dimensional volume measure without an orientation being imparted to the space. On the other hand, fixing a certain G-density at the same time makes it possible to give the space a particular orientation, just as does the restriction of the group G_m to the subgroup $J = 1$.

The establishment of a non-singular fundamental tensor in the case of affine geometry, and the establishment of a variable field of such tensors in the case of the general geometry G_1, enables a local distance metric to be introduced, and — when the tensor defines a positive quadratic form — a real angular metric. By the limiting process of integration a formula can be obtained for the arc length, given by the equations $\xi^\lambda = \xi^\lambda(\tau)$ ($\alpha \leqslant \tau \leqslant \beta$). This formula is given by

$$s = \int_{\tau_0}^{\tau} \left[\left| g_{\lambda\mu}(\xi(\tau)) \frac{d\xi^\lambda}{d\tau} \cdot \frac{d\xi^\mu}{d\tau} \right| \right]^{1/2} d\tau$$

and is, of course, invariant under transformations of the group G_1.

The introduction of a metric tensor also enables a new algebraic operation to be defined for tensors, viz. raising and lowering indices. This operation leads to a one-to-one ordering of certain quantities, and consequently to the identification of certain tensor quantities, although this is not always true.

In addition to tensor quantities, we have also learned (and we shall learn more still) about geometric objects whose transformation law is more involved. The question arises: what are all the possible geometric objects? This is the central problem in the theory of geometric objects, and is far from having been solved in general. In some cases (depending on the dimension of the space, the number of coordinates of the object, the class of the object), a solution has been found to this problem, at times under assumptions of analyticity, and in other cases with more modest assumptions. This book does not go into the details of this.

An interesting question concerns the interpretation of objects. This is not a straightforward matter and it is our impression that it has not yet been definitely settled. The interpretation of contravariant vectors presents no difficulty here. We owe the interpretation of covariant vectors to Schouten. Furthermore, we have given an interpretation of covariant tensors. Contravariant tensors can also be easily provided with an interpretation, a hyperquadric being defined as the envelope of a family of hyperplanes. An interpretation has also been given for p-vectors. But none has yet been provided for tensors with more indices.

The discovery of W-scalars assisted greatly in advancing the matter of interpreting tensor densities. A completely lucid solution of this problem none has still to be worked out.

At this stage, half-way through this book, the reader has undoubtedly discovered by personal experience the extent to which mastery of tensor calculus is dependent on the introduction of an appropriate set of symbols. Only notation allowing far-reaching mechanization of calculation is capable of overcoming all the difficulties involved in tensor calculus and of mastering the problems in this calculus which will subsequently appear for solution.

Part II

The Analysis of Tensors

CHAPTER V

PARALLEL DISPLACEMENT

59. The Derivative of a Vector Field Along a Curve. The Object of Connection. Suppose that a curve C with equations

$$\xi^{\nu} = \xi^{\nu}(\tau) \qquad (\alpha \leqslant \tau \leqslant \beta; \nu = 1, 2, \ldots, n) \tag{5.1}$$

is given in an n-dimensional Euclidean space and that a vector field is given along the curve C, i.e. suppose that a (contravariant) vector v^{ν} is attached at each point $p(\tau)$ of the curve C. The derivative of this field at the point p is defined as follows.

Consider two vectors: one at the point $p(\tau)$, and the other at a point $p(\tau+h)$. They will be denoted by the abbreviations $v(\tau)$ and $v(\tau+h)$ omitting any explicit mention of the point of attachment.

Even though the vectors $v(\tau)$ and $v(\tau+h)$ have different points of attachment, their difference can be formed since $v(\tau+h)$ can be displaced parallelly to the point of attachment $p(\tau)$ (the concept of parallelism in Euclidean spaces does exist). Then, using the parallelogram law we can construct the vector

$$v(\tau+h)-v(\tau).$$

This vector is divided by h, i.e. it is multiplied by the scalar factor $1/h$, to give the vector

$$\frac{v(\tau+h)-v(\tau)}{h}. \tag{5.2}$$

We now have to pass to the limit. In the case under discussion, this is simple. Since the variable vector (5.2) (i.e. variable in h) now has a fixed point of attachment $p(\tau)$, a limit vector will exist if the end of the variable vector (5.2) approaches a limiting position. If this limiting vector does exist, we denote it by

$$dv/d\tau \tag{5.3}$$

and we call it the *derivative of the field* $v(\tau)$ *at the point* $p(\tau)$; in fact one should speak of the derivative of the field at the point τ since the vector (5.3) depends not only on the given field $v(\tau)$ in the neighbourhood of $p(\tau)$ but also on the parametrization of the field. Indeed it can be shown—this is a standard part of every elementary course of vector analysis—that if the parameter τ is replaced on the curve C by a new parameter τ' related to τ by

$$\tau = \varphi(\tau'),$$

then, without changing the vector field along the curve, another derivative vector is obtained, viz.

$$\frac{dv}{d\tau'} = \frac{dv}{d\tau}\varphi'(\tau'),$$

the analogue of the theorem concerning the derivative of a composite function for ordinary functions, which shows that the derivative of a vector field depends on the parametrization of the field, and not only on the field itself.

If the derivative $dv/d\tau$ exists at every point on the curve, it constitutes a new field along C, parametrized with the same parameter. We shall call this the *derivative field* or the *derivative of the field.*

This field can, of course, itself be differentiated in turn to yield the second derivative of the field

$$d^2v/d\tau^2.$$

Derivatives of higher orders are obtained by following the same procedure.

This procedure can be generalized and emulated in the E_n spaces, since it is well known that the concept of parallelism of straight lines exists in these spaces. It is true that the spaces E_n do not possess metrics and that any two arbitrary vectors cannot be compared in the sense of length. Nevertheless, the ratio of two parallel vectors $\underset{1}{v}$ and $\underset{2}{v}$ can be defined as the scalar coefficient in the formula

$$\underset{2}{v} = \alpha \underset{1}{v}, \tag{5.4}$$

as it can casily be verified that this coefficient is an invariant under transformations of the group G_a. This ratio is also an invariant in spaces based on the group G_1 for two linearly independent vectors attached at the same point p, but in E_n spaces relations of the type (5.4) are also meaningful for vectors which are not attached at the same point (as is the case here) and hence the parallelogram rule is applicable here.

This procedure cannot be used in spaces more general than E_n because we do not know what it means to carry out a parallel displacement of a vector $v(\tau+h)$ to a point $p(\tau)$. We shall see below that such a definition calls for certain additional assumptions which endow the space with a certain new geometric object which is not given *a priori.*

The concept of the derivative of a vector field can be extended, without much difficulty, to fields of covariant vectors as well. Both concepts have the following fundamental properties, whose proofs are left to the reader:

$$\frac{dv}{d\tau} = \frac{dv}{d\tau'} \cdot \frac{d\tau'}{d\tau},$$

$$\frac{d(\alpha v)}{d\tau} = \alpha \frac{dv}{d\tau} + \frac{d\alpha}{d\tau} v, \tag{5.5}$$

$$\frac{d(u+v)}{d\tau} = \frac{du}{d\tau} + \frac{dv}{d\tau}.$$

Moreover, the Leibniz formula

$$\frac{d(uv)}{d\tau} = \frac{du}{d\tau} v + u \frac{dv}{d\tau} \tag{5.6}$$

holds for the scalar product of vectors u and v in Euclidean spaces (the scalar product is not defined in general affine spaces). In three-dimensional Euclidean spaces the Leibniz formula is also satisfied for the vector product $[u, v]$:

$$\frac{d[u, v]}{d\tau} = \left[\frac{du}{d\tau}, v\right] + \left[u, \frac{dv}{d\tau}\right]. \tag{5.7}$$

For Euclidean spaces relation (5.6) implies a very important relation for vector fields of constant length

$$|v| = \text{const},$$

viz.

$$\frac{dv}{d\tau} v = 0,$$

which means that the derivative vector is perpendicular to v, i.e. is zero or is perpendicular in the strict sense.

We now attempt to generalize the concept of the derivative of a vector field in spaces more general than E_n and to introduce the concept of the derivative of a tensor field in such a way that the laws given above remain valid and, moreover, so that the derivative of a field of any quantity is a field of the same sort of quantity, i.e. a quantity that transforms in the same way under transition to a new allowable coordinate system.

Since a vector is an object with n coordinates, we first discuss the problem of the derivative for simpler fields, for example a field of densities having one coordinate.

Suppose that along the (parametrized) curve C we are given a density field

$$\mathfrak{v} = \mathfrak{v}(\tau)$$

of weight -1, whose transformation law, as we recall, is

$$\mathfrak{v}' = J\mathfrak{v}. \tag{5.8}$$

REMARK. Strictly speaking, we should write $(J)_\tau$ instead of J here to indicate clearly that the value of the Jacobian J has to be calculated at that point in space which coincides with the point on the curve C corresponding to the value τ of the parameter.

We now form the ordinary density derivative

$$d\mathfrak{v}/d\tau \tag{5.9}$$

and enquire whether it is a density: the answer turns out to be in the negative. We have

$$\frac{d\mathfrak{v}'}{d\tau} = J\frac{d\mathfrak{v}}{d\tau} + \frac{dJ}{d\tau}\mathfrak{v}.$$

It is clear from this formula that the ordinary density derivative (5.9) will be a density only when the transformation law for the coordinates is such that J is constant along the entire curve. This is true for E_n, of course, but not for more general spaces.

It would plainly be pointless to define the derivative of a density as an ordinary derivative because such a derivative would not be invariant in character.

This example reveals a new difficulty encountered when defining the absolute derivative of geometric objects.

We now try to calculate the ordinary derivative of a vector field $v^\lambda(\tau)$, given along a curve C. For Euclidean space it is easily proved that, in the system (λ), the derivative vector $dv/d\tau$ has coordinates

$$dv^\lambda/d\tau \qquad (\lambda = 1, 2, \ldots, n).$$

Passing to the coordinate system (λ'), we calculate

$$dv^{\lambda'}/d\tau$$

The formula for the derivative of a product yields

$$\frac{dv^{\lambda'}}{d\tau} = \frac{d}{d\tau}(A_{\lambda}^{\lambda'})v^{\lambda} + A_{\lambda}^{\lambda'}\frac{dv^{\lambda}}{d\tau}.$$

But

$$\frac{d}{d\tau}(A_{\lambda}^{\lambda'}) = \partial_{\mu}(A_{\lambda}^{\lambda'})\frac{d\xi^{\mu}}{d\tau},$$

so we finally have

$$\frac{dv^{\lambda'}}{d\tau} = A_{\lambda}^{\lambda'}\frac{dv^{\lambda}}{d\tau} + v^{\lambda}(\partial_{\mu}A_{\lambda}^{\lambda'})\frac{d\xi^{\mu}}{d\tau}. \tag{5.10}$$

If the second term on the right-hand side were absent, this formula would state that the ordinary derivative of the field $v(\tau)$ is again a contravariant vector, in keeping with our postulate above. For E_n in which the $A_{\lambda}^{\lambda'}$'s are constant, i.e. when $\partial_{\mu}A_{\lambda}^{\lambda'} = 0$, this is indeed the case, but in more general spaces the second non-zero term on the right-hand side of (5.10) makes it impossible to take the object $dv^{\lambda}/d\tau$ as a derivative vector.

We now return to the example of the density $\mathfrak{v}$ given along the curve C. In order to define the so-called *absolute derivative* of this field, which is again a density, we attempt to overcome the difficulty mentioned above. To this end we try adding to the ordinary derivative $d\mathfrak{v}/d\tau$ as simple as possible a term in order to obtain an invariant expression which in the given case is a density. This additional "balancing" term can be expected to depend on the value of the field $\mathfrak{v}$ itself. The simplest function of the variable $\mathfrak{v}$ is, of course, a homogeneous linear function $\omega\mathfrak{v}$. We do not make any assumption regarding the nature of the coefficient ω for the moment. We shall presently see that the nature of this coefficient makes itself clear following the condition that the expression

$$\frac{D\mathfrak{v}}{d\tau} \overset{\text{def}}{=} \frac{d\mathfrak{v}}{d\tau} + \omega\mathfrak{v} \tag{5.11}$$

be a density of weight -1.

Accordingly, if we make the assumption

$$\frac{D\mathfrak{v}'}{d\tau} = \frac{d\mathfrak{v}'}{d\tau} + \omega'\mathfrak{v}',$$

then we must also have

$$\frac{D\mathfrak{v}'}{d\tau} = J\frac{D\mathfrak{v}}{d\tau}.$$

We therefore arrive at the relation

$$J\left(\frac{d\mathfrak{v}}{d\tau}+\omega\mathfrak{v}\right) = J\frac{d\mathfrak{v}}{d\tau}+\mathfrak{v}\frac{dJ}{d\tau}+\omega' J\mathfrak{v}$$

or, after simplification and simple rearrangement,

$$J\mathfrak{v}\omega = J\mathfrak{v}\frac{d\ln|J|}{d\tau}+J\mathfrak{v}\omega'.$$

Dividing this relation by $J \neq 0$, we thus have

$$\mathfrak{v}\omega' = \mathfrak{v}\omega-\mathfrak{v}\frac{d\ln|J|}{d\tau}. \tag{5.12}$$

But since, naturally, the coefficient ω is required not to depend on the field $\mathfrak{v}$ itself, i.e. relation (5.11) is required to hold for all fields $\mathfrak{v}$, we can cancel $\mathfrak{v}$ from both sides of equation (5.12). Accordingly, we finally have

$$\omega' = \omega-\frac{d\ln|J|}{d\tau}. \tag{5.13}$$

We therefore see that upon transition from the coordinate system (λ) to (λ'), the coefficient ω does not remain unchanged but is subject to the transformation expressed by formula (5.13). It is not difficult to verify that (5.13) has the group property, and hence represents a geometric object, of class two, since second derivatives of the functions defining the transformation $(\lambda) \to (\lambda')$ appear when the Jacobian J is differentiated.

It is thus turns out that if along C we take a field of geometric objects with the transformation law (5.13), which is obeyed by all possible density fields $\mathfrak{v}$, we can use equation (5.11) to define the so-called absolute derivative $D\mathfrak{v}/d\tau$ of the field $\mathfrak{v}$. This derivative is an object of the same type as $\mathfrak{v}$, namely a density. The absolute derivative is described as being defined in terms of the ordinary derivative, in terms of the field $\mathfrak{v}$, and in terms of an auxiliary second-class object ω which is independent of $\mathfrak{v}$.

Using the auxiliary object ω, we can now define the absolute derivative of a density field of an arbitrary weight $-r$

$$\mathfrak{v}' = \varepsilon|J|^r\mathfrak{v}. \tag{5.14}$$

We make the assumption

$$\frac{D\mathfrak{v}}{d\tau} = \frac{d\mathfrak{v}}{d\tau}+r\mathfrak{v}\omega,$$

where ϱ is a constant scalar factor and ω is the object mentioned above which obeys the transformation law (5.13).

At the same time, we require that $D\mathfrak{v}/d\tau$ also be a density of weight $-r$. Thus, on the one hand we have

$$\frac{D\mathfrak{v}'}{d\tau} = \varepsilon|J|^r \frac{D\mathfrak{v}}{d\tau}; \tag{5.15}$$

while on the other hand

$$\frac{D\mathfrak{v}'}{d\tau} = \frac{d\mathfrak{v}'}{d\tau} + \varrho\mathfrak{v}'\omega'.$$

When we compare these relations having regard to the transformation laws (5.14) and (5.15), we obtain $\varrho = r$, so that we finally arrive at the formula

$$\frac{D\mathfrak{v}}{d\tau} = \frac{d\mathfrak{v}}{d\tau} + r\mathfrak{v}\omega. \tag{5.16}$$

Note that in the special case of the scalar field ($r = 0$) the formula above reduces to the ordinary derivative of the scalar field.

The absolute derivative so defined can be shown to possess the properties:

1. $$\frac{D(\underset{1}{\mathfrak{v}}+\underset{2}{\mathfrak{v}})}{d\tau} = \frac{D\underset{1}{\mathfrak{v}}}{d\tau} + \frac{D\underset{2}{\mathfrak{v}}}{d\tau},$$

if $\underset{1}{\mathfrak{v}}$ and $\underset{2}{\mathfrak{v}}$ are two density fields of the same weight $-r$, and

2. $$\frac{D(\underset{1}{\mathfrak{v}}\underset{2}{\mathfrak{v}})}{d\tau} = \underset{1}{\mathfrak{v}}\frac{D\underset{2}{\mathfrak{v}}}{d\tau} + \underset{2}{\mathfrak{v}}\frac{D\underset{1}{\mathfrak{v}}}{d\tau},$$

where $\underset{1}{\mathfrak{v}}$ and $\underset{2}{\mathfrak{v}}$ are two fields of arbitrary weights. Indeed, we have

$$\frac{D(\underset{1}{\mathfrak{v}}+\underset{2}{\mathfrak{v}})}{d\tau} = \frac{d(\underset{1}{\mathfrak{v}}+\underset{2}{\mathfrak{v}})}{d\tau} + r(\underset{1}{\mathfrak{v}}+\underset{2}{\mathfrak{v}})\omega = \frac{d\underset{1}{\mathfrak{v}}}{d\tau} + \frac{d\underset{2}{\mathfrak{v}}}{d\tau} + r\underset{1}{\mathfrak{v}}\omega + r\underset{2}{\mathfrak{v}}\omega$$

$$= \frac{d\underset{1}{\mathfrak{v}}}{d\tau} + r\underset{1}{\mathfrak{v}}\omega + \frac{d\underset{2}{\mathfrak{v}}}{d\tau} + r\underset{2}{\mathfrak{v}}\omega = \frac{D\underset{1}{\mathfrak{v}}}{d\tau} + \frac{D\underset{2}{\mathfrak{a}}}{d\tau}.$$

Similarly,

$$\frac{D(\underset{1}{\mathfrak{v}}\underset{2}{\mathfrak{v}})}{d\tau} = \frac{d(\underset{1}{\mathfrak{v}}\underset{2}{\mathfrak{v}})}{d\tau} + (r_1+r_2)\underset{1}{\mathfrak{v}}\underset{2}{\mathfrak{v}}\omega = \frac{d\underset{1}{\mathfrak{v}}}{d\tau}\underset{2}{\mathfrak{v}} + \frac{d\underset{2}{\mathfrak{v}}}{d\tau} + r_1\underset{1}{\mathfrak{v}}\underset{2}{\mathfrak{v}}\omega + r_2\underset{1}{\mathfrak{v}}\underset{2}{\mathfrak{v}}\omega$$

$$= \underset{2}{\mathfrak{v}}\left\{\frac{d\underset{1}{\mathfrak{v}}}{d\tau} + r_1\underset{1}{\mathfrak{v}}\omega\right\} + \underset{1}{\mathfrak{v}}\left\{\frac{d\underset{2}{\mathfrak{v}}}{d\tau} + r_2\underset{2}{\mathfrak{v}}\omega\right\} = \underset{2}{\mathfrak{v}}\frac{D\underset{1}{\mathfrak{v}}}{d\tau} + \underset{1}{\mathfrak{v}}\frac{D\underset{2}{\mathfrak{v}}}{d\tau},$$

which was to be proved.

Note, however, that when the parameter on the curve C is changed

$$\tau = \varphi(\bar{\tau})$$

the object ω must be replaced by the object

$$\bar{\omega} = \omega \frac{d\tau}{d\bar{\tau}},$$

in order that the property

$$\frac{D\mathfrak{v}}{d\bar{\tau}} = \frac{D\mathfrak{v}}{d\tau} \cdot \frac{d\tau}{d\bar{\tau}}$$

may be preserved.

Let us now turn to the case of the vector field

$$v^\lambda(\tau).$$

As we found when examining the transformation law derived in (5.10) the ordinary derivative $dv^\lambda/d\tau$ does not represent a vector field as we would wish. In defining an absolute derivative $Dv^\lambda/d\tau$ which is itself a field of vectors (contravariant), we make use of an idea similar to that used in the example above. In order to obtain $Dv^\lambda/d\tau$ we try adding to the coordinates of the ordinary derivative $dv^\lambda/d\tau$ as simple a term as possible which hopefully "compensates" for the improper way in which $dv^\lambda/d\tau$ transforms, with the result that $Dv^\lambda/d\tau$ is then a vector field. As simple a term as possible means a term that is linear in the field v^λ itself and also linear in $d\xi^\lambda/d\tau$; the latter, as can be easily shown, represents a field of contravariant vectors tangent to C. In fact if we go over to the coordinate system (λ'), the equation of C can be written as

$$\xi^{\lambda'} = \xi^{\lambda'}(\tau) = \varphi^{\lambda'}[\xi^1(\tau), \ldots, \xi^n(\tau)].$$

Thus, by the theorem on the differentiation of composite functions we have

$$\frac{d\xi^{\lambda'}}{d\tau} = \frac{\partial \varphi^{\lambda'}}{\partial \xi^\lambda} \cdot \frac{d\xi^\lambda}{d\tau} = A_\lambda^{\lambda'} \frac{d\xi^\lambda}{d\tau},$$

which shows that $d\xi^\lambda/d\tau$ is indeed a contravariant vector. The statement that this is a vector lying on a tangent to the curve C is justified by the fact that, in the special case when the space is E_n, the end of the vector $d\xi^\lambda/d\tau$ attached at a point p on the curve C does indeed lie on a tangent drawn to C at p. Of course, when the space is more general, the statement that a vector lies on a tangent to C cannot be taken in any literal sense because, firstly, the usual interpretation of vector breaks down and, secondly, the concept of tangent has not yet been defined.

Accordingly, we write

$$\frac{Dv^\lambda}{d\tau} = \frac{dv^\lambda}{d\tau} + \Gamma^\lambda_{\mu\nu} v^\nu \frac{d\xi^\mu}{d\tau} \qquad (\lambda = 1, 2, \dots, n). \tag{5.17}$$

NOTE. Since we have assumed linearity with respect to the field v itself and with respect to the vector $d\xi/d\tau$ which is tangent to C, double summation appears in formula (5.17) and the coefficients of the sum must therefore have two indices. The third index λ derives from the fact that we have n relations (5.17), i.e. as many as the vector has coordinates.

The condition that the expression (5.17) represents a contravariant vector, together with the transformation law (5.10) previously derived for the first component of $d^\lambda v/d\tau$, leads to the transformation law for the coefficients

$$\Gamma^\lambda_{\mu\nu}. \tag{5.18}$$

By equation (5.10) and by virtue of the fact that v^λ and $d\xi^\lambda/d\tau$ are vectors, we have on the one hand

$$\frac{Dv^{\lambda'}}{d\tau} = \frac{dv^{\lambda'}}{d\tau} + \Gamma^{\lambda'}_{\mu'\nu'} v^{\nu'} \frac{d\xi^{\mu'}}{d\tau} = A^{\lambda'}_\lambda \frac{dv^\lambda}{d\tau} + (\partial_\mu A^{\lambda'}_\lambda) v^\lambda \frac{d\xi^\mu}{d\tau} + \Gamma^{\lambda'}_{\mu'\nu'} A^{\nu'}_\nu v^\nu A^{\mu'}_\mu \frac{d\xi^\mu}{d\tau}.$$

On the other hand, however, we have

$$\frac{Dv^{\lambda'}}{d\tau} = A^{\lambda'}_\lambda \frac{Dv^\lambda}{d\tau} = A^{\lambda'}_\lambda \frac{dv^\lambda}{d\tau} + A^{\lambda'}_\lambda \Gamma^\lambda_{\mu\nu} v^\nu \frac{d\xi^\mu}{d\tau},$$

whence, comparing the two, we have

$$A^{\lambda'}_\lambda \frac{dv^\lambda}{d\tau} + (\partial_\mu A^{\lambda'}_\lambda) v^\lambda \frac{d\xi^\mu}{d\tau} + \Gamma^{\lambda'}_{\mu'\nu'} A^{\nu'}_{\nu}{}^{\mu'}_{\mu} v^\nu \frac{d\xi^\mu}{d\tau} = A^{\lambda'}_\lambda \frac{dv^\lambda}{d\tau} + A^{\lambda'}_\lambda \Gamma^\lambda_{\mu\nu} v^\nu \frac{d\xi^\mu}{d\tau}$$

or, more concisely,

$$\Gamma^{\lambda'}_{\mu'\nu'} A^{\nu'}_{\nu}{}^{\mu'}_{\mu} v^\nu \frac{d\xi^\mu}{d\tau} = A^{\lambda'}_\lambda \Gamma^\nu_{\mu\nu} v^\nu \frac{d\xi^\mu}{d\tau} - (\partial_\mu A^{\lambda'}_\lambda) v^\lambda \frac{d\xi^\mu}{d\tau}.$$

Since the coefficients $\Gamma^\lambda_{\mu\nu}$ do not depend on the field v^λ — if they did the condition of linearity with respect to v would not be satisfied — the n relations above must hold for all fields v. Thus, by replacing v^λ by $A^\lambda_{\nu'} v^{\nu'}$ on the right-hand side, we obtain

$$v^{\nu'} \Gamma^{\lambda'}_{\mu'\nu'} A^{\mu'}_\mu \frac{d\xi^\mu}{d\tau} = v^{\nu'} A^{\lambda'}_{\lambda}{}^{\nu}_{\nu'} \Gamma^\lambda_{\mu\nu} \frac{d\xi^\mu}{d\tau} - v^{\nu'} A^\nu_{\nu'} (\partial_\mu A^{\lambda'}_\nu) \frac{d\xi^\mu}{d\tau}$$

and cancelling $v^{\nu'}$ on both sides, we have

$$\Gamma^{\lambda'}_{\mu'\nu'} A^{\mu'}_{\mu} \frac{d\xi^{\mu}}{d\tau} = A^{\lambda'\nu}_{\lambda\,\nu'} \Gamma^{\lambda}_{\mu\nu} \frac{d\xi^{\mu}}{d\tau} - A^{\nu}_{\nu'} (\partial_{\mu} A^{\lambda'}_{\nu}) \frac{d\xi^{\mu}}{d\tau}.$$

With the substitution

$$t^{\mu} = \frac{d\xi^{\mu}}{d\tau}, \qquad t^{\mu'} = \frac{d\xi^{\mu'}}{d\tau} = A^{\mu'}_{\mu} t^{\mu},$$

the relations can be rewritten

$$\Gamma^{\lambda'}_{\mu'\nu'} t^{\mu'} = A^{\lambda'\nu}_{\lambda\,\nu'} \Gamma^{\lambda}_{\mu\nu} t^{\mu} - A^{\nu}_{\nu'} t^{\mu} \partial_{\mu} A^{\lambda'}_{\nu}$$

or

$$\Gamma^{\lambda'}_{\mu'\nu'} t^{\mu'} = (A^{\lambda'\nu\,\mu}_{\lambda\,\nu'\mu'} \Gamma^{\lambda}_{\mu\nu} - A^{\nu\,\mu}_{\nu'\mu'} \partial_{\mu} A^{\lambda'}_{\nu}) t^{\mu'}.$$

These relations will certainly hold provided we have

$$\Gamma^{\lambda'}_{\mu'\nu'} = A^{\lambda'\mu\,\nu}_{\lambda\,\mu'\nu'} \Gamma^{\lambda}_{\mu\nu} - A^{\mu\,\nu}_{\mu'\nu'} \partial_{\mu} A^{\lambda'}_{\mu} \qquad (\lambda', \mu', \nu' = 1', 2', \ldots, n'). \tag{5.19}$$

These relations constitute the *transformation law* for the coefficients (5.18). As is shown by (5.19), or more precisely by the second term on the right-hand side of this formula, the transformation law for the coordinates $\Gamma^{\lambda}_{\mu\nu}$ is not of a tensor character in the general case.

We now show that this law possesses the group property and it will then follow that the coefficients (5.18) constitute a geometric object of class two. Before doing this, we recast the law in a form which is easier to remember.

We have

$$\begin{aligned} -A^{\mu\,\nu}_{\mu'\nu'} \partial_{\mu} A^{\lambda'}_{\nu} &= -A^{\nu}_{\nu'} A^{\mu}_{\mu'} \partial_{\mu} A^{\lambda'}_{\nu} = -A^{\nu}_{\nu'} \partial_{\mu'} A^{\lambda'}_{\nu} \\ &= -\partial_{\mu'} (A^{\nu}_{\nu'} A^{\lambda'}_{\nu}) + A^{\lambda'}_{\nu} \partial_{\mu'} A^{\nu}_{\nu'} \\ &= -\partial_{\mu'} A^{\lambda'}_{\nu'} + A^{\lambda'}_{\nu} \partial_{\mu'} A^{\nu}_{\nu'} = A^{\lambda'}_{\nu} \partial_{\mu'} A^{\nu}_{\nu'}. \end{aligned}$$

Consequently, this leads to

$$\Gamma^{\lambda'}_{\mu'\nu'} = A^{\lambda'\mu\,\nu}_{\lambda\,\mu'\nu'} \Gamma^{\lambda}_{\mu\nu} + A^{\lambda'}_{\nu} \partial_{\mu'} A^{\nu}_{\nu'}. \tag{5.20}$$

If we write down similar relations for the transition from the coordinate system (λ') to (λ''), we find that

$$\Gamma^{\lambda''}_{\mu''\nu''} = A^{\lambda''\mu'\,\nu'}_{\lambda'\,\mu''\nu''} \Gamma^{\lambda'}_{\mu'\nu'} + A^{\lambda''}_{\nu'} \partial_{\mu''} A^{\nu'}_{\nu''},$$

which, substituting from (5.20), yields

$$\begin{aligned} \Gamma^{\lambda''}_{\mu''\nu''} &= A^{\lambda''\mu'\,\nu'}_{\lambda'\,\mu''\nu''} (A^{\lambda'\mu\,\nu}_{\lambda\,\mu'\nu'} \Gamma^{\lambda}_{\mu\nu} + A^{\lambda'}_{\nu} \partial_{\mu'} A^{\nu}_{\nu'}) + A^{\lambda''}_{\nu'} \partial_{\mu''} A^{\nu'}_{\nu''} \\ &= A^{\lambda''\mu'\,\nu'\,\lambda'\mu\,\nu}_{\lambda'\,\mu''\nu''\lambda\,\mu'\nu'} \Gamma^{\lambda}_{\mu\nu} + A^{\lambda''\mu'\,\nu'\,\lambda'}_{\lambda'\,\mu''\nu''\nu} \partial_{\mu'} A^{\nu}_{\nu'} + A^{\lambda''}_{\nu'} \partial_{\mu''} A^{\nu'}_{\nu''} \end{aligned}$$

$$
\begin{aligned}
&= A^{\lambda''\mu}_{\lambda}{}^{\nu}_{\mu''\nu''}\Gamma^{\lambda}_{\mu\nu} + A^{\lambda''\nu'}_{\nu}{}_{\nu''}\partial_{\mu''}A^{\nu}_{\nu'} + A^{\lambda''}_{\nu'}\partial_{\mu''}A^{\nu'}_{\nu''} \\
&= A^{\lambda''\mu}_{\lambda}{}^{\nu}_{\mu''\nu''}\Gamma^{\lambda}_{\mu\nu} + A^{\lambda''}_{\nu}\partial_{\mu''}(A^{\nu}_{\nu'}A^{\nu'}_{\nu''}) - A^{\lambda''\nu}_{\nu}{}_{\nu'}\partial_{\mu''}A^{\nu'}_{\nu''} + A^{\lambda''}_{\nu'}\partial_{\mu''}A^{\nu'}_{\nu''} \\
&= A^{\lambda''\mu}_{\lambda}{}^{\nu}_{\mu''\nu''}\Gamma^{\lambda}_{\mu\nu} + A^{\lambda''}_{\nu}\partial_{\mu''}A^{\nu}_{\nu''} - A^{\lambda''}_{\nu'}\partial_{\mu''}A^{\nu'}_{\nu''} + A^{\lambda''}_{\nu'}\partial_{\mu''}A^{\nu'}_{\nu''} \\
&= A^{\lambda''\mu}_{\lambda}{}^{\nu}_{\mu''\nu''}\Gamma^{\lambda}_{\mu\nu} + A^{\lambda''}_{\nu}\partial_{\mu''}A^{\nu}_{\nu''},
\end{aligned}
$$

thus establishing the group property.

An object with coordinates $\Gamma^{\lambda}_{\mu\nu}$ — which in our case is defined along the curve C, or in other words the $\Gamma^{\lambda}_{\mu\nu}$ are given as functions of the parameter τ — will be called an *object of parallel displacement*. The coordinates $\Gamma^{\lambda}_{\mu\nu}$ are also themselves known as parameters of parallel (linear) displacement. P. K. Rashevskii refers to them as *Koeffitsienty svyaznostii* (coefficients of connectivity), or *coefficients of connection*. If this object is defined throughout space, i.e. if an n-dimensional field of objects Γ is defined, the field itself enables one to define the absolute derivative $Dv/d\tau$, as defined by equations (5.17), along C with respect to the parameter τ possible for every curve C and for every field of contravariant vectors v; this derivative is again a vector field.

It is not meaningful to say that the object Γ is zero, for the relationships

$$
\Gamma^{\lambda}_{\mu\nu} \overset{*}{=} 0 \qquad (\lambda, \mu, \nu = 1, 2, \ldots, n) \tag{5.21}
$$

which hold in the coordinate system (λ) are not invariant, unless we restrict the group to the affine subgroup. In this case, however, we have

$$
\partial_{\mu'}A^{\nu}_{\nu'} = 0
$$

and the transformation law (5.20) reduces to

$$
\Gamma^{\lambda'}_{\mu'\nu'} = A^{\lambda'\mu}_{\lambda}{}^{\nu}_{\mu'\nu'}\Gamma^{\lambda}_{\mu\nu},
$$

i.e. the object Γ (of class two) becomes in E_n a tensor or an object of class one.

Relationships (5.21) are not invariant, it is true, but one can nevertheless ask what kind of object must Γ be for there to exist a coordinate system (λ) in which relations (5.21) are satisfied at every point in space. This problem will be solved on p. 215. If is possible, however, to pose a more local problem, viz. whether for a given object Γ and a given point p of space there exists a coordinate system (λ) such that the relationships

$$
(\Gamma^{\lambda}_{\mu\nu})_p \overset{*}{=} 0,
$$

hold in that system, i.e. such that these equalities hold only at the point p. It turns out that if the object $\Gamma^{\lambda}_{\mu\nu}$ is symmetric, the answer to this question is always in the affirmative. A coordinate system in which relationships (5.21) hold is called a system of *normal or geodesic coordinates* at the point p.

Let us see what form the transformation law (5.20) takes when non-holonomic systems are admitted to the discussion.

If by (λ) we denote an arbitrary (holonomic) coordinate system while by (K) we denote any arbitrary system which in general is not holonomic, then (5.19) gives

$$\Gamma^{L}_{MN} = A^{L\mu\ \nu}_{\lambda MN}\Gamma^{\lambda}_{\mu\nu} - A^{\mu\ \nu}_{MN}\partial_{\mu}A^{L}_{\nu}. \tag{5.22}$$

Following convention (4.32), we can write these relationships more simply as

$$\Gamma^{L}_{MN} = A^{L\mu\ \nu}_{\lambda MN}\Gamma^{\lambda}_{\mu\nu} - A^{\nu}_{N}\partial_{M}A^{L}_{\nu}. \tag{5.23}$$

When the second term on the right-hand side of equation (5.23) is transformed according to the rule

$$0 = \partial_{M}A^{L}_{N} = \partial_{M}(A^{L}_{\nu}A^{\nu}_{N}) = A^{\nu}_{N}\partial_{M}A^{L}_{\nu} + A^{L}_{\nu}\partial_{M}A^{\nu}_{N}$$

relations (5.22) can be rewritten as

$$\Gamma^{L}_{MN} = A^{L\mu\ \nu}_{\lambda MN}\Gamma^{\lambda}_{\mu\nu} + A^{L}_{\nu}\partial_{M}A^{\nu}_{N}, \tag{5.24}$$

which is rather similar to form (5.20) valid for holonomic systems. On the other hand, we have by (4.34)

$$\Gamma^{L}_{MN} = A^{L\mu\ \nu}_{\lambda MN}\Gamma^{\lambda}_{\mu\nu} - A^{\mu}_{M}\partial_{N}A^{L}_{\mu} - 2\Omega_{MN}{}^{L}. \tag{5.25}$$

using equation (5.25), it can be checked that the transformation law

$$\Gamma^{L'}_{M'N'} = A^{L'M\ N}_{L\ M'N'}\Gamma^{L}_{MN} + A^{L'}_{N}\partial_{M}A^{N}_{N'} \tag{5.26}$$

is valid for any two systems (K) and (K'), whether or not they are holonomic.

60. The Object Γ_{μ}. We saw that for tensors, or densities, the operation of contraction always led to tensors, or tensor densities. A question arises as to whether for more general geometric objects contraction leads to new objects, and if so, what kind of objects. In particular, two contractions can be effected in the object $\Gamma^{\lambda}_{\mu\nu}$: contraction over the indices λ, μ or over λ, ν. It turns out that this leads to two new one-index objects.

We first introduce the definitions

$$\Gamma_{\mu} \overset{\text{def}}{=} \Gamma^{\lambda}_{\mu\lambda}, \qquad \Lambda_{\nu} \overset{\text{def}}{=} \Gamma^{\lambda}_{\lambda\nu}. \tag{5.27}$$

Next we work out the transformation law for Γ_μ and Λ_μ. By (5.20) we have

$$\begin{aligned}\Gamma_{\mu'} &= \Gamma^{\lambda'}_{\mu'\lambda'} = A^{\lambda'\mu\,\nu}_{\lambda\,\mu'\nu'}\Gamma^{\lambda}_{\mu\nu} + A^{\lambda'}_{\nu}\partial_{\mu'}A^{\nu}_{\lambda'} = A^{\mu}_{\mu'}\Gamma^{\lambda}_{\mu\lambda} + A^{\lambda'}_{\nu}\partial_{\mu'}A^{\nu}_{\lambda'}\\ &= A^{\mu}_{\mu'}\Gamma_\mu - \partial_{\mu'}\ln|J|.\end{aligned}$$

It is easily checked that the transformation law

$$\Gamma_{\mu'} = A^{\mu}_{\mu'}\Gamma_\mu - \partial_{\mu'}\ln|J| \tag{5.28}$$

possesses the group property and thus defines a geometric object of class two. When the group of transformations is restricted to be the generalized unimodular group G_u ($J = \text{const}$), the object Γ_μ reduces to a covariant vector (although the object $\Gamma^{\lambda}_{\mu\nu}$ does not reduce to a tensor for the group G_u). Similarly, by contraction over the indices λ' and μ' in equation (5.20) we obtain

$$\Lambda_{\nu'} = \Gamma^{\lambda'}_{\lambda'\nu'} = A^{\lambda'\mu\,\nu}_{\lambda\,\lambda'\nu'}\Gamma^{\lambda}_{\mu\nu} + A^{\lambda'}_{\nu}\partial_{\lambda'}A^{\nu}_{\nu'} = A^{\nu}_{\nu'}\Lambda_\nu + \partial_\nu A^{\nu}_{\nu'},$$

which, by virtue of equation (1.41) from p. 28, yields

$$\Lambda_{\nu'} = A^{\nu}_{\nu'}\Lambda_\nu - \partial_{\nu'}\ln|J|. \tag{5.29}$$

We thus obtain exactly the same transformation law for the object Λ_μ as we obtain for the object Γ_μ.

Now consider what results from the contraction of Γ_μ with the vector field $d\xi^\mu/d\tau$ tangent to the curve C. We introduce the notation

$$\overline{\omega} \stackrel{\text{def}}{=} \Gamma_\mu \frac{d\xi^\mu}{d\tau} \tag{5.30}$$

and we check whether or not this is a geometric object. (Note that if Γ_μ were a vector, we could say *a priori* that $\overline{\omega}$ was a scalar, but Γ_μ is not a vector under the general group of transformations.)

We accordingly write

$$\begin{aligned}\overline{\omega}' &= \Gamma_{\mu'}\frac{d\xi^{\mu'}}{d\tau} = [A^{\mu}_{\mu'}\Gamma_\mu - \partial_{\mu'}\ln|J|]A^{\mu'}_{\lambda}\frac{d\xi^\lambda}{d\tau}\\ &= A^{\mu\,\mu'}_{\mu'\lambda}\Gamma_\mu\frac{d\xi^\lambda}{d\tau} - A^{\mu'}_{\lambda}\frac{d\xi^\lambda}{d\tau}\partial_{\mu'}\ln|J| \\ &= \Gamma_\lambda\frac{d\xi^\lambda}{d\tau} - \frac{d\xi^\lambda}{d\tau}\partial_\lambda\ln|J| = \overline{\omega} - \frac{d[\ln|J|]}{d\tau},\end{aligned} \tag{5.31}$$

since

$$\frac{d[\ln|J|]}{d\tau} = \partial_\lambda(\ln|J|)\frac{d\xi^\lambda}{d\tau}.$$

We thus find that in a straightforward manner, formula (5.30) can be applied to the object to generate a one-coordinate object having the same transformation law as the object ω on p. 167; ω was the object which served to define the absolute derivative of a density field along the curve C.

61. The Symmetric and Antisymmetric Parts of the Object of Connection. We now split the object $\Gamma^{\lambda}_{\mu\nu}$ into a sum of symmetric and antisymmetric parts

$$\Gamma^{\lambda}_{\mu\nu} = \Gamma^{\lambda}_{(\mu\nu)} + \Gamma^{\lambda}_{[\mu\nu]} \tag{5.32}$$

and discuss how each of these parts transforms separately.

Confining ourselves for the moment to holonomic systems, we have

$$\begin{aligned} \Gamma^{\lambda'}_{(\mu'\nu')} &= A^{\lambda'\mu\ \nu}_{\lambda(\mu'\nu')}\Gamma^{\lambda}_{\mu\nu} + A^{\lambda'}_{\lambda}\partial_{(\mu'}A^{\lambda}_{\nu')} \\ &= A^{\lambda'\mu\ \nu}_{\lambda\ \mu'\nu'}\Gamma^{\lambda}_{(\mu\nu)} + A^{\lambda'}_{\lambda}\partial_{\mu'}A^{\lambda}_{\nu'}, \end{aligned} \tag{5.33}$$

since

$$\partial_{(\mu'}A^{\lambda}_{\nu')} = \partial_{\mu'}A^{\lambda}_{\nu'}.$$

We see from equation (5.33) that the $\Gamma^{\lambda}_{(\mu\nu)}$'s transform according to the same law that governs the $\Gamma^{\lambda}_{\mu\nu}$.

On the other hand,

$$\Gamma^{\lambda'}_{[\mu'\nu']} = A^{\lambda'\mu\ \nu}_{\lambda[\mu'\nu']}\Gamma^{\lambda}_{\mu\nu} + A^{\lambda'}_{\lambda}\partial_{[\mu'}A^{\lambda}_{\nu']} = A^{\lambda'\mu\ \nu}_{\lambda\ \mu'\nu'}\Gamma^{\lambda}_{[\mu\nu]}, \tag{5.34}$$

since $\partial_{[\mu'}A^{\lambda}_{\mu']} = 0$; it is therefore plain that $\Gamma^{\lambda}_{[\mu\nu]}$ is a tensor, in fact a three-index tensor antisymmetric in the inferior indices.

Passing now to non-nolonomic systems, following Schouten we introduce the expression

$$S_{MN}{}^{K} \overset{\text{def}}{=} \Gamma^{K}_{[MN]} + \Omega_{MN}{}^{K}, \tag{5.35}$$

where $\Omega_{MN}{}^{K}$ is the object of anholonomicity. By virtue of equations (5.26) and (4.37) we infer that

$$\begin{aligned} S_{M'N'}{}^{K'} &= \Gamma^{K'}_{[M'N']} + \Omega_{M'N'}{}^{K'} \\ &= A^{K'M\ N}_{K[M'N']}\Gamma^{K}_{MN} + A^{K'}_{K}\partial_{[M'}A^{K}_{N']} + A^{K'M\ N}_{K\ M'N'}\Omega_{MN}{}^{K} - A^{K}_{[M'}\partial_{N']}A^{K'}_{K}. \end{aligned}$$

But

$$A^{M\ N}_{[M'N']}\Gamma^{K}_{MN} = A^{M\ N}_{M'N'}\Gamma^{K}_{[MN]}$$

and

$$\begin{aligned} -A^{K}_{[M'}\partial_{N']}A^{K'}_{K} + A^{K'}_{K}\partial_{[M'}A^{K}_{N']} &= \tfrac{1}{2}\{-A^{K}_{M'}\partial_{N'}A^{K'}_{K} + A^{K}_{N'}\partial_{M'}A^{K'}_{K} + A^{K'}_{K}\partial_{M'}A^{K}_{N'} - A^{K'}_{K}\partial_{N'}A^{K}_{M'}\} \\ &= \tfrac{1}{2}\{-\partial_{N'}(A^{K'}_{K}A^{K}_{M'}) + \partial_{M'}(A^{K'}_{K}A^{K}_{N'})\} \\ &= \tfrac{1}{2}\{-\partial_{N'}A^{K'}_{M'} + \partial_{M'}A^{K'}_{N'}\} = 0. \end{aligned}$$

This implies that the expression $S_{NM}{}^{K}$ is a tensor (torsion tensor), and in fact that it is antisymmetric with respect to the inferior indices.

DEFINITION. If the tensor S is zero

$$S_{MN}{}^{K} = 0 \tag{5.36}$$

the object Γ is said to be *symmetric*.

If the tensor S is such that there exists a vector field S_μ with the property that

$$S_{\mu\nu}{}^{\lambda} = S_{[\mu} A^{\lambda}_{\nu]}, \tag{5.37}$$

the object is said to be *semi-symmetric* (1).

We now introduce the notation

$$\overset{s}{\Gamma}{}^{L}_{MN} \overset{\text{def}}{=} \Gamma^{L}_{(MN)} - \Omega_{MN}{}^{L}. \tag{5.38}$$

and we next attempt to find the transformation law for $\overset{s}{\Gamma}$. We have

$$\begin{aligned}
\overset{s}{\Gamma}{}^{L'}_{M'N'} &= \Gamma^{L'}_{(M'N')} - \Omega_{M'N'}{}^{L'} \\
&= A^{L'M\,N}_{L(M'N')} \Gamma^{L}_{MN} + A^{L'}_{L} \partial_{(M'} A^{L}_{N')} - A^{L'M\,N}_{L\,M'N'} \Omega_{MN}{}^{L} + A^{L}_{[M'} \partial_{N']} A^{L'}_{L} \\
&= A^{L'M\,N}_{L\,M'N'} \overset{s}{\Gamma}{}^{L}_{MN} + \tfrac{1}{2}\{A^{L'}_{L} \partial_{M'} A^{L}_{N'} + A^{L'}_{L} \partial_{N'} A^{L}_{M'} + A^{L}_{M'} \partial_{N'} A^{L'}_{L} - A^{L}_{N'} \partial_{M'} A^{L'}_{L}\}.
\end{aligned}$$

The second and third terms in the brackets cancel since when we add them we get $\partial_{N'} A^{L'}_{M'} = 0$ and for the same reasons the fourth term is equal to the first. We therefore have finally

$$\overset{s}{\Gamma}{}^{L'}_{M'N'} = A^{L'M\,N}_{L\,M'N'} \overset{s}{\Gamma}{}^{L}_{MN} + A^{L'}_{L} \partial_{M'} A^{L}_{N'}, \tag{5.39}$$

which shows that the transformation law for $\overset{s}{\Gamma}$ is the same as that for Γ.

DEFINITION. The object $\overset{s}{\Gamma}$ defined by equations (5.38) is called the *symmetric part of the object* Γ.

This definition is justified by the fact that if we form the tensor S for $\overset{s}{\Gamma}$, it turns out to be equal to zero. This is because

$$\overset{s}{\Gamma}{}^{L}_{[MN]} + \Omega_{MN}{}^{L} = \Gamma^{\ L}_{[(MN)]} - \Omega_{MN}{}^{L} + \Omega_{MN}{}^{L} = 0,$$

since the result of symmetrization and antisymmetrization of the same group of indices is always zero.

(1) After J. A. Schouten.

A space provided with an object $\Gamma^{\lambda}_{\mu\nu}$ will be denoted by L_n, while a space furnished with the symmetric object $\Gamma(S = 0)$ will be denoted by A_n, and the object Γ will be called an *affine connection.*

The geometric interpretation of the tensor $S^{\lambda}_{\mu\nu}$ will be given later on (cf. p. 221).

62. Other Ways of Introducing the Object of Parallel Displacement. We now return to the object $\Gamma^{\lambda}_{\mu\nu}$ and present a different line of geometric reasoning leading up to this object, reasoning which appeared later (1918) [66] than did the concept of covariant derivative (Christoffel implicitly in 1869 and Ricci and Levi-Civita in explicit form in 1901). The representation which I shall give is my own, and is in fact modelled on the representation due to Schouten (1918).

Suppose that along a curve C parametrized by parameter τ we are given a field $v^{\lambda}(\tau)$ of contravariant vectors; our objective is to define the absolute derivative $Dv^{\lambda}/d\tau$ of the field in such a way that it is itself a field of contravariant vectors. The difficulty is that we do not know how to subtract vectors $v^{\lambda}(\tau+h)$ and $v^{\lambda}(\tau)$, since as they have different points of attachment $p(\tau+h)$ and $p(\tau)$.

However, if we knew some way of effecting the parallel displacement of a vector attached at one point $\underset{0}{\tau}$ to some other point on the curve C, we could then formulate the concept of the derivative of the field at the point $p(\underset{0}{\tau})$ in the following manner. Let $\bar{v}^{\lambda}(\tau)$ denote a vector attached at the point τ; this vector arises from the parallel displacement of $v^{\lambda}(\underset{0}{\tau})$ from $p(\underset{0}{\tau})$ to $p(\tau)$. We form the difference $v^{\lambda}(\tau)-\bar{v}^{\lambda}(\tau)$ of the vectors attached at the same point $p(\tau)$, divide this difference by $h = \tau-\underset{0}{\tau}$, and then pass to the limit as $h \to 0$; the limiting vector could be called the absolute derivative of the field at the point $p(\underset{0}{\tau})$ with respect to the parameter τ

$$\frac{Dv^{\lambda}}{d\tau} = \lim_{\tau\to\underset{0}{\tau}} \frac{v^{\lambda}(\tau)-\bar{v}^{\lambda}(\tau)}{\tau-\underset{0}{\tau}}.$$

In this way, the concept of absolute derivative is made to depend on the concept of parallel displacement of a given vector at one point on a curve C to other points on C.

As we pointed out earlier, the idea of defining parallel vectors at neighbouring points as vectors which have identical coordinates breaks down because such a definition is not invariant.

A field of parallel vectors would have to be such that the value of the field at one point $p(\underset{0}{\tau})$ determine the whole field.

This property suggests the postulate that this field be the solution of a vectorial first-order differential equation or, equivalently that the coordinates of this field should be the integrals of a system of first-order differential equations. The simplest such system is a system of linear equations, i.e. a system of the form

$$\frac{dv^{\lambda}}{d\tau}+\gamma^{\lambda}_{\mu}(\tau)v^{\mu} = \zeta^{\lambda}(\tau) \qquad (\lambda = 1, 2, \ldots, n), \tag{5.40}$$

where γ^{λ}_{μ} and $\zeta^{\lambda}_{\cdot}$ are given coefficients depending on the parameter τ. Let us see what transformations have to be carried out on the coefficients γ and ζ in the transition from the original coordinate system (λ) to a new one (λ') for equations (5.40) to remain valid.

If they are to hold, we must have

$$\frac{dv^{\lambda'}}{d\tau}+\gamma^{\lambda'}_{\mu'}(\tau)v^{\mu'} = \zeta^{\lambda'}(\tau),$$

whence, on using the relations

$$v^{\lambda'} = A^{\lambda'}_{\lambda} v^{\lambda}$$

by differentating them, we must have

$$\frac{dA^{\lambda'}_{\lambda}}{d\tau} v^{\lambda}+A^{\lambda'}_{\lambda}[\zeta^{\lambda}-\gamma^{\lambda}_{\mu}v^{\mu}]+\gamma^{\lambda'}_{\mu'}A^{\mu'}_{\mu}v^{\mu} = \zeta^{\lambda'}.$$

These relationships are satisfied if we assume that

$$\zeta^{\lambda'} = A^{\lambda'}_{\lambda}\zeta^{\lambda} \tag{5.41}$$

and

$$v^{\lambda}\frac{dA^{\lambda'}_{\lambda}}{d\tau}-A^{\lambda'}_{\lambda}\gamma^{\lambda}_{\mu}v^{\mu}+\gamma^{\lambda'}_{\mu'}A^{\mu'}_{\mu}v^{\mu} = 0. \tag{5.42}$$

Equation (5.41) expresses the fact that it is sufficient to take any vector field for $\zeta^{\lambda}(\tau)$.

We rearrange equation (5.42) by interchanging the dummy indices and bringing the factor v^{μ} out in front of the braces

$$v^{\mu}\left\{\frac{dA^{\lambda'}_{\mu}}{d\tau}-A^{\lambda'}_{\lambda}\gamma^{\lambda}_{\mu}+\gamma^{\lambda'}_{\mu'}A^{\mu'}_{\mu}\right\} = 0,$$

and this again is satisfied if we have

$$\frac{dA^{\lambda'}_{\mu}}{d\tau} - A^{\lambda'}_{\lambda}\gamma^{\lambda}_{\mu} + \gamma^{\lambda'}_{\mu'}A^{\mu'}_{\mu} = 0 \quad (\mu = 1, 2, \ldots, n;\ \lambda' = 1', 2', \ldots, n'). \tag{5.43}$$

Relations (5.43) take a different form when they are multiplied by $A^{\mu}_{\nu'}$ and the first two terms are transferred to the right-hand side. We then have

$$\gamma^{\lambda'}_{\nu'} = A^{\lambda'\mu}_{\lambda\ \nu'}\gamma^{\lambda}_{\mu} - A^{\mu}_{\nu'}\frac{dA^{\lambda'}_{\mu}}{d\tau} \quad (\lambda', \nu' = 1', 2', \ldots, n'). \tag{5.44}$$

It is not difficult to verify that these equations constitute the transformation law for a two-index geometric object of class two which, for a general group of transformations ($A^{\lambda'}_{\mu}$ not constant), is not a mixed tensor.

If along the curve C we have a field of objects γ^{λ}_{ν} whose coordinates transform according to the law (5.41) and if we take a field of zero vectors $\zeta^{\lambda} = 0$ (the simplest assumption), we will have a set of equations of the type (5.40). This set of equations will enable us to displace along the curve C any arbitrary (contravariant) vectors which are parallel in the following sense.

If at any point $p(\underset{0}{\tau})$ of C we take an arbitrary (contravariant) vector $\underset{0}{v}^{\lambda}$, the only integral $v^{\lambda}(\tau)$ of the set of equations (5.40) which satisfies the initial conditions

$$v^{\lambda}(\underset{0}{\tau}) = \underset{0}{v}^{\lambda} \quad (\lambda = 1, 2, \ldots, n) \tag{5.45}$$

is regarded as a field of vectors generated by the parallel displacement of $\underset{0}{v}^{\lambda}$ along the curve C.

This definition, as is readily seen, implies the following properties.

1. If a vector $\underset{2}{v}^{\lambda}$ at a point $p(\underset{2}{\tau})$ arises from the parallel displacement (along C) of a vector $\underset{1}{v}^{\lambda}$ attached at the point $p(\underset{1}{\tau})$, the vector $\underset{1}{v}^{\lambda}$ is in turn generated by the parallel displacement of $\underset{2}{v}^{\lambda}$ from the point $p(\underset{2}{\tau})$ to $p(\underset{1}{\tau})$.

2. A null vector displaced to any point on a curve always remains a null vector.

3. Every vector can undergo parallel displacement from any point to any other point.

4. If at a point $p(\underset{1}{\tau})$ we are given two vectors $\underset{1}{v}^{\lambda}$ and $\underset{2}{v}^{\lambda}$ such that

$\underset{2}{v}^\lambda = \varrho \underset{1}{v}^\lambda$ and if we displace them parallelly along C to the point $p(\underset{2}{\tau})$, the relation $\underset{2}{v} = \varrho \underset{1}{v}$ remains valid after the displacement.

5. If at a point $p(\underset{1}{\tau})$ we are given two linearly independent vectors $\underset{1}{v}$ and $\underset{2}{v}$ and we displace them parallelly to the point $p(\underset{2}{\tau})$, we still have linearly independent vectors after the displacement.

Property 5 holds for any number of independent vectors $\underset{1}{v}, \ldots, \underset{p}{v}$ $(p \leqslant n)$.

Property 1 follows from the uniqueness of the integral of the system of equations (5.40) for the given initial conditions. Property 2 emerges from the homogeneity of equation (5.40), i.e. the assumption that $\zeta^\lambda = 0$. Property 3 stems from the existence of an integral for any arbitrary initial conditions. Property 4 is an immediate consequence of the fact that the system of equations

$$\frac{dv^\lambda}{d\tau} + \gamma^\lambda_\mu(\tau) v^\mu = 0 \quad (\lambda = 1, 2, \ldots, n)$$

is linear and homogeneous. Property 5 can easily be proved by using properties 1 and 4, arguing by contradiction. For $p \geqslant 3$ the reasoning is again fairly straightforward.

Remark. In the space E_n, where the $A^{\lambda'}_\mu$'s are constants and the object γ^λ_μ reduces to a tensor, it can be assumed that $\gamma^\lambda_\mu = 0$ so that parallel displacement reduces in this case to ordinary parallel displacement. In more general spaces, parallel displacement using γ^λ_μ is a new idea, which cannot be defined without bringing in the object γ.

If the curve C is closed and if from some point $\underset{0}{p}$ on the curve the vector $\underset{0}{v}$ is displaced parallelly along C until it return to the initial point $\underset{0}{p}$, then we in general obtain a vector $\underset{1}{v}^\lambda$ which differs from the original vector $\underset{0}{v}^\lambda$. This phenomenon will be discussed more rigorously later; it is related to a certain new quantity called the *curvature tensor*.

The object γ^λ_μ, used to effect the parallel displacement of vectors, was defined for points on a curve C, along which we displaced vectors parallelly. It would be troublesome if we had to define separate independent objects γ^λ_μ for each curve C in space. The question arises as to whether or not the object γ^λ_μ can be defined throughout space as functions of the coordinates ξ^ν

in such a way that for any particular curve

$$\xi^\nu = f^\nu(\tau)$$

the object $\gamma^\lambda_\mu(\tau)$ can be calculated from the equations

$$\gamma^\lambda_\mu(\tau) = \gamma^\lambda_\mu[f^1(\tau), \ldots, f^n(\tau)].$$

We now give an answer to this question, which turns out to be in the negative. In fact, suppose that the functions $\gamma^\lambda_\mu(\xi^\nu)$ depend only on the point ξ^ν. In view of this, since the value of the first term on the right-hand side of the transformation law (5.44) depends only on the point in space and on the transformations of the coordinates at the point, the second term on the right-hand side would also have to depend only on the values of $A^{\lambda'}_\lambda$, $A^\lambda_{\lambda'}$, $\partial_\mu A^{\lambda'}_\lambda$ at the point under consideration. In other words, for all curves passing through the given point, the value of the term

$$A^\mu_{\lambda'} \frac{dA^{\lambda'}_\mu}{d\tau}$$

would have to be constant for a given transformation. Hence, it further follows that the value of the term

$$A^\mu_{\nu'} \partial_\lambda (A^{\lambda'}_\mu) \frac{df^\lambda}{d\tau} \tag{5.46}$$

would have to be the same for all $df^\lambda/d\tau$ (for fixed λ', ν'). Since the form (5.46) is linear, this is possible only when all coefficients

$$A^\mu_{\nu'} \partial_\lambda A^{\lambda'}_\mu = 0 \qquad (\lambda = 1, 2, \ldots, n;\ \lambda', \nu' = 1', 2', \ldots, n'),$$

which now easily leads us to the conclusion that

$$\partial_\lambda A^{\lambda'}_\mu = 0. \tag{5.47}$$

Since these relationships would have to hold for each point, we would have to restrict the group of transformations to the affine group.

Extension of the object $\gamma^\lambda_\mu(\tau)$ to $\gamma^\lambda_\mu(\xi)$ is thus not feasible. We therefore try to find a three-index object

$$\Gamma^\lambda_{\mu\nu}(\xi) \tag{5.48}$$

with the property that its contraction with the vector field tangent to the curve C yields γ^λ_μ. To be more precise, we are looking for an object (5.48) with the property

$$\Gamma^\lambda_{\mu\nu}[f^\varrho(\tau)] \frac{df^\mu}{d\tau} = \gamma^\lambda_\nu(\tau). \tag{5.49}$$

Substituting from equation (5.49) into equation (5.44), we obtain

$$\Gamma^{\lambda'}_{\mu'\nu'}\frac{df^{\mu'}}{d\tau} = \Gamma^{\lambda}_{\mu\nu}\frac{df^{\mu}}{d\tau}A^{\lambda'\nu}_{\lambda\ \nu'} - A^{\nu}_{\nu'}(\partial_\lambda A^{\lambda'}_{\nu})\frac{df^{\lambda}}{d\tau}$$

or, on changing the dummy index,

$$\frac{df^{\mu}}{d\tau}\Gamma^{\lambda'}_{\mu'\nu'}A^{\mu'}_{\mu} = \{\Gamma^{\lambda}_{\mu\nu}A^{\lambda'\nu}_{\lambda\ \nu'} - A^{\nu}_{\nu'}\partial_\mu A^{\lambda'}_{\nu}\}\frac{df^{\mu}}{d\tau}.$$

Since these relations are to hold for all curves C, i.e. for any arbitrary fields $df^{\mu}/d\tau$, it follows that

$$\Gamma^{\lambda'}_{\mu'\nu'}A^{\mu'}_{\mu} = \Gamma^{\lambda}_{\mu\nu}A^{\lambda'\nu}_{\lambda\ \nu'} - A^{\nu}_{\nu'}\partial_\mu A^{\lambda'}_{\nu}$$

or, after multiplication by $A^{\mu}_{\varrho'}$,

$$\Gamma^{\lambda'}_{\varrho'\nu'} = \Gamma^{\lambda}_{\mu\nu}A^{\lambda'\nu\ \mu}_{\lambda\ \nu'\varrho'} - A^{\nu\ \mu}_{\nu'\varrho'}\partial_\mu A^{\lambda'}_{\nu} \qquad (\lambda', \nu', \varrho' = 1', 2', \ldots, n').$$

We see that this is the same law as that given on p. 172 (cf. equation (5.19)), which we arrived at by a different route.

63. Parallel Displacement in Riemannian Space. We now sketch out the way in which Levi-Civita, in his famous paper, arrived at the concept of parallelism in general metric, but curved spaces [134]. Levi-Civita considers an n-dimensional Riemannian space V_n, i.e. a space endowed with a field of metric tensors $g_{\lambda\mu}(\xi^{\nu})$. Next, he embeds this space in an Euclidean metric space R_m of higher dimension $\left(n+1 \leqslant m \leqslant \binom{n+1}{2}\right)$. Embedded in this way, the space becomes a subspace in R_m and the vectors of V_n are now vectors of R_m, tangent to that subspace. We now set up in V_n a curve C and a vector $\underset{0}{v}$ at a certain point on that curve. Along C we form a one-parameter family of n-dimensional planes tangent to V_n. Such a family envelops a certain surface developable in R_n. After development, the curve C goes over into C' and the vector $\underset{0}{v}$ into $\underset{0}{v}'$. In R_n we consider a field of vectors v along the curve C' parallel to (and the same length as) $\underset{0}{v}'$ (the concept of parallelism exists because R_n is an Euclidean space!). Next, R_n is "transvected" along C back onto V_n, each vector of the field v going over into a vector lying in a plane tangent to V_n; Levi-Civita refers to this vector field as a field of parallel vectors along C. This concept of parallelism preserves the angles between parallel vectors as well as the lengths. The

Levi-Civita method for the parallel displacement of vectors is defined uniquely, i.e. the metric tensor $g_{\lambda\mu}$ of the space induces an object $\Gamma^{\lambda}_{\mu\nu}$ of parallel displacement in a unique manner.

There is still another geometric interpretation of the parallel displacement of vectors along a curve. We give it in the special case of a surface V_2 embedded in a three-dimensional Euclidean space R_3. Suppose that a regular curve C lies on $V_2 \subset R_3$. We fix a point p_0 on C and at this point attach a contravariant vector $\underset{0}{v}$ tangent to V_2. We next take some other point p on C. In order to define uniquely the vector v, tangent to V_2 at p, formed by the parallel displacement of the vector $\underset{0}{v}$ from the point p_0 along C, we use a certain procedure of the integral type. Let k denote an arbitrary natural number. On the arc C between the points p_0 and p we choose k intermediate points $p_1, \ldots, p_k$ in an order dictated by the orientation of the arc C from p_0 to p. Let $\underset{1}{v}$ denote a vector attached at p_1 which is equivalent to v (equivalence does exist in R_3). Next, let $\underset{1}{v'}$ denote the orthogonal projection of v onto the plane tangent to V_2 at the point p_1. We repeat this operation denoting by $\underset{2}{v}$ a vector which is attached at p_2 and equivalent to the vector $\underset{1}{v'}$, and by $\underset{2}{v'}$ the orthogonal projection of the vector $\underset{2}{v}$ onto the plane tangent to V_2 at the point p_2. After $k+1$ such steps we arrive at the vector $\underset{k+1}{v'}$ attached at p and tangent to V_2. It can be proved that in the limit, when $k \to \infty$ and the lattice of intermediate points on C has a diameter tending to zero, the vector $\underset{k+1}{v'}$ has a limiting position v, which is called the *vector formed by the parallel displacement of the vector* $\underset{0}{v}$ *from a point* p_0 *to a point* p *along* C. The vector v depends only on the choice of the points p_0 and p, the arc C linking p with p_0, and on the choice of the vector $\underset{0}{v}$. This construction is equivalent to the original definition due to Levi-Civita.

In a non-metric space we do not have *a priori* any rules for making an invariant choice of a particular one from an infinite number of objects of parallel displacement. The Levi-Civita geometric method of parallel displacement of vectors becomes very clear in the case when the two-dimensional space V_2 is embedded in the space R_3. Consider, for example, a two-dimensional sphere and follow the parallel displacement of the vector v

along a parallel C; at the initial point the vector v is tangent to the parallel. The one-parameter family of planes tangent to the sphere along C in this case envelops a circular cone whose apex lies on a straight line perpendicular to the plane of C and passing through the centre of the sphere. This cone degenerates into a cylinder when the parallel C is a great circle. Suppose that this cone is unwrapped onto the plane intersecting it along the generatrix which passes through the point on the parallel C corresponding to the initial position of the vector v. Next, the vector v is moved parallelly to the point which corresponds to the final position and lies at the other end of the edge of the cut. When the cone is rewrapped back onto the sphere, we obtain the end position of the displaced vector which will no longer be tangent to the parallel C. See Fig. 4 (This is taken from the book by

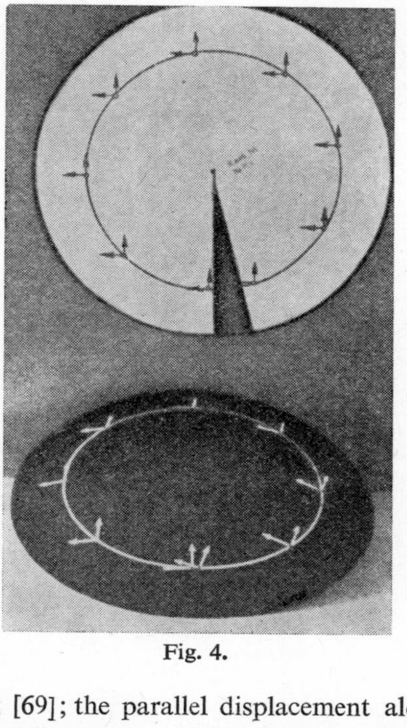

Fig. 4.

Schouten and Struik [69]; the parallel displacement along C of a vector which in the initial position lay on the generatrix of the cone is also shown in this illustration).

Let us carry out the calculations for this example. We first endow our spherical surface with geographical coordinates, with φ denoting the geographical latitude and λ the geographical longitude. Let the circle C correspond to the angle φ_0, where $0 \leqslant \varphi_0 < \pi/2$, and the value $\lambda = 0$, to the initial position of the vector v. If v is to be tangent to C, its curvilinear coordinates will be $(\varrho, 0)$ e.g. $(1, 0)$. In order to find the coordinates of v in its final position, we must know the coordinates of the object of parallel displacement, i.e. the $\Gamma^{\nu}_{\lambda\mu}$, which—as we shall see further on—are equal to the so-called *Christoffel symbols* and in the case of our example are

$$\begin{aligned} &\Gamma^1_{11} = 0, \quad \Gamma^1_{12} = \Gamma^1_{21} = 0, \quad \Gamma^1_{22} = \sin\varphi\cos\varphi, \\ &\Gamma^2_{11} = 0, \quad \Gamma^2_{12} = \Gamma^2_{21} = -\tan\varphi, \quad \Gamma^2_{22} = 0. \end{aligned} \tag{5.50}$$

The system of differential equations for the coordinates v^μ of the parallel displacement vector will be

$$\frac{dv^\mu}{dt} + \Gamma^\mu_{\varrho\sigma} v^\sigma \frac{d\xi^\varrho}{dt} = 0 \qquad (\mu = 1, 2). \tag{5.51}$$

Moreover, in our case the equation of the circle C is

$$\xi^1 = \varphi = \varphi_0, \qquad \xi^2 = \lambda = t,$$

whence

$$\frac{d\xi^1}{dt} = 0, \qquad \frac{d\xi^2}{dt} = 1,$$

and the system of equations (5.51) assumes the form

$$\frac{dv^1}{dt} + \Gamma^1_{2\sigma} v^\sigma = 0, \qquad \frac{dv^2}{dt} + \Gamma^2_{2\sigma} v^\sigma = 0.$$

Thus, by virtue of equations (5.50) and of the fact that $\varphi = \varphi_0 = \text{const}$ along C, we arrive at

$$\begin{aligned} &\frac{dv^1}{dt} + \sin\varphi_0 \cos\varphi_0 v^2 = 0, \\ &\frac{dv^2}{dt} - \tan\varphi_0 v^1 = 0. \end{aligned} \tag{5.52}$$

Integration of the foregoing system of equations is simple and leads to

$$\begin{aligned} v^1 &= A\sin\alpha t + B\cos\alpha t \\ v^2 &= \frac{1}{\cos\varphi_0}[-A\cos\alpha t + B\sin\alpha t] \end{aligned} \qquad (\alpha = \sin\varphi_0). \tag{5.53}$$

Using the initial conditions

$$(v^1)_0 = 1, \quad (v^2)_0 = 0, \tag{5.54}$$

we obtain for the constants of integration A, B the formulae

$$A = 0, \quad B = 1,$$

so that

$$v^1 = \cos \alpha t,$$

$$v^2 = \frac{\sin \alpha t}{\cos \varphi_0}. \tag{5.55}$$

The end position of the vector v corresponds to the values $\lambda = t = 2\pi$, i.e. after parallel cyclical displacement around C the coordinates of v are

$$(v^1)_{2\pi} = \cos[2\pi \sin \varphi_0], \quad (v^2)_{2\pi} = \frac{\sin[2\pi \sin \varphi_0]}{\cos \varphi_0}. \tag{5.56}$$

The question now is, when do we return to the initial position? The equalities

$$(v^1)_{2\pi} = (v^1)_0, \quad (v^2)_{2\pi} = (v^2)_0$$

lead to

$$\sin[2\pi \sin \varphi_0] = 0, \quad \cos[2\pi \sin \varphi_0] = 1,$$

whence

$$2\pi \sin \varphi_0 = 0, \quad \text{or} \quad \varphi_0 = 0.$$

It thus follows that only when the parallel is an equatorial circle (great circle) do we return to the initial position. Otherwise, the vectors $(v)_{2\pi}$ and $(v)_0$ form a certain angle θ which is given by the formula

$$\cos \theta = \frac{g_{\lambda\mu}(v^\lambda)_0(v^\mu)_{2\pi}}{|(v)_0|\,|(v)_{2\pi}|}.$$

The coefficients of the metric tensor $g_{\lambda\mu}$ in our example are

$$g_{11} = R^2, \quad g_{12} = g_{21} = 0, \quad g_{22} = R^2 \cos^2 \xi^1 = R^2 \cos^2 \varphi_0.$$

It is also easily calculated that

$$|v| = R.$$

From this we have

$$\begin{aligned} g_{\mu\lambda}(v^\lambda)_0(v^\mu)_{2\pi} &= g_{11}(v^1)_0(v^1)_{2\pi} + g_{22}(v^2)_0(v^2)_{2\pi} \\ &= g_{11}(v^1)_{2\pi} = R^2 \cos[2\pi \sin \varphi_0]. \end{aligned}$$

Inserting these values into the formula for $\cos\theta$, we have

$$\cos\theta = \cos[2\pi \sin \varphi_0],$$

i.e. the angle θ depends on the geographical latitude φ_0. If $\varphi_0 = 0$, the angle $\theta = 0$. When $\varphi_0 \to \frac{1}{2}\pi$, then

$$\cos[2\pi \sin\varphi_0] \to 1,$$

whence $\cos\theta \to 1$, which implies that $\theta \to 0$.

The maximum angle θ is attained when $2\pi \sin\varphi_0 = \pi$, that is $\sin\varphi_0 = \frac{1}{2}$, and we then have $\theta = \pi$. Thus, after one complete cycle, we have a vector whose direction is reversed relative to its initial position.

Note also that if the vector v is displaced parallelly from a point p along the closed curve C back to p, the angle which it makes with the initial position depends only on the curve C, and not on the choice of the vectors v. This follows from the fact that angles do not change under parallel displacement of vectors.

64. Another Way of Introducing the Object $\Gamma^\lambda_{\mu\nu}$. We give another simple, albeit unrigorous, method of arriving at the object Γ. Suppose that at a point $p(\xi^\nu)$ we are given an arbitrary contravariant vector v. Consider an "infinitely close" neighbouring point $p(\xi^\nu + d\xi^\nu)$, set up a system of n^3 numbers $\Gamma^\lambda_{\mu\nu}$, and use them at the point $p(\xi^\nu + d\xi^\nu)$ form the vector

$$v^\nu - \Gamma^\nu_{\mu\lambda} v^\lambda d\xi^\mu.$$

This vector is said to be *parallel* to the vector v^ν at the point $p(\xi^\nu)$. This formula enables vectors v^ν to be propagated "parallelly" in the "infinitely close neighbourhood" of the point $p(\xi^\nu)$. However, this formula must not be applied to finite distances. To reach an exact definition, the field of objects $\Gamma^\lambda_{\mu\nu}$ must be defined and the limiting process carried out along the given curve C which comes from the point p. Naturally, when this method is made more precise we arrive at the expected system of equations

$$\frac{dv^\lambda}{d\tau} + \Gamma^\lambda_{\mu\nu} v^\nu \frac{d\xi^\mu}{d\tau} = 0 \quad (\lambda = 1, 2, \ldots, n) \tag{5.57}$$

and the transformation law for the coordinates of the object $\Gamma^\lambda_{\mu\nu}$.

The two processes — the definition of the absolute derivative of a vector field along the curve C and the definition of a field of parallel vectors along C — have each led us to concepts of geometric objects with the same transformation law. It is therefore to be expected that a close relationship between these concepts exist, and this is indeed the case. If we take the same object $\Gamma^\lambda_{\mu\nu}$ in both definitions, the formulae derived lead to the following result:

If the absolute derivative $Dv^\lambda/d\tau$ of a field of vectors v^λ is identically zero (along C), the field $v^\lambda(\tau)$ is a field of parallel vectors, and the converse is also true.

65. A System Moving Geodesically Along a Line. At some point p on the curve C we take n linearly independent vectors $\underset{i}{v}^\nu$ $(i = 1, 2, \ldots, n)$. Moving each of these parallelly, we obtain n independent fields (cf. (2.36)) of vectors along the entire curve C.

Each field $w^\nu(\tau)$ can therefore be represented in the form

$$w^\nu = \overset{i}{\alpha}\underset{i}{v}^\nu \qquad (\nu = 1, 2, \ldots, n),$$

where the $\overset{i}{\alpha}$ are scalar fields. Let (k) denote a system which is defined by the system of vectors $\underset{i}{v}$ and which is local (along C). If this is the case, we have

$$w^k \overset{*}{=} \overset{k}{\alpha}. \tag{5.58}$$

We can write

$$\frac{Dw^\nu}{d\tau} = \frac{D\overset{i}{\alpha}}{d\tau}\underset{i}{v}^\nu + \overset{i}{\alpha}\frac{D\underset{i}{v}^\nu}{d\tau},$$

but

$$\frac{D\overset{i}{\alpha}}{d\tau} = \frac{d\overset{i}{\alpha}}{d\tau},$$

since the $\overset{i}{\alpha}$ are scalar and

$$\frac{D\underset{i}{v}^\nu}{d\tau} = 0 \qquad (i = 1, 2, \ldots, n)$$

from the definition of the fields $\underset{i}{v}$. Accordingly, we have

$$\frac{Dw^\nu}{d\tau} = \frac{d\overset{i}{\alpha}}{d\tau}\underset{i}{v}^\nu.$$

In the system (k) we can write $(v_i^k \overset{*}{=} \delta_i^k)$.

$$\frac{Dw^k}{d\tau} = \frac{d\overset{k}{\alpha}}{d\tau}. \tag{5.59}$$

On the other hand, equation (5.58) implies

$$\frac{dw^k}{d\tau} \overset{*}{=} \frac{d\overset{k}{\alpha}}{d\tau}; \tag{5.60}$$

by comparison of equations (5.59) and (5.60) we deduce that

$$\frac{Dw^k}{d\tau} \overset{*}{=} \frac{dw^k}{d\tau}, \tag{5.61}$$

i.e. that in the system (k) the absolute derivative reduces to an ordinary derivative.

A system of vectors $\underset{i}{v}^\nu$ with the properties

$$|\underset{i}{v}^\nu| \neq 0$$

and

$$\frac{D\underset{i}{v}^\nu}{d\tau} = 0 \quad (i = 1, \ldots, n)$$

is called a *system moving geodesically along the curve C*. We thus have the result, which can be generalized to fields of any quantities that the absolute derivative of a field reduces to an ordinary derivative in a geodesically moving system.

This was the idea behind the method of Schouten (1918) [66] concerning the concept of a (pseudo) parallel displacement of vectors along curves.

Henceforth, in the case of fields of other quantities we shall deal only with the concept of absolute derivative, and the concept of parallel fields will emerge automatically when the absolute derivative in set equal to zero.

66. The Covariant Derivative. Suppose now that we are given a field of vectors v^λ not only along a curve C, but throughout space, or perhaps in some n-dimensional part of it

$$v^\lambda = v^\lambda(\xi^\nu). \tag{5.62}$$

Equations (5.57) can then be written in the form

$$(\partial_\mu v^\lambda)\frac{d\xi^\mu}{d\tau} + \Gamma^\lambda_{\mu\nu} v^\nu \frac{d\xi^\mu}{d\tau} = 0 \quad (\lambda = 1, 2, \ldots, n), \tag{5.63}$$

or

$$[\partial_\mu v^\lambda + \Gamma^\lambda_{\mu\nu} v^\nu]\frac{d\xi^\mu}{d\tau} = 0. \tag{5.64}$$

Now consider the expression in brackets, which we abbreviate by setting

$$\nabla_\mu v^\lambda \overset{\text{def}}{=} \partial_\mu v^\lambda + \Gamma^\lambda_{\mu\nu} v^\nu. \tag{5.65}$$

This expression is defined throughout space since both the vector field

(5.62), and the field of objects (5.48) are defined at every point of the space. If we calculate $\nabla_{\mu'} v^{\lambda'}$ and apply the transformation law for the coordinates of the object Γ, we can easily verify that

$$\nabla_{\mu'} v^{\lambda'} = A^{\mu\ \lambda'}_{\mu'\lambda} \nabla_\mu v^\lambda, \tag{5.66}$$

i.e. that $\nabla_\mu v^\lambda$ is a field of tensors (mixed).

The tensor $\nabla_\mu v^\lambda$ is called the *covariant derivative* of the field v^λ. This name derives from the fact that the index μ which the field v^λ gains when we form the tensor (5.65) is a covariant index.

With the aid of the covariant derivative the absolute derivative of a vector field along a curve C can be written in a very simple form, viz.

$$\frac{Dv^\lambda}{d\tau} = \nabla_\mu v^\lambda \frac{d\xi^\mu}{d\tau} = \frac{d\xi^\mu}{d\tau} \nabla_\mu v^\lambda, \tag{5.67}$$

so that we have the following symbolical operator equation

$$\frac{D}{d\tau} = \frac{d\xi^\mu}{d\tau} \nabla_\mu . \tag{5.68}$$

67. The Absolute Derivative of Other Quantities. Hitherto, we have defined the absolute derivative for densities and for fields of contravariant vectors. The question now is how to define the absolute derivative for other kinds of quantities, i.e. covariant vectors, tensors of higher valences, and tensor densities. We now address ourselves to this problem. Without limiting the generality of the discussion in any way, we solve the problem for covariant derivatives and the concept of absolute derivative is then established automatically because, as it turns out, the symbolical validity of (5.68) is preserved for quantities of all kinds.

Consider a field of covariant vectors

$$u_\lambda(\xi^\nu). \tag{5.69}$$

It can easily be seen that the term $\partial_\mu u_\lambda$ does not transform like a covariant tensor under the transition to a new system (λ'), and so we consider (as on p. 190) the expression

$$\partial_\mu u_\lambda + \Lambda^\nu_{\mu\lambda} u_\nu$$

where we assume that the coefficients $\Lambda^\nu_{\mu\lambda}$ transform in such a way that this expression is a tensor.

Let us now see what transformation law for the coefficients Λ follows from this assumption.

On the one hand, we have

$$\partial_{\mu'} u_{\lambda'} + \Lambda^{\nu'}_{\mu'\lambda'} u_{\nu'} = A^{\mu\ \lambda}_{\mu'\lambda'}[\partial_\mu u_\lambda + \Lambda^{\nu}_{\mu\lambda} u_\nu].$$

But on the other hand

$$\begin{aligned}\partial_{\mu'} u_{\lambda'} + \Lambda^{\nu'}_{\mu'\lambda'} u_{\nu'} &= A^{\mu}_{\mu'}\partial_\mu(u_\lambda A^{\lambda}_{\lambda'}) + \Lambda^{\nu'}_{\mu'\lambda'} u_\nu A^{\nu}_{\nu'} \\ &= A^{\mu\ \lambda}_{\mu'\lambda'}\partial_\mu u_\lambda + A^{\mu}_{\mu'} u_\lambda \partial_\mu A^{\lambda}_{\lambda'} + A^{\nu}_{\nu'} u_\nu \Lambda^{\nu'}_{\mu'\lambda'},\end{aligned}$$

whence, on comparing these two equations, we obtain

$$\Lambda^{\nu'}_{\mu'\lambda'} A^{\nu}_{\nu'} u_\nu + A^{\mu\ \lambda}_{\mu'\lambda'}\partial_\mu u_\lambda + A^{\mu}_{\mu'} u_\lambda \partial_\mu A^{\lambda}_{\lambda'} = A^{\mu\ \lambda}_{\mu'\lambda'}\partial_\mu u_\lambda + A^{\mu\ \lambda}_{\mu'\lambda'}\Lambda^{\nu}_{\mu\lambda} u_\nu$$

or, upon simplification and change of the dummy index

$$u_\nu A^{\nu}_{\nu'}\Lambda^{\nu'}_{\mu'\lambda'} = u_\nu\{A^{\mu\ \lambda}_{\mu'\lambda'}\Lambda^{\nu}_{\mu\lambda} - A^{\mu}_{\mu'}\partial_\mu A^{\nu}_{\lambda'}\}.$$

Since the foregoing relation is to hold for all fields u_ν, we must have

$$A^{\nu}_{\nu'}\Lambda^{\nu'}_{\mu'\lambda'} = A^{\mu\ \lambda}_{\mu'\lambda'}\Lambda^{\nu}_{\mu\lambda} - A^{\mu}_{\mu'}\partial_\mu A^{\nu}_{\lambda'}$$

or, multiplying both sides by $A^{\varrho'}_{\nu}$

$$\Lambda^{\varrho'}_{\mu'\lambda'} = A^{\varrho'\mu\ \lambda}_{\nu'\mu'\lambda'}\Lambda^{\nu}_{\mu\lambda} - A^{\mu\ \varrho'}_{\mu'\nu}\partial_\mu A^{\nu}_{\lambda'},$$

or, finally, on replacing the index ϱ' by ν' and substituting $\partial_{\mu'}$ for $A^{\mu}_{\mu'}\partial_\mu$,

$$\Lambda^{\nu'}_{\mu'\lambda'} = A^{\nu'\mu\ \lambda}_{\nu\ \mu'\lambda'}\Lambda^{\nu}_{\mu\lambda} - A^{\nu'}_{\nu}\partial_{\mu'} A^{\nu}_{\lambda'}. \tag{5.70}$$

Comparison of this transformation law for the coefficients Λ with that for the coordinates of the object Γ reveals a strong similarity. The only difference is that the second term on the right-hand side here has a minus sign, whereas there it has a plus sign.

Now consider in general the transformation law for the object

$$\alpha\Gamma^{\lambda}_{\mu\nu}. \tag{5.71}$$

where α is an arbitrary scalar. We have

$$\begin{aligned}\alpha\Gamma^{\lambda'}_{\mu'\nu'} &= \alpha\{A^{\lambda'\mu\ \nu}_{\lambda\ \mu'\nu'}\Gamma^{\lambda}_{\mu\nu} + A^{\lambda'}_{\lambda}\partial_{\mu'} A^{\lambda}_{\nu'}\} \\ &= A^{\lambda'\mu\ \nu}_{\lambda\ \mu'\nu'}(\alpha\Gamma^{\lambda}_{\mu\nu}) + \alpha A^{\lambda'}_{\lambda}\partial_{\mu'} A^{\lambda}_{\nu'}.\end{aligned} \tag{5.72}$$

This implies in particular that the object $\Lambda^{\nu}_{\mu\lambda}$ has the same transformation law as $-\Gamma^{\nu}_{\mu\lambda}$.

We could therefore simply put

$$\Lambda^{\nu}_{\mu\lambda} \overset{\text{def}}{=} -\Gamma^{\nu}_{\mu\lambda}. \tag{5.73}$$

However, the point is that hypothesis (5.73) should follow as a consequence of some other rational postulate. Now let us require that the concept of absolute (or covariant) derivative obey the *Leibniz rule for differentiation of a product*, i.e. that

$$D(ab) = (Da)b + a(Db), \tag{5.74}$$

where a, b are arbitrary quantities, and the abbrieviated symbol D appears in place of $D/d\tau$.

If, moreover, we require the derivative Da to be a quantity of the same kind as a, it follows from these two assumptions that the law (5.74) must also hold for transvection. Thus we must have in particular

$$D(u_\lambda v^\lambda) = (Du_\lambda)v^\lambda + u_\lambda Dv^\lambda.$$

Now denote the scalar $u_\lambda v^\lambda$ for short by

$$\sigma \stackrel{\text{def}}{=} u_\lambda v^\lambda.$$

We thus have

$$\begin{aligned}\nabla_\mu \sigma &= (\nabla_\mu u_\lambda)v^\lambda + u\nabla_\mu v^\lambda = (\partial_\mu u_\lambda + \Lambda^{\nu}_{\mu\lambda} u_\nu)v^\lambda + u_\lambda(\partial_\mu v^\lambda + \Gamma^{\lambda}_{\mu\nu} v^\nu)\\ &= \partial_\mu \sigma + \Lambda^{\nu}_{\mu\lambda} u_\nu v^\lambda + \Gamma^{\lambda}_{\mu\nu} u_\lambda v^\nu = \partial_\mu \sigma + \Lambda^{\nu}_{\mu\lambda} u_\nu v^\lambda + \Gamma^{\nu}_{\mu\lambda} u_\nu v^\lambda\\ &= \partial_\mu \sigma + (\Lambda^{\nu}_{\mu\lambda} + \Gamma^{\nu}_{\mu\lambda}) u_\nu v^\lambda.\end{aligned}$$

Now since $\partial_\mu \sigma$ is a covariant vector, we can reasonably assume that

$$\nabla_\mu \sigma = \partial_\mu \sigma \quad \text{for all scalars } \sigma, \tag{5.75}$$

and the last relation then reduces to the equality

$$(\Lambda^{\nu}_{\mu\lambda} + \Gamma^{\nu}_{\mu\lambda}) u_\nu v^\lambda = 0.$$

Since this is to hold for all fields u_ν and v^λ, this implies the relation

$$\Lambda^{\nu}_{\mu\lambda} + \Gamma^{\nu}_{\mu\lambda} = 0 \quad (\lambda, \mu, \nu = 1, 2, \ldots, n), \tag{5.76}$$

i.e., relation (5.73).

The problem of the covariant derivative for covariant vectors thus is resolved and we have the formula

$$\nabla_\mu u_\lambda = \partial_\mu u_\lambda - \Gamma^{\nu}_{\mu\lambda} u_\nu, \tag{5.77}$$

where the object Γ is the same one as that used to define the covariant derivative (5.65) of contravariant vectors.

We now go on to consider tensors of higher valence. First, we assume that the tensor $b^{\lambda\mu}$ is the product of two vector fields

$$b^{\lambda\mu} \stackrel{\text{def}}{=} u^\lambda v^\mu.$$

We must, by (5.74), have

$$\nabla_\nu b^{\lambda\mu} = (\nabla_\nu u^\lambda)v^\mu + u^\lambda \nabla_\nu v^\mu,$$

which further implies

$$\nabla_\nu b^{\lambda\mu} = (\partial_\nu u^\lambda + \Gamma^\lambda_{\nu\varrho} u^\varrho) v^\mu + u^\lambda(\partial_\nu v^\mu + \Gamma^\mu_{\nu\varrho} v^\varrho) = \partial_\nu (u^\lambda v^\mu) + \Gamma^\lambda_{\nu\varrho} u^\varrho v^\mu + \Gamma^\mu_{\nu\varrho} u^\lambda v^\varrho$$
$$= \partial_\nu b^{\lambda\mu} + \Gamma^\lambda_{\nu\varrho} b^{\varrho\mu} + \Gamma^\mu_{\nu\varrho} b^{\lambda\varrho}.$$

It is thus seen that the covariant derivative of the tensor b is defined in terms of an ordinary derivative ∂_ν and additional expressions which are linear in b (just as in the case of vectors), with only the coordinates of the object Γ appearing as coefficients.

The formula given above for the covariant derivative of the tensor b can also be written in the following form, which incidentally is easier to remember,

$$\nabla_\mu b^{\lambda\nu} = \partial_\mu b^{\lambda\nu} + \Gamma^{\lambda\nu}_{\mu\varrho\sigma} b^{\varrho\sigma}, \tag{5.78}$$

where

$$\Gamma^{\lambda\nu}_{\mu\varrho\sigma} \stackrel{\text{def}}{=} \Gamma^\lambda_{\mu\varrho} \delta^\nu_\sigma + \Gamma^\nu_{\mu\sigma} \delta^\lambda_\varrho. \tag{5.79}$$

In order to have a universal formula for the covariant derivative of any two-index contravariant tensor, and not only for those which are products of vectors, formula (5.78) is taken by definition to be valid in general for all tensors $b^{\lambda\mu}$.

By completely analogous reasoning we arrive at the following general formula for the covariant derivative of any tensor field $T^{\lambda_1\ldots\lambda_p}{}_{\mu_1\ldots\mu_q}$:

$$\nabla_\nu T^{\lambda_1\ldots\lambda_p}{}_{\mu_1\ldots\mu_q} = \partial_\nu T^{\lambda_1\ldots\lambda_p}{}_{\mu_1\ldots\mu_q} +$$
$$+ \Gamma^{\lambda_1\ldots\lambda_p}_{\nu\varrho_1\ldots\varrho_p} T^{\varrho_1\ldots\varrho_p}{}_{\mu_1\ldots\mu_q} - \Gamma^{\sigma_1\ldots\sigma_q}{}_{\nu\mu_1\ldots\mu_q} T^{\lambda_1\ldots\lambda_p}{}_{\sigma_1\ldots\sigma_q}, \tag{5.80}$$

where we have in general set

$$\Gamma^{\alpha_1\ldots\alpha_m}_{\nu\beta_1\ldots\beta_m} \stackrel{\text{def}}{=} \sum_{i=1}^{m} \Gamma^{\alpha_i}_{\nu\beta_i} \delta^{\alpha_1\ldots\alpha_{i-1}\alpha_{i+1}\ldots\alpha_m}_{\beta_1\ldots\beta_{i-1}\beta_{i+1}\ldots\beta_m}, \qquad m \geqslant 2. \tag{5.81}$$

Thus, as can be checked, the covariant derivative of a tensor of valence (p, q) is a tensor of valence $(p, q+1)$. For instance, for $p = 3$ and $q = 2$ formula (5.80) becomes

$$\nabla_\nu T^{\lambda\mu\varrho}{}_{\sigma\tau} = \partial_\nu T^{\lambda\mu\varrho}{}_{\sigma\tau} + \Gamma^\lambda_{\nu\omega} T^{\omega\mu\varrho}{}_{\sigma\tau} + \Gamma^\mu_{\nu\omega} T^{\lambda\omega\varrho}{}_{\sigma\tau}$$
$$+ \Gamma^\varrho_{\nu\omega} T^{\lambda\mu\omega}{}_{\sigma\tau} - \Gamma^\omega_{\nu\sigma} T^{\lambda\mu\varrho}{}_{\omega\tau} - \Gamma^\omega_{\nu\tau} T^{\lambda\mu\varrho}{}_{\sigma\omega}.$$

Next, we adopt a general definition for the absolute derivative

$$\frac{DT}{d\tau} \stackrel{\text{def}}{=} \frac{d\xi^\mu}{d\tau} \nabla_\mu T \tag{5.82}$$

whence it follows that the absolute derivative (along a curve C with respect to the parameter τ) of an arbitrary tensor field is once again a tensor field of the same valence.

We therefore find that the same object has enabled us to define the covariant, and consequently the absolute, derivative for all tensor fields.

We still have to resolve the matter of the covariant derivative for a density.

On p. 169 we saw that the absolute derivative of a density of weight -1 can be defined by the formula

$$\frac{D\mathfrak{v}}{d\tau} = \frac{d\mathfrak{v}}{d\tau} + \omega\mathfrak{v}, \tag{5.83}$$

where ω is an object with the transformation law

$$\omega' = \omega - \frac{d\ln|J|}{d\tau} \tag{5.84}$$

Moreover, we found further on that if we set

$$\omega = \Gamma_\mu \frac{d\xi^\mu}{d\tau} \tag{5.85}$$

we obtain an object with exactly the same transformation law. Hence if we assume that

$$\nabla_\mu \mathfrak{v} = \partial_\mu \mathfrak{v} + \Gamma_\mu \mathfrak{v} \tag{5.86}$$

then, we first obtain — as can easily be verified — a vector density of weight -1 and, what is more, the quantity

$$\nabla_\mu \mathfrak{v} \cdot \frac{d\xi^\mu}{d\tau} \tag{5.87}$$

turns out to be a density of weight -1, which could be taken to be the absolute derivative $D\mathfrak{v}/d\tau$. Our objective, however, is to obtain formula (5.86) as the logical outcome of previous assumptions and formulae. In order to achieve this, we start from the now established fact that if

$$w^{\lambda_1 \dots \lambda_n} \tag{5.88}$$

is an n-vector, then

$$w^{1 \dots n} \tag{5.89}$$

represents a density of weight -1. We already have an established formula for the covariant derivative of the n-vector (5.88), viz. formula (5.80). Applying this formula to the n-vector (5.88), we obtain

$$\nabla_\nu w^{\lambda_1 \dots \lambda_n} = \partial_\nu w^{\lambda_1 \dots \lambda_n} + \Gamma^{\lambda_1}_{\nu\varrho} w^{\varrho\lambda_2 \dots \lambda_n} + \dots + \Gamma^{\lambda_n}_{\nu\varrho} w^{\lambda_1 \dots \lambda_{n-1}\varrho} .$$

Hence, in particular for $\lambda_k = k$ we have

$$\nabla_\nu w^{1\ldots n} = \partial_\nu w^{1\ldots n} + \Gamma^1_{\nu\varrho} w^{\varrho 2\ldots n} + \ldots + \Gamma^n_{\nu\varrho} w^{1\ldots(n-1)\varrho}.$$

However, since each coordinate of the n-vector w is zero whenever at least two indices have the same value, we obtain

$$\begin{aligned}\nabla_\nu w^{1\ldots n} &= \partial_\nu w^{1\ldots n} + \Gamma^1_{\nu 1} w^{1\ldots n} + \ldots + \Gamma^n_{\nu n} w^{1\ldots n}\\ &= \partial_\nu w^{1\ldots n} + \Gamma^\lambda_{\lambda\nu} w^{1\ldots n}\\ &= \partial_\nu w^{1\ldots n} + \Gamma_\nu w^{1\ldots n},\end{aligned}$$

and since each density of weight -1 can be regarded as a fundamental coordinate of a corresponding n-vector, formula (5.86) is now proved.

In a similar manner we derive a formula for the covariant derivative of a density with weight $+1$

$$\nabla_\nu \mathfrak{v} = \partial_\nu \mathfrak{v} - \Gamma_\nu \mathfrak{v}. \tag{5.90}$$

Suppose now that we are given a W-density of weight $-r$

$$\mathfrak{v}' = |J|^r \mathfrak{v}.$$

We set

$$\nabla_\nu \mathfrak{v} = \partial_\nu \mathfrak{v} + \omega_\nu \mathfrak{v}$$

and take $\nabla_\nu \mathfrak{v}$ to be a covariant W-density of weight $-r$, i.e. we assume that

$$\nabla_{\nu'} \mathfrak{v}' = |J|^r A^\nu_{\nu'} \nabla_\nu \mathfrak{v}.$$

We have

$$\begin{aligned}\partial_\nu \mathfrak{v}' &= \partial_\nu(|J|^r \mathfrak{v}) = |J|^r \partial_\nu \mathfrak{v} + r|J|^{r-1} \partial_\nu J \operatorname{sgn} J \cdot \mathfrak{v}\\ &= |J|^r \partial_\nu \mathfrak{v} + |J|^r r \partial_\nu \ln|J| \cdot \mathfrak{v}.\end{aligned}$$

Consequently,

$$\nabla_{\nu'} \mathfrak{v}' = A^\nu_{\nu'} \partial_\nu \mathfrak{v}' + \omega_{\nu'} \mathfrak{v}' = |J|^r \{A^\nu_{\nu'} \partial_\nu \mathfrak{v} + r A^\nu_{\nu'} \partial_\nu \ln|J| \cdot \mathfrak{v} + \omega_{\nu'} \mathfrak{v}\},$$

whence, by comparison, we obtain

$$A^\nu_{\nu'}(\partial_\nu \mathfrak{v} + \omega_\nu \mathfrak{v}) = A^\nu_{\nu'}\{\partial_\nu \mathfrak{v} + r\partial_\nu \ln|J| \cdot \mathfrak{v}\} + \omega_{\nu'} \mathfrak{v}.$$

From this, on simplifying and cancelling $\mathfrak{v}$ we get

$$\omega_{\nu'} = A^\nu_{\nu'} \omega_\nu - r\partial_{\nu'} \ln|J|.$$

The object ω_ν thus has the same transformation law as the object $r\Gamma_\nu$.

Hence putting

$$\nabla_\nu \mathfrak{v} \stackrel{\text{def}}{=} \partial_\nu \mathfrak{v} + r\Gamma_\nu \mathfrak{v}, \tag{5.91}$$

we obtain a general definition for the covariant derivative of a W-density.

This definition now leads us to the conclusion that the covariant derivative of a W-scalar reduces to an ordinary derivative. Indeed, let $\tilde{\mathfrak{s}}$ be a W-density of weight $+1$, and let $\mathfrak{v}$ be an ordinary density of weight -1, and set

$$\tilde{\sigma} \stackrel{\text{def}}{=} \tilde{\mathfrak{s}}\mathfrak{v}.$$

Since

$$\tilde{\sigma}' = \tilde{\mathfrak{s}}'\mathfrak{v}' = |J|^{-1}\tilde{\mathfrak{s}}J\mathfrak{v} = \operatorname{sgn} J \cdot \tilde{\mathfrak{s}}\mathfrak{v} = \operatorname{sgn} \mathrm{J} \cdot \tilde{\sigma},$$

it follows that $\tilde{\sigma}$ is a W-scalar. By virtue of these formulae and the general Leibniz rule applied to the symbol ∇_μ, we have

$$\begin{aligned}\nabla_\mu \tilde{\sigma} &= (\nabla_\mu \tilde{\mathfrak{s}})\mathfrak{v} + \tilde{\mathfrak{s}}\nabla_\mu \mathfrak{v} = (\partial_\mu \tilde{\mathfrak{s}} + \Gamma_\mu \tilde{\mathfrak{s}})\mathfrak{v} + \tilde{\mathfrak{s}}(\partial_\mu \mathfrak{v} - \Gamma_\mu \mathfrak{v}) \\ &= \mathfrak{v}_\mu \partial \tilde{\mathfrak{s}} + \tilde{\mathfrak{s}}\partial_\mu \mathfrak{v} + \Gamma_\mu \tilde{\mathfrak{s}}\mathfrak{v} - \Gamma_\mu \mathfrak{v}\tilde{\mathfrak{s}} = \partial_\mu(\tilde{\mathfrak{s}}\mathfrak{v}) = \partial_\mu \tilde{\sigma}.\end{aligned}$$

Now let $\mathfrak{w}$ be a G-density of weight $-r$, i.e. let

$$\mathfrak{w}' = |J|^r \operatorname{sgn} J \cdot \mathfrak{w}.$$

Next, we set

$$\nabla_\mu \mathfrak{w} = \partial_\mu \mathfrak{w} + \eta_\mu \mathfrak{w}$$

and assume that $\nabla_\mu \mathfrak{w}$ is a G-density of weight $-r$, i.e. that

$$\nabla_{\mu'} \mathfrak{w}' = \operatorname{sgn} J \cdot |J|^r A^\mu_{\mu'} \nabla_\mu \mathfrak{w}.$$

We have

$$\begin{aligned}\partial_\mu \mathfrak{w}' &= \operatorname{sgn} J\{r|J|^{r-1} \operatorname{sgn} J \cdot \partial_\mu J \cdot \mathfrak{w} + |J|^r \partial_\mu \mathfrak{w}\} \\ &= \operatorname{sgn} J\{r|J|^r \partial_\mu \ln |J| \cdot \mathfrak{w} + |J|^r \partial_\mu \mathfrak{w}\} \\ &= \operatorname{sgn} J \cdot |J|^r \{r\partial_\mu \ln |J| \cdot \mathfrak{w} + \partial_\mu \mathfrak{w}\}.\end{aligned}$$

Hence

$$\begin{aligned}\nabla_{\mu'} \mathfrak{w}' &= \partial_{\mu'} \mathfrak{w}' + \eta_{\mu'} \mathfrak{w}' \\ &= A^\mu_{\mu'} \operatorname{sgn} J \cdot |J|^r \{r\partial_\mu \ln |J| + \partial_\mu \mathfrak{w}\} + \eta_{\mu'} |J|^r \operatorname{sgn} J \cdot \mathfrak{w} \\ &= \operatorname{sgn} J \cdot |J|^r A^\mu_{\mu'} (\partial_\mu \mathfrak{w} + \eta_\mu \mathfrak{w}),\end{aligned}$$

and so on dividing both sides by $\operatorname{sgn} J \cdot |J|^r \mathfrak{w}$, we obtain after simplification

$$\eta_{\mu'} = A^\mu_{\mu'} \eta_\mu - r\partial_{\mu'} \ln |J|, \tag{5.92}$$

in other words, the same transformation law as that for ω_μ. Therefore,

The covariant derivative for both a W-density and a G-density is given by the same formula (5.91), *and it represents a W-density or a G-density, as the case may be.*

Since each tensor density $\mathfrak{T}$ can be regarded as the product of a density and a tensor, the formulae above can be used to derive a formula for the covariant derivative of tensor *W*- or *G*-densities of weight $-r$.

Suppose that

$$\mathfrak{T} = \mathfrak{v}T,$$

where $\mathfrak{T}$ is a density, and T is a tensor. Since we must have

$$\nabla_\mu \mathfrak{T} = (\nabla_\mu \mathfrak{v})T + \mathfrak{v}\nabla_\mu T,$$

we have, explicitly,

$$\begin{aligned} \nabla_\mu \mathfrak{T}^{\lambda_1\dots\lambda_p}{}_{\mu_1\dots\mu_q} &= (\partial_\mu \mathfrak{v} + r\Gamma_\mu \mathfrak{v})\, T^{\lambda_1\dots\lambda_p}{}_{\mu_1\dots\mu_q} + \mathfrak{v}\{\partial_\mu T^{\lambda_1\dots\lambda_p}{}_{\mu_1\dots\mu_q} + \\ &\quad + \Gamma_{\mu\varrho_1\dots\varrho_p}^{\lambda_1\dots\lambda_p} T^{\varrho_1\dots\varrho_p}{}_{\mu_1\dots\mu_q} - \Gamma_{\mu\mu_1\dots\mu_q}^{\sigma_1\dots\sigma_q} T^{\lambda_1\dots\lambda_p}{}_{\sigma_1\dots\sigma_q}\} \\ &= \partial_\mu\{\mathfrak{v}T^{\lambda_1\dots\lambda_p}{}_{\mu_1\dots\mu_q}\} + \Gamma_{\mu\varrho_1\dots\varrho_p}^{\lambda_1\dots\lambda_p} \mathfrak{v} T^{\varrho_1\dots\varrho_p}{}_{\mu_1\dots\mu_q} - \\ &\quad - \Gamma_{\mu\mu_1\dots\mu_q}^{\sigma_1\dots\sigma_q} \mathfrak{v} T^{\lambda_1\dots\lambda_p}{}_{\sigma_1\dots\sigma_q} + r\Gamma_\mu T\mathfrak{v}^{\lambda_1\dots\lambda_p}{}_{\mu_1\dots\mu_q} \\ &= \partial_\mu \mathfrak{T}^{\lambda_1\dots\lambda_p}{}_{\mu_1\dots\mu_q} + \Gamma_{\mu\varrho_1\dots\varrho_p}^{\lambda_1\dots\lambda_p} \mathfrak{T}^{\varrho_1\dots\varrho_p}{}_{\mu_1\dots\mu_q} - \\ &\quad - \Gamma_{\mu\mu_1\dots\mu_q}^{\sigma_1\dots\sigma_q} \mathfrak{T}^{\lambda_1\dots\lambda_p}{}_{\sigma_1\dots\sigma_q} + r\Gamma_\mu \mathfrak{T}^{\lambda_1\dots\lambda_p}{}_{\mu_1\dots\mu_q}. \end{aligned} \tag{5.93}$$

This is the most general formula and it contains the special cases: the covariant derivative of scalars, *W*-scalars, *W*-densities, *G*-densities, tensors of any valence, and finally, tensor densities.

Thus, for instance, for $p = 2$, $q = 2$, $r = -2$, formula (5.93) for the covariant derivative of the quantity $\mathfrak{T}$ takes on the following special form

$$\begin{aligned} \nabla_\mu \mathfrak{T}^{\lambda\nu}{}_{\varrho\sigma\tau} = \partial_\mu \mathfrak{T}^{\lambda\nu}{}_{\varrho\sigma\tau} + \Gamma_{\mu\omega}^{\lambda} \mathfrak{T}^{\omega\nu}{}_{\varrho\sigma\tau} + \Gamma_{\mu\omega}^{\nu} \mathfrak{T}^{\lambda\omega}{}_{\varrho\sigma\tau} - \Gamma_{\mu\varrho}^{\omega} \mathfrak{T}^{\lambda\nu}{}_{\omega\sigma\tau} - \\ - \Gamma_{\mu\sigma}^{\omega} \mathfrak{T}^{\lambda\nu}{}_{\varrho\omega\tau} - \Gamma_{\mu\tau}^{\omega} \mathfrak{T}^{\lambda\nu}{}_{\varrho\sigma\omega} - 2\Gamma_\mu \mathfrak{T}^{\lambda\nu}{}_{\varrho\sigma\tau}. \end{aligned}$$

68. The Derivative of the Unit Tensor. In particular, we now calculate the covariant derivative of a field of unit tensors A_ν^λ. We have

$$\nabla_\mu A_\nu^\lambda = \partial_\mu A_\nu^\lambda + \Gamma_{\mu\varrho}^\lambda A_\nu^\varrho - \Gamma_{\mu\nu}^\varrho A_\varrho^\lambda = 0 + \Gamma_{\mu\nu}^\lambda - \Gamma_{\mu\nu}^\lambda = 0. \tag{5.94}$$

We therefore see that the covariant derivative of a field of unit tensors is zero and hence the absolute derivative of a field of unit tensors along a curve C is identically zero.

As another example we calculate the covariant derivative of the *n*-vector density $\varepsilon_{\lambda_1\dots\lambda_n}$ which, as we know, has weight $r = 1$. By the formula given above, we then have

$$\nabla_\mu \varepsilon_{\lambda_1\ldots\lambda_n} = \partial_\mu \varepsilon_{\lambda_1\ldots\lambda_n} - \Gamma^{\varrho}_{\mu\lambda_1} \varepsilon_{\varrho\lambda_2\ldots\lambda_n} - \ldots - \Gamma^{\varrho}_{\mu\lambda_n} \varepsilon_{\lambda_1\ldots\lambda_{n-1}\varrho} + \Gamma_\mu \varepsilon_{\lambda_1\ldots\lambda_n}$$

$$= 0 - \Gamma^{\lambda_1}_{\mu\lambda_1} \varepsilon_{\lambda_1\ldots\lambda_n} - \ldots - \Gamma^{\lambda_n}_{\mu\lambda_n} \varepsilon_{\lambda_1\ldots\lambda_n} + \Gamma_\mu \varepsilon_{\lambda_1\ldots\lambda_n}$$

(N.B. no summation is intended in terms containing three identical indices!).

If in the terms with a minus sign we take the density $\varepsilon_{\lambda_1\ldots\lambda_n}$ out in front of the brackets, we are left inside the brackets with

$$-\sum_{\lambda_1} \Gamma^{\lambda_1}_{\mu\lambda_1} = -\Gamma_\mu$$

and we therefore finally arrive at

$$\nabla_\mu \varepsilon_{\lambda_1\ldots\lambda_n} = 0. \tag{5.95}$$

Similarly, we can show that

$$\nabla_\mu \varepsilon^{\lambda_1\ldots\lambda_n} = 0. \tag{5.96}$$

The question now is whether it is possible to define a covariant or absolute derivative for geometric objects other than quantities (e.g. for an object of parallel displacement). This problem will not be taken up in this textbook.

69. Geodesics of the Space L_n. Suppose that we are given a space L_n, i.e. an n-dimensional space X_n based on the group G_1 with the object Γ of parallel displacement.

In the special case when $L_n = E_n$, straight lines have the property that if at any point we take a tangent vector, which, of course, lies on the line itself, and displace it parallelly along the line to another point, it remains on that line, i.e. is still a tangent vector. This property will be used to generalize the concept of geodesics in the spaces L_n.

DEFINITION. A curve C in a space L_n is called a *geodesic* if it has the following property: if at any point p on C we take an arbitrary vector t tangent to C and displace it parallelly along C to any other point q, we always get a vector tangent to C[(1)].

In other words, curve C is a geodesic if there exists along it a field of

---

(1) This definition expresses the fact that a geodesic is a straight line in the sense formulated long ago by Euclid as a line with the property that we do not change direction as we move along it. Note that the above definition is not metric. The object Γ does not yet determine the metric, and so it is meaningless to say that a geodesic is the shortest of all possible lines joining two points.

tangent vectors t^ν such that

$$\frac{Dt^\nu}{d\tau} \equiv 0. \tag{5.97}$$

We now write down in analytic and explicit form the condition that the curve

$$\xi^\nu = \xi^\nu(\tau) \tag{5.98}$$

be a geodesic in the light of the definition above. Since the field of vectors $t^\nu(\tau)$ is to be tangent to the curve C, it follows that there exists a scalar factor $\varrho(\tau) \neq 0$ such that

$$t^\nu = \varrho(\tau)\frac{d\xi^\nu}{d\tau}. \tag{5.99}$$

We now write out the condition

$$\frac{Dt^\nu}{d\tau} = 0,$$

in full.

We obtain consecutively

$$\begin{aligned}
\nabla_\mu t^\nu \cdot \frac{d\xi^\mu}{d\tau} &= 0 \qquad (\nu = 1, 2, \ldots, n), \\
(\partial_\mu t^\nu + \Gamma^\nu_{\mu\lambda} t^\lambda)\frac{d\xi^\mu}{d\tau} &= 0, \\
\frac{d\xi^\mu}{d\tau}\partial_\mu t^\nu + \Gamma^\nu_{\mu\lambda} t^\lambda \frac{d\xi^\mu}{d\tau} &= 0, \\
\frac{dt^\nu}{d\tau} + \Gamma^\nu_{\mu\lambda} t^\lambda \frac{d\xi^\mu}{d\tau} &= 0, \\
\varrho'(\tau)\frac{d\xi^\nu}{d\tau} + \varrho(\tau)\frac{d^2\xi^\nu}{d\tau^2} + \varrho(\tau)\Gamma^\nu_{\mu\lambda}\frac{d\xi^\lambda}{d\tau}\cdot\frac{d\xi^\mu}{d\tau} &= 0,
\end{aligned} \tag{5.100}$$

and, finally,

$$\frac{d^2\xi^\nu}{d\tau^2} + \Gamma^\nu_{\mu\lambda}\frac{d\xi^\lambda}{d\tau}\cdot\frac{d\xi^\mu}{d\tau} = -\frac{d\ln|\varrho|}{d\tau}\cdot\frac{d\xi^\nu}{d\tau} \qquad (\nu = 1, 2, \ldots, n). \tag{5.101}$$

Naturally, the variables ξ^ν in the coefficients $\Gamma^\nu_{\mu\lambda}$ must be replaced by the functions $\xi^\nu(\tau)$ which determine the curve C.

This is the analytic condition that C be a geodesic. The form of this condition is inconvenient because these equations contain an unknown function $\varrho(\tau)$; we therefore put these equations into a different form.

This is accomplished by replacing the parameter τ. Namely, we set $\sigma = \varphi(\tau)$, where φ is a function of class C^2, and on introducing the parameter σ in place of τ, we obtain

$$\frac{d\xi^\nu}{d\tau} = \frac{d\xi^\nu}{d\sigma} \cdot \frac{d\sigma}{d\tau} = \varphi'(\tau)\frac{d\xi^\nu}{d\sigma}$$

and

$$\frac{d^2\xi^\nu}{d\tau^2} = \frac{d}{d\tau}\left(\frac{d\xi^\nu}{d\tau}\right) = \frac{d}{d\tau}\left[\varphi' \frac{d\xi^\nu}{d\sigma}\right] = \varphi'' \frac{d\xi^\nu}{d\sigma} + \varphi'^2 \frac{d^2\xi^\nu}{d\sigma^2},$$

which, on substitution into (5.101), yields

$$\varphi'' \frac{d\xi^\nu}{d\sigma} + \varphi'^2 \frac{d^2\xi^\nu}{d\sigma^2} + \Gamma^\nu_{\mu\lambda}\varphi'^2 \frac{d\xi^\lambda}{d\sigma} \cdot \frac{d\xi^\mu}{d\sigma} = -\frac{\varphi'\varrho'}{\varrho} \cdot \frac{d\xi^\nu}{d\sigma},$$

that is,

$$\varphi'^2\left[\frac{d^2\xi^\nu}{d\sigma^2} + \Gamma^\nu_{\mu\lambda}\frac{d\xi^\lambda}{d\sigma} \cdot \frac{d\xi^\mu}{d\sigma}\right] = -\left(\varphi'' + \frac{\varrho'}{\varrho}\varphi'\right)\frac{d\xi^\nu}{d\sigma}$$

or

$$\frac{d^2\xi^\nu}{d\sigma^2} + \Gamma^\nu_{\mu\lambda}\frac{d\xi^\lambda}{d\sigma} \cdot \frac{d\xi^\nu}{d\sigma} = -\left(\frac{\varphi''}{\varphi'^2} + \frac{\varrho'}{\varrho\varphi'}\right)\frac{d\xi^\nu}{d\sigma}. \tag{5.102}$$

The system of equations (5.102) assumes the simplest form if the right-hand side is set equal to zero, i.e. if the function φ (replacing the original parameter τ with σ) is chosen so that

$$\frac{\varphi''}{\varphi'} = -\frac{\varrho'}{\varrho},$$

or

$$\varphi = \int \frac{c\,d\tau}{\varrho(\tau)}.$$

Hence, if we take

$$\sigma = c_1 + c_2 \int \frac{d\tau}{\varrho(\tau)}, \tag{5.103}$$

then for the new parameter σ we obtain the equations of the geodesics in the form

$$\frac{d^2\xi^\nu}{d\sigma^2} + \Gamma^\nu_{\mu\lambda}\frac{d\xi^\lambda}{d\sigma} \cdot \frac{d\xi^\mu}{d\sigma} = 0 \quad (\nu = 1, 2, \ldots, n). \tag{5.104}$$

This is a system of ordinary differential equations of the second order, which is in normal form (solved for the second derivative) and which does not explicitly contain the functions $\xi^\nu(\sigma)$ which we are looking for; however, it is not a system of linear equations since the first derivatives are present in the second degree, while the functions sought, which appear in the arguments Γ, do not occur linearly.

The parameter σ, for which the equations of the geodesics have the simple form (5.104), is known as the *natural parameter*. It is defined up to affine transformations: in fact, assuming that τ is already a natural parameter, we have $\frac{\varrho'}{\varrho} = 0$, that is, $\varrho(\tau) = \text{const}$, whence by virtue of equation (5.103) it follows that $\sigma = c_1 + c_2\tau$.

With the natural parameter σ, the system of equations (5.104) can be written as

$$\frac{D\left(\dfrac{d\xi^\nu}{d\sigma}\right)}{d\sigma} = 0, \tag{5.105}$$

which implies that the field of vectors $d\xi^\nu/d\sigma$ tangent to the geodesic has a zero absolute derivative.

REMARK. Not every field of vectors tangent to a geodesic constitutes a field of parallel vectors!

It is well known from the theory of second-order differential equations that the system of equations (5.104) has the property that the initial conditions

$$\left.\begin{aligned} \xi^\nu(0) &= \underset{0}{\xi^\nu} \\ \left(\frac{d\xi^\nu}{d\sigma}\right)_0 &= \underset{0}{m^\nu} \end{aligned}\right\} \quad (\nu = 1, 2, \ldots, n), \tag{5.106}$$

determine locally a unique solution $\xi^\nu(\sigma)$, provided the functions $\Gamma^\nu_{\mu\lambda}(\xi)$ are of class C^1. This means that through each point and for each direction there exists exactly one geodesic passing through that point and having the tangent at that point with the prescribed direction $\underset{0}{m^\nu}$.

Solving equations (5.104) and writing the solution in the form

$$\xi^\nu = \psi^\nu(\sigma; \underset{0}{\xi^\nu}, \underset{0}{m^\nu}) \tag{5.107}$$

we obtain the equations of geodesics in an explicit form, where the parameter σ here is the natural parameter for all geodesics.

If the object Γ is such that there exists a coordinate system (λ) with the property that

$$\Gamma^\lambda_{\mu\nu} \overset{*}{=} 0 \quad (\lambda, \mu, \nu = 1, 2, \ldots, n), \tag{5.108}$$

then the equations of the geodesics in this coordinate system take the form

$$\xi^\nu = \overset{\nu}{c}\sigma + \overset{\nu}{d} \qquad (\nu = 1, 2, \ldots, n), \tag{5.109}$$

where $\overset{\nu}{c}$, $\overset{\nu}{d}$ are constants, and so the equations are linear. In another coordinate system (λ'), however, these equations will in general no longer be linear. In affine spaces E_n, where Γ reduces to a tensor and equations (5.108) are invariant, geodesics are simply straight lines.

Since the natural parameter is defined up to affine transformations (and not translations), one cannot speak of the length of a geodesic line; it is, however, possible to define an invariant on any given geodesic, viz. the ratio of two pieces of that geodesic, i.e. a number s equal to $(\sigma_4 - \sigma_3)/(\sigma_2 - \sigma_1)$, if σ_1 is the value of the natural parameter for each of the four points p_1, where p_1p_2 specifies one piece of the geodesic, and p_3p_4 the other.

When

$$\Gamma^\nu_{\mu\lambda} = \Gamma^\nu_{(\mu\lambda)} + \Gamma^\nu_{[\mu\lambda]}$$

is inserted in equation (5.104), the result is

$$\frac{d^2\xi^\nu}{d\sigma^2} + \Gamma^\nu_{(\mu\lambda)} \frac{d\xi^\lambda}{d\sigma} \cdot \frac{d\xi^\mu}{d\sigma} + \Gamma^\nu_{[\mu\lambda]} \frac{d\xi^\lambda}{d\sigma} \cdot \frac{d\xi^\mu}{d\sigma} = 0.$$

But

$$\Gamma^\lambda_{[\mu\lambda]} \frac{d\xi^\lambda}{d\sigma} \cdot \frac{d\xi^\mu}{d\sigma} = 0,$$

since the factor $\Gamma^\nu_{[\mu\lambda]}$ is antisymmetric with respect to the indices μ, λ, and the factor $\frac{d\xi^\lambda}{d\sigma} \cdot \frac{d\xi^\mu}{d\sigma}$ is symmetric with respect to those indices. The sum is thus zero, and the equations of geodesics may therefore, be written as

$$\frac{d^2\xi^\nu}{d\sigma^2} + \Gamma^\nu_{(\mu\lambda)} \frac{d\xi^\lambda}{d\sigma} \cdot \frac{d\xi^\mu}{d\sigma} = 0, \tag{5.110}$$

which shows that the geodesics depend only on the symmetric part of the object Γ and not on the tensor $S_{\mu\lambda}{}^\nu$.

One may ask when two different objects $\underset{1}{\Gamma}{}^\nu_{\lambda\mu}$ and $\underset{2}{\Gamma}{}^\nu_{\lambda\mu}$ define the same system of geodesics. The answer to this question has been given by Weyl [153], and is expressed by a simple relationship. The necessary and sufficient condition for this is that the difference between the objects,

$$\underset{2}{\Gamma}{}^\lambda_{(\mu\nu)} - \underset{1}{\Gamma}{}^\lambda_{(\mu\nu)}, \tag{5.111}$$

which can be shown to be a tensor, be of the form

$$\underset{2}{\Gamma}^{\lambda}_{(\mu\nu)} - \underset{1}{\Gamma}^{\lambda}_{(\mu\nu)} = A^{\lambda}_{\mu} p_{\nu} + A^{\lambda}_{\nu} p_{\mu}, \tag{5.112}$$

where p_{ν} is a field of covariant vectors.

70. Affine Parameter on a Curve. It is well known from the elements of classical differential geometry that in Euclidean space a preferred parameter, defined up to translations, known as an arc can always be introduced on a sufficiently regular curve C. Similarly, we have seen that a certain preferred parameter, defined up to affine transformations, which we called a *natural parameter*, can also be introduced on geodesics. We now show that a certain preferred parameter, called an *affine parameter*, can be introduced for a arbitrary (sufficiently regular) curve C by means of a given field of objects $\Gamma^{\nu}_{\mu\lambda}$ of parallel displacement; this parameter is defined up to affine transformations and coincides with the natural parameter in the special case, when C is a geodesic.

To construct this parameter, we start with an arbitrary *a priori* parameter τ in terms of which the curve

$$\xi^{\nu} = \xi^{\nu}(\tau)$$

is given. Next, we define along C a sequence of vector fields (contravariant) by means of the recursion formula

$$\begin{aligned} \underset{1}{t}^{\nu}(\tau) &= \frac{d\xi^{\nu}}{d\tau} \qquad (\nu = 1, 2, \ldots, n), \\ \underset{k+1}{t}^{\nu}(\tau) &= \frac{D\underset{k}{t}^{\nu}}{d\tau} \qquad (k = 1, 2, \ldots). \end{aligned} \tag{5.113}$$

Let p denote the only natural number determined uniquely by the condition that the vectors

$$\underset{1}{t}, \ldots, \underset{p}{t} \tag{5.114}$$

are linearly indepedent of each other, while the vector $\underset{p+1}{t}$ is a linear combination of the vectors (5.114).

REMARK. It may happen, of course, that the number p depends on the point τ of the curve C, but we further assume here that p is constant along the whole or certainly part of the curve C.

We now introduce the p-vector

$$v^{\lambda_1 \dots \lambda_p} \overset{\text{def}}{=} \underset{1}{t}^{[\lambda_1} \dots \underset{p}{t}^{\lambda_p]} . \tag{5.115}$$

The direction of this p-vector is called the *osculating p-direction* of the curve C at the point in question. If in particular C is a geodesic, we have $p = 1$, this follows from the definition of a geodesic

$$\frac{D\left(\dfrac{d\xi^\nu}{d\tau}\right)}{d\tau} = \alpha \frac{d\xi^\nu}{d\tau} .$$

We calculate

$$\frac{Dv^{\lambda_1 \dots \lambda_p}}{d\tau} .$$

By the now familiar rules of absolute differentiation, we have

$$\frac{Dv^{\lambda_1 \dots \lambda_p}}{d\tau} = \sum_{i=1}^{p} \underset{1}{t}^{[\lambda_1} \dots \underset{i-1}{t}^{\lambda_{i-1}} \frac{D\underset{i}{t}^{\lambda_i}}{d\tau} \underset{i+1}{t}^{\lambda_{i+1}} \dots \underset{p}{t}^{\lambda_p]}$$

$$= \sum_{i=1}^{p} \underset{1}{t}^{[\lambda_1} \dots \underset{i-1}{t}^{\lambda_{i-1}} \underset{i+1}{t}^{\lambda_i} \underset{i+1}{t}^{\lambda_{i+1}} \dots \underset{p}{t}^{\lambda_p]},$$

and we see see that in the sum above all the components for $i = 1, 2, \dots$ $\dots, p-1$ are zero, leaving only the last component so that we have

$$\frac{Dv^{\lambda_1 \dots \lambda_p}}{d\tau} = \underset{1}{t}^{[\lambda_1} \dots \underset{p-1}{t}^{\lambda_{p-1}} \frac{D}{d\tau} \underset{p}{t}^{\lambda_p]} .$$

However, it follows from the definition of p that

$$\frac{D\underset{p}{t}^\nu}{d\tau} = \sum_{i=1}^{p} \overset{i}{\alpha} \underset{i}{t}^\nu$$

and, consequently,

$$\frac{Dv^{\lambda_1 \dots \lambda_p}}{d\tau} = \underset{1}{t}^{[\lambda_1} \dots \underset{p-1}{t}^{\lambda_{p-1}} \sum_{i=1}^{p} \overset{i}{\alpha} \underset{i}{t}^{\lambda_p]} = \sum_{i=1}^{p} \overset{i}{\alpha} \underset{1}{t}^{[\lambda_1} \dots \underset{p-1}{t}^{\lambda_{p-1}} \underset{i}{t}^{\lambda_p]}.$$

Again in the sum above all components for $i = 1, 2, \dots, p-1$ are zero, and we finally have

$$\frac{Dv^{\lambda_1 \dots \lambda_p}}{d\tau} = \overset{p}{\alpha} \underset{1}{t}^{[\lambda_1} \dots \underset{p}{t}^{\lambda_p]} = \overset{p}{\alpha} v^{\lambda_1 \dots \lambda_p}, \tag{5.116}$$

which means that the absolute derivative of the p-vector $v^{\lambda_1 \dots \lambda_p}$ differs from the p-vector itself only by the scalar factor $\overset{p}{\alpha}$.

We now attempt to introduce a new parameter

$$\sigma = \sigma(\tau)$$

in such a way that we have

$$\frac{Dv^{\lambda_1 \dots \lambda_p}}{d\sigma} \equiv 0. \tag{5.117}$$

To this end, we first note that the sequence (5.113) of vectors depends on the choice of the curve parameter. We therefore denote the corresponding sequence of vectors for the parameter σ by $\underset{1}{s}, \underset{2}{s}, \dots, \underset{p}{s}$ to identify it as the sequence associated with σ.

The following relationships exist between the sequences of vectors $\underset{i}{t}$ and $\underset{i}{s}$

$$\underset{1}{s} = \frac{d\xi}{d\sigma} = \frac{d\xi}{d\tau} \cdot \frac{d\tau}{d\sigma} = \underset{1}{t} \frac{d\tau}{d\sigma},$$

$$\underset{2}{s} = \frac{D\underset{1}{s}}{d\sigma} = \frac{D\left(\underset{1}{t} \dfrac{d\tau}{d\sigma}\right)}{d\sigma} = \underset{1}{t} \frac{d^2\tau}{d\sigma^2} + \frac{d\tau}{d\sigma} \cdot \frac{D\underset{1}{t}}{d\sigma}$$

$$= \underset{1}{t} \frac{d^2\tau}{d\sigma^2} + \left(\frac{d\tau}{d\sigma}\right)^2 \frac{D\underset{1}{t}}{d\tau} = \underset{2}{t} \left(\frac{d\tau}{d\sigma}\right)^2 + \underset{1}{t} \frac{d^2\tau}{d\sigma^2}.$$

By mathematical induction it can be shown that in general we have

$$\underset{j}{s} = \underset{j}{t} \left(\frac{d\tau}{d\sigma}\right)^j + \sum_{i=1}^{j-1} \underset{ji}{\beta} \overset{i}{t} \qquad (j = 2, 3, \dots, p),$$

where the coefficients $\underset{j}{\overset{i}{\beta}}$ are rational functions of the derivatives $d\tau/d\sigma$, $d^2\tau/d^2\sigma, \dots, d^j\tau/d\sigma^j$.

In view of the above, we have

$$\frac{D\bar{v}^{\lambda_1 \dots \lambda_p}}{d\sigma} = \frac{D\underset{1}{s}^{[\lambda_1} \dots \underset{p}{s}^{\lambda_p]}}{d\sigma} = \frac{d\tau}{d\sigma} \cdot \frac{D\underset{1}{s}^{[\lambda_1} \dots \underset{p}{s}^{\lambda_p]}}{d\tau} = \varrho \frac{D\underset{1}{t}^{[\lambda_1}\varrho \underset{2}{t}^{\lambda_2}\varrho^2 \dots \underset{p}{t}^{\lambda_p]}\varrho^p}{d\tau},$$

where $\bar{v}^{\lambda_1 \dots \lambda_p} \overset{\text{def}}{=} \underset{1}{s}^{[\lambda_1} \dots \underset{p}{s}^{\lambda_p]}$ and we have, for short, set

$$\varrho \overset{\text{def}}{=} \frac{d\tau}{d\sigma}.$$

Thus, we have

$$\frac{D\bar{v}^{\lambda_1\ldots\lambda_p}}{d\sigma} = \varrho\frac{D\left[\varrho^{\frac{p(p+1)}{2}} \underset{1}{t}^{[\lambda_1} \ldots \underset{p}{t}^{\lambda_p]}\right]}{d\tau}$$

$$= \varrho^{1+\binom{p+1}{2}} \frac{D\underset{1}{t}^{[\lambda_1} \ldots \underset{p}{t}^{\lambda_p]}}{d\tau} + \varrho \underset{1}{t}^{[\lambda_1} \ldots \underset{p}{t}^{\lambda_p]} \frac{d}{d\tau}\left\{\varrho^{\frac{p(p+1)}{2}}\right\}$$

$$= \varrho^{1+\binom{p+1}{2}} \overset{p}{\alpha} v^{\lambda_1\ldots\lambda_p} \ldots + \varrho v^{\lambda_1\ldots\lambda_p} \frac{p(p+1)}{2} \varrho^{\binom{p+1}{2}-1} \frac{d\varrho}{d\tau}$$

$$= v^{\lambda_1\ldots\lambda_p} \varrho^{\binom{p+1}{2}} \left\{\varrho\overset{p}{\alpha} + \frac{p(p+1)}{2} \cdot \frac{d\varrho}{d\tau}\right\}.$$

Since relationship (5.117) is to hold, ϱ must be such that

$$\varrho\overset{p}{\alpha} + \frac{p(p+1)}{2} \cdot \frac{d\varrho}{d\tau} = 0.$$

Hence, we have

$$\frac{d\varrho}{\varrho} = -\frac{2}{p(p+1)} \overset{p}{\alpha} d\tau,$$

from which we get

$$\varrho = \frac{1}{C} \exp\left[-\frac{2}{p(p+1)} \int \overset{p}{\alpha}(\tau) d\tau\right],$$

or

$$\frac{1}{\varrho} = C \exp\left[\frac{2}{p(p+1)} \int \overset{p}{\alpha} d\tau\right],$$

whence

$$\frac{d\sigma}{d\tau} = C \exp\left[\frac{2}{p(p+1)} \int \overset{p}{\alpha} d\tau\right]$$

and, finally

$$\sigma = \int C \exp\left[\frac{2}{p(p+1)} \int \overset{p}{\alpha} d\tau\right] d\tau. \tag{5.118}$$

With the notation

$$f(\tau) \overset{\text{def}}{=} \int_{\tau_0}^{\tau} \exp\left[\frac{2}{p(p+1)} \int_{\tau_0}^{u} \overset{p}{\alpha}(v) dv\right] du \tag{5.119}$$

we can write

$$\sigma = C_1 f(\tau) + C_2, \tag{5.120}$$

where C_1 and C_2 are arbitrary constants (of integration). If we had $\overset{p}{\alpha}(\tau) \equiv 0$, then $f(\tau) = \tau - \tau_0$ and we thus see that the new preferred parameter σ is a linear function of the original i.e. we see that the preferred parameter, called the *affine arc of the curve C*, is defined up to inhomogeneous linear transformations (Hlavatý [32]).

Now consider the special case $p = 1$, which means that the curve C is a geodesic. We then have

$$f(\tau) = \int_{\tau_0}^{\tau} \exp\left[\int_{\tau_0}^{u} \alpha(v)\,dv\right] du,$$

where

$$\frac{D\left[\frac{d\xi^\nu}{d\tau}\right]}{d\tau} = \alpha \frac{d\xi^\nu}{d\tau}.$$

The parameter σ found from the relationship

$$\sigma = C_1 f(\tau) + C_2$$

is, of course, the natural parameter for the geodesic.

71. Locally Geodesic (Normal) Coordinates. We have seen that if in a particular coordinate system (λ) we have

$$\Gamma^\nu_{\mu\lambda} \overset{*}{=} 0 \tag{5.121}$$

at every point, the equations of the geodesic are particularly simple, indeed, they are linear equations. In the space E_n this property is enjoyed by so-called Cartesian systems. In a later chapter we shall become acquainted with a more extensive class of spaces, based on a more general pseudogroup of transformations, in which coordinate systems with property (5.121) exist. A more modest problem can, however, be posed. We fix a point p in a space L_n and ask whether or not there exists a coordinate system (λ') with the property that

$$(\Gamma^{\nu'}_{\mu'\lambda'}) \overset{*}{=} 0 \quad (\lambda', \mu', \nu' = 1', 2', \dots, n'). \tag{5.122}$$

It turns out that when $L_n = A_n$, that is when the object of parallel displacement is symmetric, then with some assumptions as to the regularity of the coordinates $\Gamma^\nu_{\lambda\mu}$ of the object it is possible to prove the existence of a coordinate system (λ') for which relationships (5.122) hold. Such a coordinate system is said to be *locally geodesic* at point p, whereas the coordinates $\xi^{\lambda'}$ themselves are called *normal* (at the point p) (cf. p. 211).

This concept together with this name for it were introduced by Veblen [151]. Geodesic coordinates in various spaces have been the subject of a large number of works. Proofs of the existence of local geodesic systems for the most part leave much to be desired as far as rigour is concerned. The most correct proof can be found in a paper [156] by J. H. C. Whitehead who was the first to prove the existence of convex neighbourhoods in general spaces, i.e. neighbourhoods such that any two points in them can be joined by a single arc of a geodesic lying entirely within the given neighbourhood.

It seems to us that the proof of the existence of local geodesic systems can be presented without any assumptions as to the regularity of the coordinates of the object $\Gamma^{\nu}_{\mu\lambda}$, as we demonstrate below. In the meantime, most authors, such as Schouten for example, assume the functions $\Gamma^{\nu}_{\mu\lambda}$ to be analytic, which in our opinion is a superfluous. Not until we work with normal Riemannian coordinates (Section 72) does it becomes necessary to assume that the functions $\Gamma^{\nu}_{\mu\lambda}$ are analytic.

We now denote the coordinates of the point p in the system (λ) by $\underset{0}{\xi}^{\lambda}$ and we write

$$\alpha^{\lambda}_{\mu\nu} \overset{\text{def}}{=} \tfrac{1}{2}\Gamma^{\lambda}_{\mu\nu}(\underset{0}{\xi}).$$

Next, let us consider the following transformation:

$$\xi^{\lambda'} = \delta^{\lambda'}_{\lambda}\{\xi^{\lambda} - \underset{0}{\xi}^{\lambda} + \alpha^{\lambda}_{\mu\nu}(\xi^{\mu} - \underset{0}{\xi}^{\mu})(\xi^{\nu} - \underset{0}{\xi}^{\nu})\}. \tag{5.123}$$

The question we now ask whether this transformation is locally invertible We have

$$\begin{aligned} A^{\lambda'}_{\varrho} &= \frac{\partial \xi^{\lambda'}}{\partial \xi^{\varrho}} = \delta^{\lambda'}_{\lambda}\{\delta^{\lambda}_{\varrho} + \alpha^{\lambda}_{\mu\nu}[\delta^{\mu}_{\varrho}(\xi^{\nu} - \underset{0}{\xi}^{\nu}) + \delta^{\nu}_{\varrho}(\xi^{\mu} - \underset{0}{\xi}^{\mu})]\} \\ &= \delta^{\lambda'}_{\varrho} + \delta^{\lambda'}_{\lambda}\{\alpha^{\lambda}_{\varrho\nu}(\xi^{\nu} - \underset{0}{\xi}^{\nu}) + \alpha^{\lambda}_{\mu\varrho}(\xi^{\mu} - \underset{0}{\xi}^{\mu})\}, \end{aligned} \tag{5.124}$$

whence

$$(A^{\lambda'}_{\varrho})_p = \delta^{\lambda'}_{\varrho}. \tag{5.125}$$

It thus follows that

$$(J)_p = |(A^{\lambda'}_{\varrho})_p| = \begin{vmatrix} 1 & 0 \ldots 0 \\ 0 & 1 \ldots 0 \\ \cdot & \cdots\cdots \\ 0 & 0 \ldots 1 \end{vmatrix} = 1,$$

and, hence, in a certain neighbourhood of the point p we have $J \neq 0$, i.e. transformation (5.123) is allowable. Let us now calculate $(\Gamma^{\lambda'}_{\mu'\nu'})_p$.

On the one hand, we have

$$\Gamma^{\lambda'}_{\mu'\nu'} = \Gamma^{\lambda}_{\mu\nu} A^{\lambda'\mu\ \nu}_{\lambda\ \mu'\nu'} - A^{\mu\ \nu}_{\mu'\nu'}\partial_{\mu} A^{\lambda'}_{\nu}.$$

On the other hand, differentiating (5.124), which is permissible by virtue of assumption (5.123), we have

$$\partial_\mu A_\nu^{\lambda'} = \frac{\partial}{\partial \xi^\mu} \{\delta_\nu^{\lambda'} + \delta_\lambda^{\lambda'} [\alpha_{\nu\varkappa}^\lambda (\xi^\varkappa - \overset{0}{\xi^\varkappa}) + \alpha_{\varkappa\nu}^\lambda (\xi^\varkappa - \overset{0}{\xi^\varkappa})]\}$$

$$= \delta_\lambda^{\lambda'} [\alpha_{\nu\varkappa}^\lambda \delta_\mu^\varkappa + \alpha_{\varkappa\nu}^\lambda \delta_\mu^\varkappa] = \delta_\lambda^{\lambda'} [\alpha_{\nu\mu}^\lambda + \alpha_{\mu\nu}^\lambda].$$

Now the object of parallel displacement is by assumption symmetric ($L_n = A_n$), and thus $\alpha_{\nu\mu}^\lambda = \alpha_{\mu\nu}^\lambda$. Hence

$$(\partial_\mu A_\nu^{\lambda'})_p = \delta_\lambda^{\lambda'} (\Gamma_{\mu\nu}^\lambda)_p. \tag{5.126}$$

Note, moreover, that similar formulae

$$(A_{\mu'}^\mu)_p = \delta_{\mu'}^\mu \tag{5.127}$$

hold for the inverse transformation of (5.123), as can be easily shown. When we substitute from (5.125), (5.126) and (5.127), in the formula

$$(\Gamma_{\mu'\nu'}^{\lambda'})_p = \{\Gamma_{\mu\nu}^\lambda A_{\lambda\ \mu'\nu'}^{\lambda'\mu\ \nu} - A_{\mu'\nu'}^{\mu\ \nu} \partial_\mu A_\nu^{\lambda'}\}_p,$$

we obtain

$$(\Gamma_{\mu'\nu'}^{\lambda'}) = (\Gamma_{\mu\nu}^\lambda)_p \delta_{\lambda\ \mu'\nu'}^{\lambda'\mu\ \nu} - \delta_{\mu'\nu'\lambda}^{\mu\ \nu\ \lambda'} (\Gamma_{\mu\nu}^\lambda)_p = 0$$

and we have thus shown that the system (λ') is geodesic at the point p. Note that if we pass from the system (λ'), which is geodesic at point p, to the system (λ'') by means of an affine transformation, then in (λ'') we again get a system that is geodesic at p.

Fermi [104], the famous theoretical physicist, proved that in Riemannian space in which, as we shall see below, an object of parallel displacement can be defined uniquely by means of the metric tensor, it is possible to find for every arbitrary regular line C a coordinate system which is geodesic at every point p on C. This theorem was later generalized by Eisenhart [25] for arbitrary spaces A_n. The proofs given for these theorems by these authors, as well as that given by Levi-Civita [47], are not strictly correct, from the point of view of rigour.

Note that in coordinate systems which are geodesic along a curve C the absolute derivative of any field of quantities, defined along the curve C, reduces to the ordinary derivative of that field.

For an m-dimensional surface X_m, where $m \geqslant 2$, it is not in general possible to set up a coordinate system with the property that the coordinates of the object of parallel displacement are identically equal to zero in the entire hypersurface X_m.

72. Riemann's Normal Coordinates. A special case of geodesic coordinates at a fixed point p_0 are the *normal coordinates* which are uniquely determined (in contrast to geodesic coordinates) and which can be regarded as a generalization and modification of polar coordinates in the plane. In order to construct them, consider the set of all geodesics passing through a fixed point p_0. Let the original coordinates of the point p_0 be $\underset{0}{\xi}^\lambda$. On each geodesic we introduce the affine parameter σ, in terms of which the equations of the geodesics assume the form

$$\frac{d^2\xi^\nu}{d\sigma^2}+\Gamma^\nu_{\lambda\mu}\frac{d\xi^\lambda}{d\sigma}\frac{d\xi^\mu}{d\sigma}=0 \quad (\nu=1,2,\ldots,n). \tag{5.128}$$

We have seen that the unique choice of the parameter σ can only be made once two constants have been fixed. These constants are fixed by, for example, the two conditions

$$\begin{aligned} &\text{for } \sigma=0 \quad \text{we must have} \quad \xi^\lambda=\underset{0}{\xi}^\lambda,\\ &\text{for } \sigma=1 \quad \text{we must have} \quad \sum_\lambda(\xi^\lambda-\underset{0}{\xi}^\lambda)^2=1. \end{aligned} \tag{5.129}$$

The geodesic through p_0 is uniquely defined by two initial conditions

$$\begin{aligned} \xi^\nu(0)&=\underset{0}{\xi}^\nu,\\ \left(\frac{d\xi^\nu}{d\sigma}\right)_{\sigma=0}&=v^\nu, \end{aligned} \tag{5.130}$$

where v^ν is an *a priori* non-zero vector attached at the point p_0.

We now make the assumption that the functions $\Gamma^\nu_{\lambda\mu}(\xi)$ are analytic functions in the neighbourhood of $(\underset{0}{\xi}^\nu)$. Equations (5.128) can then be differentiated freely. After the first differentiation (with respect to σ), we obtain

$$\frac{d^3\xi^\nu}{d\sigma^3}+\frac{d\Gamma^\nu_{\lambda\mu}}{d\sigma}\frac{d\xi^\lambda}{d\sigma}\frac{d\xi^\mu}{d\sigma}+\Gamma^\nu_{\lambda\mu}\frac{d^2\xi^\lambda}{d\sigma^2}\frac{d\xi^\mu}{d\sigma}+\Gamma^\nu_{\lambda\mu}\frac{d\xi^\lambda}{d\sigma}\frac{d^2\xi^\mu}{d\sigma^2}=0$$

or

$$\frac{d^3\xi^\nu}{d\sigma^3}+(\partial_\omega\Gamma^\nu_{\lambda\mu})\frac{d\xi^\omega}{d\sigma}\frac{d\xi^\lambda}{d\sigma}\frac{d\xi^\mu}{d\sigma}+\Gamma^\nu_{\lambda\mu}\left(\frac{d^2\xi^\lambda}{d\sigma^2}\frac{d\xi^\mu}{d\sigma}+\frac{d\xi^\lambda}{d\sigma}\frac{d^2\xi^\mu}{d\sigma^2}\right)=0,$$

or, on eliminating the second derivatives from equations (5.128),

$$\frac{d^3\xi^\nu}{d\sigma^3}+\Gamma^\nu_{\omega\lambda\mu}\frac{d\xi^\omega}{d\sigma}\frac{d\xi^\lambda}{d\sigma}\frac{d\xi^\mu}{d\sigma}=0,$$

where, for brevity of notation, we have put

$$\Gamma^{\nu}_{\omega\lambda\mu} \stackrel{\text{def}}{=} \partial_{\omega}\Gamma^{\nu}_{\lambda\mu}-\Gamma^{\nu}_{\tau\omega}\Gamma^{\tau}_{\lambda\mu}-\Gamma^{\nu}_{\tau\omega}\Gamma^{\tau}_{\mu\lambda} = \partial_{\omega}\Gamma^{\nu}_{\lambda\mu}-2\Gamma^{\nu}_{\tau\omega}\Gamma^{\tau}_{\lambda\mu}, \tag{5.131}$$

which, by the symmetry of the object $\Gamma^{\nu}_{\lambda\mu}$ and the symmetry of the product $\dfrac{d\xi^{\omega}}{d\sigma}\cdot\dfrac{d\xi^{\lambda}}{d\sigma}\cdot\dfrac{d\xi^{\mu}}{d\sigma}$, may also be written as

$$\Gamma^{\nu}_{\omega\lambda\mu} = \partial_{(\omega}\Gamma^{\nu}_{\lambda\mu)}-2\Gamma^{\tau}_{(\omega\lambda}\Gamma^{\nu}_{\mu)\tau}.$$

By further differentiation and repeated elimination of the second derivatives, we arrive at

$$\frac{d^j\xi^{\nu}}{d\sigma^j}+\Gamma^{\nu}_{\lambda_j\ldots\lambda_1}\frac{d\xi^{\lambda_1}}{d\sigma}\cdots\frac{d\xi^{\lambda_j}}{d\sigma} = 0, \tag{5.132}$$

where for the coefficients of Γ we have the recursion formula

$$\Gamma^{\nu}_{\lambda_j\ldots\lambda_1} \stackrel{\text{def}}{=} \partial_{(\lambda_j}\Gamma^{\nu}_{\lambda_{j-1}\ldots\lambda_1)}-j\Gamma^{\omega}_{(\lambda_j\lambda_{j-1}}\Gamma^{\nu}_{\lambda_{j-2}\ldots\lambda_1)\omega}. \tag{5.133}$$

From these formulae we can calculate in particular the values of derivatives of all orders

$$\left(\frac{d^j\xi^{\nu}}{d\sigma^j}\right)_0 = -\Gamma^{\nu}_{\lambda_j\ldots\lambda_1}v^{\lambda_1}\ldots v^{\lambda_j}, \tag{5.134}$$

whereby for every geodesic through the point p_0 we have the expansion

$$\xi^{\nu} = \underset{0}{\xi^{\nu}}+\sigma v^{\nu}-\frac{1}{2!}\Gamma^{\nu}_{\lambda\mu}(\underset{0}{\xi})v^{\lambda}v^{\mu}\sigma^2-\frac{1}{3!}\Gamma^{\nu}_{\lambda_3\lambda_2\lambda_1}v^{\lambda_1}v^{\lambda_2}v^{\lambda_3}\sigma^3-\ldots$$

We now introduce

$$\eta^{\nu} \stackrel{\text{def}}{=} v^{\nu}\sigma$$

as new coordinates in the neighbourhood of p_0 for a point which lies on the geodesic tangent to the vector v^{ν} and the value of whose (uniquely determined) affine parameter is σ. (To avoid using the generalized Kronecker symbol, we here depart from the general rule that the new coordinates be denoted by $\xi^{\lambda'} = \delta^{\lambda'}_{\lambda}\eta^{\lambda}$.) We then have

$$\xi^{\nu} = \underset{0}{\xi^{\nu}}+\eta^{\nu}-\frac{1}{2!}\Gamma^{\nu}_{\lambda_2\lambda_1}\eta^{\lambda_1}\eta^{\lambda_2}-\frac{1}{3!}\Gamma^{\nu}_{\lambda_3\lambda_2\lambda_1}\eta^{\lambda_1}\eta^{\lambda_2}\eta^{\lambda_3}-\ldots \tag{5.135}$$

For comparatively small η^{ν}'s, this expansion will be convergent (under certain assumptions as to the object Γ) and invertible so that we are able to write

$$\eta^{\nu} = \xi^{\nu}-\underset{0}{\xi^{\nu}}+\Lambda^{\nu}_{\lambda\mu}(\underset{0}{\xi})(\xi^{\lambda}-\underset{0}{\xi^{\lambda}})(\xi^{\mu}-\underset{0}{\xi^{\mu}})+\ldots, \tag{5.136}$$

where the coefficients Λ are once again given in terms of the coefficients $\Gamma^{\nu}_{\lambda_j \ldots \lambda_1}$ by means of recursion formulae.

It can easily be shown that we have

$$\left(\frac{\partial \eta^{\nu}}{\partial \xi^{\lambda}}\right)_0 = \delta^{\nu}_{\lambda} \tag{5.137}$$

and

$$\left(\frac{\partial^2 \eta^{\nu}}{\partial \xi^{\lambda} \partial \xi^{\mu}}\right)_0 = 0, \tag{5.138}$$

from which it follows that all of the coordinates of the object Γ are zero at a point p_0 in the coordinate system η^{ν}. The (uniquely defined) coordinates of the system η^{ν} are called the *normal coordinates* at the point p_0. Note that, the coordinates η^{ν} are geodesic at the point p_0 (some authors refer to normal coordinates as Riemannian coordinates).

In normal coordinates, the equations of geodesics passing through the point p_0, although not of all geodesics in general, are linear in form

$$\eta^{\nu} = \alpha^{\nu} \sigma. \tag{5.139}$$

Some authors use the term *normal tensors* for tensors in normal coordinates, i.e. tensors attached at the point p_0, the origin of the normal coordinates. We regard this name as inappropriate; one cannot speak of normal tensors but only of the normal coordinates of a tensor.

CHAPTER VI

CURVATURE AND TORSION TENSORS

73. The Curvature Tensor. We have already seen that if the equations

$$\Gamma^{\lambda}_{\mu\nu} \overset{*}{=} 0 \quad (\lambda, \mu, \nu = 1, 2, \ldots, n) \tag{6.1}$$

hold in some coordinate system (λ), then the equations of geodesics in this system take on linear form. Equations (6.1) in a general space based on the group G_1 are not of an invariant nature. The problem then is find an invariant condition on the objects Γ for relations (6.1) to hold in certain coordinate systems (λ). Let us find that condition.

We start from an arbitrary coordinate system (λ) in the space L_n in which the $\Gamma^{\lambda}_{\mu\nu}$'s are not identically zero and we assume that there exists a coordinate system (λ') with the property that

$$\Gamma^{\lambda'}_{\mu'\nu'} \overset{*}{=} 0.$$

In this event, there exists a transformation of coordinates $\xi^{\lambda} \to \xi^{\lambda'}$ such that we have the relations

$$\Gamma^{\lambda}_{\mu\nu} = \Gamma^{\lambda'}_{\mu'\nu'} A^{\lambda\ \mu'\nu'}_{\lambda'\mu\ \nu} + A^{\lambda}_{\lambda'} \partial_{\mu} A^{\lambda'}_{\nu} = A^{\lambda}_{\lambda'} \partial_{\mu} A^{\lambda'}_{\nu},$$

whence, on multiplying both sides by $A^{\mu'}_{\lambda}$, we obtain

$$\partial_{\mu} A^{\mu'}_{\nu} = A^{\mu'}_{\lambda} \Gamma^{\lambda}_{\mu\nu}. \tag{6.2}$$

The set of equation (6.2) represents a system of partial differential equations of the second order with the $\Gamma^{\lambda}_{\mu\nu}$ as (known) coefficients. Since

$$\partial_{[\mu} A^{\mu'}_{\nu]} = 0,$$

we see at once that the necessary condition for solubility is of the form

$$A^{\mu'}_{\lambda} \Gamma^{\lambda}_{[\mu\nu]} = 0.$$

This leads to the conclusion that

$$\Gamma^{\lambda}_{[\mu\nu]} = 0, \tag{6.3}$$

i.e. that the object Γ must be symmetric, and hence that the space must be the space A_n.

Now suppose further that the condition is satisfied. Assuming that the transformation $(\lambda) \to (\lambda')$ is of class C^3, we differentiate both sides of equations (6.2) with respect to ξ^ϱ, to get

$$\partial_{\varrho\mu} A^{\mu'}_{\nu} = (\partial_\varrho A^{\mu'}_{\lambda})\Gamma^{\lambda}_{\mu\nu} + A^{\mu'}_{\lambda}\partial_\varrho \Gamma^{\lambda}_{\mu\nu}.$$

Using equation (6.2) we substitute

$$\partial_\varrho A^{\mu'}_{\lambda} = A^{\mu'}_{\sigma}\Gamma^{\sigma}_{\varrho\lambda}$$

and obtain

$$\partial_{\varrho\mu} A^{\mu'}_{\nu} = A^{\mu'}_{\sigma}\Gamma^{\sigma}_{\varrho\lambda}\Gamma^{\lambda}_{\mu\nu} + A^{\mu'}_{\lambda}\partial_\varrho\Gamma^{\lambda}_{\mu\nu}$$

or, on changing the dummy index,

$$\partial_{\varrho\mu} A^{\mu'}_{\nu} = A^{\mu'}_{\sigma}\{\Gamma^{\sigma}_{\varrho\lambda}\Gamma^{\lambda}_{\mu\nu} + \partial_\varrho\Gamma^{\sigma}_{\mu\nu}\}. \tag{6.4}$$

When we interchange the indices ϱ, μ, we obtain

$$\partial_{\mu\varrho} A^{\mu'}_{\nu} = A^{\mu'}_{\sigma}\{\Gamma^{\sigma}_{\mu\lambda}\Gamma^{\lambda}_{\varrho\nu} + \partial_\mu\Gamma^{\sigma}_{\varrho\nu}\}. \tag{6.5}$$

Since

$$\partial_{\varrho\mu} A^{\mu'}_{\nu} = \partial_{\mu\varrho} A^{\mu'}_{\nu},$$

then on subtracting equation (6.4) from equation (6.5), we get

$$A^{\mu'}_{\sigma}\{\partial_{[\varrho}\Gamma^{\sigma}_{\mu]\nu} + \Gamma^{\sigma}_{[\varrho|\lambda|}\Gamma^{\lambda}_{\mu]\nu}\} = 0$$

or, on multiplying by $A^{\lambda}_{\mu'}$ and at the same time replacing the dummy index λ by ω, we have

$$\partial_{[\varrho}\Gamma^{\lambda}_{\mu]\nu} + \Gamma^{\lambda}_{[\varrho|\omega|}\Gamma^{\omega}_{\mu]\nu} = 0 \qquad (\lambda, \mu, \nu, \varrho = 1, 2, \ldots, n). \tag{6.6}$$

It turns out that equations (6.6), together with equation (6.3), constitute conditions for the system (6.2) to be totally integrable, i.e. these conditions, conversely, are sufficient for the system (6.2) to be soluble (we ignore the question of initial conditions for the time being).

If we abbreviate the (doubled) left-hand side of equations (6.2) by setting

$$R_{\varrho\mu\nu}{}^{\lambda} \stackrel{\text{def}}{=} 2\partial_{[\varrho}\Gamma^{\lambda}_{\mu]\nu} + 2\Gamma^{\lambda}_{[\varrho|\omega|}\Gamma^{\omega}_{\mu]\nu}, \tag{6.7}$$

conditions (6.6) then assume the form

$$R_{\varrho\mu\nu}{}^{\lambda} = 0 \qquad (\lambda, \mu, \nu, \varrho = 1, 2, \ldots, n). \tag{6.8}$$

Since the system (λ) was arbitrary, conditions (6.8) are invariant, and this suggests that the object R is probably a tensor. This is indeed the case, for we have

$$\begin{aligned}
R_{\varrho'\mu'\nu'}{}^{\lambda'} &= \partial_{\varrho'}\Gamma^{\lambda'}_{\mu'\nu'} - \partial_{\mu'}\Gamma^{\lambda'}_{\varrho'\nu'} + \Gamma^{\lambda'}_{\varrho'\omega'}\Gamma^{\omega'}_{\mu'\nu'} - \Gamma^{\lambda'}_{\mu'\omega'}\Gamma^{\omega'}_{\varrho'\nu'} \\
&= A^{\varrho}_{\varrho'}\partial_{\varrho}\{\Gamma^{\lambda}_{\mu\nu}A^{\lambda'\mu\,\nu}_{\lambda\,\mu'\nu'} + A^{\lambda'}_{\lambda}\partial_{\mu'}A^{\lambda}_{\nu'}\} - A^{\mu}_{\mu'}\partial_{\mu}\{\Gamma^{\lambda}_{\varrho\nu}A^{\lambda'\varrho\,\nu}_{\lambda\,\varrho'\nu'} + A^{\lambda'}_{\lambda}\partial_{\varrho'}A^{\lambda}_{\nu'}\} + \\
&\quad + \{\Gamma^{\lambda}_{\varrho\omega}A^{\lambda'\varrho\,\omega}_{\lambda\,\varrho'\omega'} + A^{\lambda'}_{\lambda}\partial_{\varrho'}A^{\lambda}_{\omega'}\}\{\Gamma^{\sigma}_{\mu\nu}A^{\omega'\mu\,\nu}_{\sigma\,\mu'\nu'} + A^{\omega'}_{\sigma}\partial_{\mu'}A^{\sigma}_{\nu'}\} - \\
&\quad - \{\Gamma^{\lambda}_{\mu\omega}A^{\lambda'\mu\,\omega}_{\lambda\,\mu'\omega'} + A^{\lambda'}_{\lambda}\partial_{\mu'}A^{\lambda}_{\omega'}\}\{\Gamma^{\sigma}_{\varrho\nu}A^{\omega'\varrho\,\nu}_{\sigma\,\varrho'\nu'} + A^{\omega'}_{\sigma}\partial_{\varrho'}A^{\sigma}_{\nu'}\} \\
&= \underline{A^{\varrho\,\lambda'\mu\,\nu}_{\varrho'\lambda\,\mu'\nu'}\partial_{\varrho}\Gamma^{\lambda}_{\mu\nu}} + A^{\varrho\,\mu\,\nu}_{\varrho'\mu'\nu'}(\partial_{\varrho}A^{\lambda'}_{\lambda})\Gamma^{\lambda}_{\mu\nu} + \Gamma^{\lambda}_{\mu\nu}[A^{\lambda'\nu}_{\lambda\,\nu'}\partial_{\varrho'}A^{\mu}_{\mu'} + A^{\lambda'\mu}_{\lambda\,\mu'}\partial_{\varrho'}A^{\nu}_{\nu'}] - \\
&\quad - \underline{A^{\varrho\,\lambda'\mu\,\nu}_{\varrho'\lambda\,\mu'\nu'}\partial_{\mu}\Gamma^{\lambda}_{\varrho\nu}} - A^{\varrho\,\mu\,\nu}_{\varrho'\mu'\nu'}(\partial_{\mu}A^{\lambda'}_{\lambda})\Gamma^{\lambda}_{\varrho\nu} - \Gamma^{\lambda}_{\varrho\nu}[A^{\lambda'\varrho}_{\lambda\,\varrho'}\partial_{\mu'}A^{\nu}_{\nu'} + A^{\lambda'\nu}_{\lambda\,\nu'}\partial_{\mu'}A^{\varrho}_{\varrho'}] + \\
&\quad + A^{\lambda'}_{\lambda}\partial_{\varrho'\mu'}A^{\lambda}_{\nu'} + (\partial_{\varrho'}A^{\lambda'}_{\lambda})\partial_{\mu'}A^{\lambda}_{\nu'} - A^{\lambda'}_{\lambda}\partial_{\mu'\varrho'}A^{\lambda}_{\nu'} - (\partial_{\mu'}A^{\lambda'}_{\lambda})\partial_{\varrho'}A^{\lambda}_{\nu'} + \\
&\quad + \underline{A^{\lambda'\varrho\,\omega\,\omega'\mu\,\nu}_{\lambda\,\varrho'\omega'\sigma\,\mu'\nu'}\Gamma^{\lambda}_{\varrho\omega}\Gamma^{\sigma}_{\mu\nu}} + A^{\lambda'\omega'\mu\,\nu}_{\lambda\,\sigma\,\mu'\nu'}\Gamma^{\sigma}_{\mu\nu}\partial_{\varrho'}A^{\lambda}_{\omega'} + A^{\lambda'\varrho\,\omega\,\omega'}_{\lambda\,\varrho'\omega'\sigma}\Gamma^{\lambda}_{\varrho\omega}\partial_{\mu'}A^{\sigma}_{\nu'} - \\
&\quad - \underline{A^{\lambda'\mu\,\omega\,\omega'\varrho\,\nu}_{\lambda\,\mu'\omega'\sigma\,\varrho'\nu'}\Gamma^{\lambda}_{\mu\omega}\Gamma^{\sigma}_{\varrho\nu}} - A^{\lambda'\mu\,\omega\,\omega'}_{\lambda\,\mu'\omega'\sigma}\Gamma^{\lambda}_{\mu\omega}\partial_{\varrho'}A^{\sigma}_{\nu'} - A^{\lambda'\omega'\varrho\,\nu}_{\lambda\,\sigma\,\varrho'\nu'}\Gamma^{\sigma}_{\varrho\nu}\partial_{\mu'}A^{\lambda}_{\omega'} + \\
&\quad + A^{\lambda'\omega'}_{\lambda\,\sigma}(\partial_{\varrho'}A^{\lambda}_{\omega'})\partial_{\mu'}A^{\sigma}_{\nu'} - A^{\lambda'\omega'}_{\lambda\,\sigma}(\partial_{\varrho'}A^{\sigma}_{\nu'})\partial_{\mu'}A^{\lambda}_{\omega'}.
\end{aligned}$$

The terms underlined with a wavy line taken together yield $A^{\varrho\,\lambda'\mu\,\nu}_{\varrho'\lambda\,\mu'\nu'}R_{\varrho\mu\nu}{}^{\lambda}$.

It must be shown that the other terms cancel out. The terms containing third derivatives $\partial_{\varrho'\mu'}A^{\lambda}_{\nu'}$ and $\partial_{\mu'\varrho'}A^{\lambda}_{\nu'}$ obviously cancel, while the second term in row three cancels with the first of row six and, similarly, the fourth term of row three cancels with the second term of row six, since

$$\begin{aligned}
A^{\lambda'\omega'}_{\lambda\,\sigma}(\partial_{\varrho'}A^{\lambda}_{\omega'})\partial_{\mu'}A^{\sigma}_{\nu'} &= -A^{\lambda\,\omega'}_{\omega'\sigma}(\partial_{\varrho'}A^{\lambda'}_{\lambda})\partial_{\mu'}A^{\sigma}_{\nu'} = -A^{\lambda}_{\sigma}(\partial_{\varrho'}A^{\lambda'}_{\lambda})\partial_{\mu'}A^{\sigma}_{\nu'} \\
&= -(\partial_{\varrho'}A^{\lambda'}_{\sigma})\partial_{\mu'}A^{\sigma}_{\nu'} = -(\partial_{\varrho'}A^{\lambda'}_{\lambda})\partial_{\mu'}A^{\lambda}_{\nu'}.
\end{aligned}$$

Still to be dealt with are ten terms containing Γ in the first degree. The fourth term of row one cancels with the second term of row five and, similarly, the third term of row two cancels with the second term of row four and the second term of row two with the third term of row five. Thus, out of this group of ten terms, two are now left: the third in row one and the fourth in row two, and they too cancel, as can be seen when the dummy index in one of them is changed. Hence, we finally have

$$R_{\varrho'\mu'\nu'}{}^{\lambda'} = R_{\varrho\mu\nu}{}^{\lambda}A^{\lambda'\varrho\,\mu\,\nu}_{\lambda\,\varrho'\mu'\nu'}, \tag{6.9}$$

whence it follows that R is a tensor of valence $(1, 3)$. This tensor is called the *curvature tensor* of the space A_n.

Although it is true that we arrived at equations (6.8) under the assumption that the object Γ was symmetric, throughout the entire proof given above to show that $R_{\varrho\mu\nu}{}^{\lambda}$ is a tensor we did not make use of the symmetry of the object $\Gamma^{\lambda}_{\mu\nu}$.

We thus have the theorem that if in a space A_n the tensor of curvature is identically zero, then there exists a coordinate system (λ) with property (6.1). The converse theorem is clearly also true.

In Riemannian spaces, the tensor of curvature is called the *Christoffel–Riemann tensor*.

Note that the tensor R is antisymmetric with respect to the first two indices, as is immediately implied by the definition:

$$R_{(\omega\mu)\nu}{}^{\lambda} = 0. \tag{6.10}$$

74. Another Way of Introducing the Curvature Tensor. The curvature tensor can be arrived at in another way, as we demonstrate below.

It is well known that ordinary differentiation is independent of the order in which the various differentiations are carried out, i.e. that we always have

$$\partial_\lambda \partial_\mu = \partial_\mu \partial_\lambda.$$

The question arises as to whether covariant differentiation also enjoys the same property, or expressing it symbolically, whether

$$\nabla_\lambda \nabla_\mu \overset{?}{=} \nabla_\mu \nabla_\lambda.$$

To answer this question, let us consider the case of a vector field (of class C^2) v^ν and let us apply to it the operation of two-fold covariant differentiation in both orders. Note that repeated covariant differentiation is allowable because the result of differentiation of a quantity can always be differentiated, under appropriate regularity conditions on the field.

In order to get a more general result, we consider a system (K) which is not necessarily holonomic. We have

$$\begin{aligned}\nabla_J \nabla_I v^K &= \partial_J \nabla_I v^K + \Gamma^K_{JL} \nabla_I v^L - \Gamma^L_{JI} \nabla_L v^K \\ &= \partial_J[\partial_I v^K + \Gamma^K_{IM} v^M] + \Gamma^K_{JL}[\partial_I v^L - \Gamma^L_{IM} v^M] - \Gamma^L_{JI}(\partial_L v^K + \Gamma^K_{LM} v^M) \\ &= \partial_J \partial_I v^K + (\partial_J \Gamma^K_{IM}) v^M + \Gamma^K_{IM}\, \partial_J v^M + \Gamma^K_{JL}\, \partial_I v^L - \Gamma^L_{JI}\, \partial_L v^K + \\ &\quad + v^M [\Gamma^K_{JL} \Gamma^K_{IM} - \Gamma^L_{JI} \Gamma^K_{LM}].\end{aligned}$$

Writing down the analogous formula for $\nabla_I \nabla_J v^K$ and subtracting it from the formula above, we obtain

$$\nabla_{[J} \nabla_{I]} v^K = \partial_{[J}\, \partial_{I]} v^K + (\partial_{[J}\, \Gamma^K_{I]M}) v^M + \Gamma^K_{[J|L|}\, \Gamma^L_{I]M} v^M - \Gamma^L_{[JI]} \nabla_L v^K. \tag{6.11}$$

But

$$\begin{aligned}\partial_{[J}\partial_{I]}v^K &= A^{\mu}_{[J}\partial_{|\mu|}(A^{\nu}_{I]}\partial_\nu v^K)\\ &= \tfrac{1}{2}[A^{\mu}_{J}(\partial_\mu A^{\nu}_{I})\partial_\nu v^K + A^{\mu\nu}_{JI}\partial_{\mu\nu}v^K - A^{\mu}_{I}(\partial_\mu A^{\nu}_{J})\partial_\nu v^K - A^{\mu\nu}_{IJ}\partial_{\mu\nu}v^K]\\ &= \tfrac{1}{2}[\partial_J A^{\nu}_{I} - \partial_I A^{\nu}_{J}]\partial_\nu v^K = \partial_{[J}A^{\nu}_{I]}\partial_\nu v^K = -\Omega_{IJ}{}^{M}\partial_M v^K\end{aligned}$$

and, therefore, formula (6.11) takes the form

$$\nabla_{[J}\nabla_{I]}v^K = -\Omega_{JI}{}^{M}\partial_M v^K + [\partial_{[J}\Gamma^K_{I]M} + \Gamma^K_{[J|L|}\Gamma^L_{I]M}]v^M - \Gamma^L_{[JI]}\{\partial_L v^K + \Gamma^K_{LM}v^M\},$$

which, when the equality

$$\Gamma^L_{[JI]} = S_{JI}{}^{L} - \Omega_{JI}{}^{L}$$

is used, can be rewritten as

$$\begin{aligned}\nabla_{[J}\nabla_{I]}v^K = -\Omega_{JI}{}^{M}\partial_M v^K + \{\partial_{[J}\Gamma^K_{I]M} + \Gamma^K_{[J|L|}\Gamma^L_{I]M} + \Omega_{JI}{}^{L}\Gamma^K_{LM}\}v^M -\\ - S_{JI}{}^{L}\nabla_L v^K + \Omega_{JI}{}^{L}\partial_L v^K.\end{aligned}$$

With the notation

$$R_{JIM} \stackrel{\text{def}}{=} 2\partial_{[J}\Gamma^K_{I]M} + 2\Gamma^L_{[J|L|}\Gamma^L_{I]M} + 2\Gamma^K_{LM}\Omega_{JI}{}^{L} \qquad (6.12)$$

we have the coordinates of the curvature tensor in an anholonomic system and, finally, the expression (6.11) becomes

$$\nabla_{[J}\nabla_{I]}v^K = \tfrac{1}{2}R_{JIM}{}^{K}v^M - S_{JI}{}^{M}\nabla_M v^K. \qquad (6.13)$$

In exactly the same way, we obtain

$$\nabla_{[J}\nabla_{I]}w_K = -\tfrac{1}{2}R_{JIK}{}^{M}w_M - S_{JI}{}^{M}\nabla_M w_K. \qquad (6.14)$$

It is seen from the formulae above that covariant differentiation is not in general commutative. For symmetric objects $\Gamma (S = 0)$, these formulae simplify to

$$\nabla_{[J}\nabla_{I]}v^K = \tfrac{1}{2}R_{JIM}{}^{K}v^M, \quad \nabla_{[J}\nabla_{I]}w_K = -\tfrac{1}{2}R_{JIK}{}^{M}w_M. \qquad (6.15)$$

The right-hand sides of relations (6.15) reduce to zero when the curvature tensor is zero and in this special case covariant differentiation is commutative.

75. Two-Fold Covariant Differentiation of a Density. We now calculate the quantity $\nabla_{[J}\nabla_{I]}\mathfrak{v}$, where $\mathfrak{v}$ is a density of weight $-r$. Accordingly, we have

$$\nabla_I\mathfrak{v} = \partial_I\mathfrak{v} + r\Gamma_I\mathfrak{v}$$

and, subsequently

$$\nabla_J \nabla_I \mathfrak{v} = \partial_J[\nabla_I \mathfrak{v}] - \Gamma^K_{JI} \nabla_K \mathfrak{v} + r\Gamma_J \nabla_I \mathfrak{v} = \partial_J[\partial_1 \mathfrak{v} + r\Gamma_I \mathfrak{v}] - \Gamma^K_{JI} \nabla_K \mathfrak{v} + r\Gamma_J \nabla_I \mathfrak{v}.$$

From this it follows further that

$$\begin{aligned}\nabla_{[J} \nabla_{I]} \mathfrak{v} &= \partial_{[J} \partial_{I]} \mathfrak{v} + r\mathfrak{v}\partial_{[J} \Gamma_{I]} - \Gamma^K_{[JI]} \nabla_K \mathfrak{v} + r\Gamma_{[J} \partial_{I]} \mathfrak{v} + r\Gamma_{[I} \partial_{J]} \mathfrak{v} \\ &= \partial_{[J} \partial_{I]} \mathfrak{v} + r\mathfrak{v}\partial_{[J} \Gamma_{I]} - \Gamma^K_{[JI]} \nabla_K \mathfrak{v}.\end{aligned}$$

If on the right-hand side we set

$$\partial_{[J} \partial_{I]} \mathfrak{v} = -\Omega_{JI}{}^K \partial_K \mathfrak{v}, \qquad \Gamma^K_{[JI]} = S_{JI}{}^K - \Omega_{JI}{}^K$$

this can be written

$$\begin{aligned}\nabla_{[J} \nabla_{I]} \mathfrak{v} &= r\mathfrak{v}\partial_{[J} \Gamma_{I]} - \Omega_{JI}{}^K \partial_K \mathfrak{v} - S_{JI}{}^K \nabla_K \mathfrak{v} + \Omega_{JI}{}^K (\partial_K \mathfrak{v} + r\mathfrak{v}\Gamma_K) \\ &= r\mathfrak{v}\partial_{[J} \Gamma_{I]} - S_{JI}{}^K \nabla_K \mathfrak{v} + r\mathfrak{v}\Omega_{JI}{}^K \Gamma_K. \end{aligned} \tag{6.16}$$

In the expression

$$R_{JIK}{}^L = \partial_J \Gamma^L_{IK} - \partial_I \Gamma^L_{JK} + \Gamma^L_{JM} \Gamma^M_{IK} - \Gamma^L_{IM} \Gamma^M_{JK} + 2\Gamma^L_{MK} \Omega_{JI}{}^K$$

we contract the indices K and L and put

$$V_{JI} \stackrel{\text{def}}{=} R_{JIK}{}^K. \tag{6.17}$$

We then obtain

$$V_{JI} = 2\partial_{[J} \Gamma_{I]} + 2\Gamma_M \Omega_{JI}{}^M. \tag{6.18}$$

Substituting from this into (6.16), we have

$$\nabla_{[J} \nabla_{I]} \mathfrak{v} = \tfrac{1}{2} r\mathfrak{v} V_{JI} - S_{JI}{}^K \nabla_K \mathfrak{v}. \tag{6.19}$$

For the symmetric object Γ we obtain the simple formula

$$\nabla_{[J} \nabla_{I]} \mathfrak{v} = \tfrac{1}{2} r\mathfrak{v} V_{JI}, \tag{6.20}$$

which for holonomic systems goes over into

$$\nabla_{[\mu} \nabla_{\nu]} \mathfrak{v} = r\mathfrak{v}\partial_{[\mu} \Gamma_{\nu]}. \tag{6.21}$$

If $\mathfrak{v}$ is a scalar ($\mathfrak{v} = \sigma$), that is, $r = 0$, formula (6.19) yields

$$\nabla_{[\mu} \nabla_{\nu]} \sigma = -S_{\mu\nu}{}^\lambda \partial_\lambda \sigma. \tag{6.22}$$

For the symmetric object Γ, on the other hand, we simply get

$$\nabla_{[\mu} \nabla_{\nu]} \sigma = 0. \tag{6.23}$$

76. The Torsion Tensor and its Interpretation. We now give the geometric interpretation of the tensor $S_{\mu\nu}{}^\lambda$. Imagine at the point $\underset{0}{p}$ two infinitesimal vectors $\underset{1}{v}{}^\lambda dt$ and $\underset{2}{v}{}^\lambda dt$ which are linearly independent.

Let $\underset{1}{p}$ denote the terminal point of the vector $\underset{1}{v}\,dt$ [1] and $\underset{2}{p}$ the terminal point of the vector $\underset{2}{v}\,dt$. We now displace the vector $\underset{2}{v}\,dt$ parallelly to the point $\underset{1}{p}$ to obtain, disregarding infinitesimals of higher order, the vector

$$\underset{2}{\bar{v}}^{\nu}\,dt = \underset{2}{v}^{\nu}\,dt - \Gamma^{\nu}_{\mu\lambda}\underset{2}{v}^{\lambda}\,dt\underset{1}{v}^{\mu}\,dt = (\underset{2}{v}^{\nu} - \Gamma^{\nu}_{\mu\lambda}\underset{2}{v}^{\lambda}\underset{1}{v}^{\mu}\,dt)\,dt\,.$$

Similarly, if vector $\underset{1}{v}\,dt$ is subjected to a parallel displacement to the point $\underset{2}{p}$, we obtain vector $\underset{1}{\bar{v}}\,dt$ according to the formula

$$\underset{1}{\bar{v}}^{\nu}\,dt = (\underset{1}{v}^{\nu} - \Gamma^{\nu}_{\mu\lambda}\underset{1}{v}^{\lambda}\underset{2}{v}^{\mu}\,dt)\,dt\,.$$

If $r dt$ denotes a vector whose terminal point coincides with the terminal point of the vector $\underset{1}{\bar{v}}\,dt$, and whose origin coincides with the origin of the vector $\underset{2}{\bar{v}}\,dt$ (Fig. 5), we have

$$(\underset{1}{v} + \underset{2}{\bar{v}} + r - \underset{1}{\bar{v}} - \underset{2}{v})\,dt = 0\,,$$

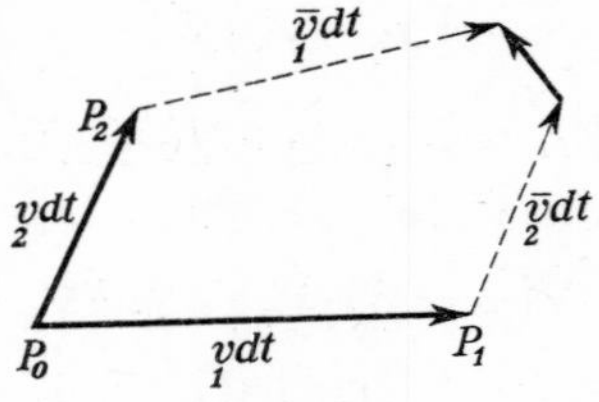

Fig. 5.

and so, making use of the formulae above, we arrive at

$$r^{\nu} = \underset{1}{v}^{\nu} - \Gamma^{\nu}_{\mu\lambda}\underset{1}{v}^{\lambda}\underset{2}{v}^{\mu}\,dt + \underset{2}{v}^{\nu} - \underset{1}{v}^{\nu} - \underset{2}{v}^{\nu} + \Gamma^{\nu}_{\mu\lambda}\underset{2}{v}^{\lambda}\underset{1}{v}^{\mu}\,dt = 2\Gamma^{\nu}_{[\mu\lambda]}\underset{2}{v}^{\lambda}\underset{1}{v}^{\mu}\,dt$$

$$= 2S_{\mu\lambda}{}^{\nu}\underset{2}{v}^{\lambda}\underset{1}{v}^{\mu}\,dt = 2S_{\mu\lambda}{}^{\nu}\underset{2}{v}^{[\lambda}\underset{1}{v}^{\mu]}\,dt\,. \qquad (6.24)$$

---

[1] It is true that the concept of the terminal point of a (contravariant) vector is not invariant in a space based on the general group of transformations. Nevertheless, in the discussion which follows because the vectors under consideration are "infinitesimal", owing to the introduction of the factor dt, the terminal point of a vector may be taken to mean a point whose coordinates are such that the differences between the coordinates of the terminal point and the point of attachment are equal to the coordinates of the vector. It can be showns that this rounding off does not affect the final result of the limiting transition $dt \to 0$.

Thus, if the object Γ is symmetric ($S = 0$), the vectors $\underset{1}{v}\,dt$, $\underset{2}{v}\,dt$, $\underset{1}{\overline{v}}\,dt$ and $\underset{2}{\overline{v}}\,dt$ form an infinitesimal parallelogram ($r\,dt = 0$). Accordingly, the tensor $S_{\mu\lambda}{}^{\nu}$ is called a *torsion tensor* by some authors (e.g. Cartan).

One geometric interpretation of the torsion tensor has been given by Cartan [97]. Another, simpler one, which we expound here is due to Bompiani [94].

Suppose that a vector v^{ν} is given at the point p with coordinates ξ^{ν}. Displacing it parallelly to the point $\underset{1}{p}$ with coordinates $\xi^{\nu}+u^{\nu}dt$, we obtain a vector with coordinates

$$v^{\nu}-\Gamma^{\nu}_{\mu\lambda}v^{\lambda}u^{\mu}dt. \tag{6.25}$$

If the object Γ is not symmetric, i.e. when

$$\Gamma^{\nu}_{\mu\lambda} = \Gamma^{\nu}_{(\mu\lambda)}+S_{\mu\lambda}{}^{\nu} \qquad (S_{\mu\lambda}{}^{\nu} \neq 0),$$

the vector (6.25) can be constructed, as shown by Bompiani, by means of the following two consecutive operations. From the vector v^{ν} (attached at the point p) by means of a rotation (infinitesimal) we go over to the vector $\overline{v}^{\nu}$ (also attached at p, of course) defined by the formula

$$\overline{v}^{\nu} = v^{\nu}-S_{(\mu\lambda)}{}^{\nu}v^{\lambda}u^{\mu}dt,$$

and displace the vector $\overline{v}^{\nu}$ parallelly to the terminal point of the vector $u^{\nu}dt$ by means of the object $\Gamma^{\nu}_{(\mu\lambda)}$ of parallel displacement. We then have

$$\overline{v}^{\nu}-\Gamma^{\nu}_{(\mu\lambda)}\overline{v}^{\lambda}u^{\mu}dt,$$

which yields

$$v'-S_{\mu\lambda}{}^{\nu}v^{\lambda}u^{\mu}dt-\Gamma^{\nu}_{(\mu\lambda)}[v^{\lambda}-S_{\varrho\sigma}{}^{\lambda}v^{\sigma}u^{\varrho}dt]u^{\mu}dt$$

and, omitting infinitesimals of higher orders, we finally obtain

$$v^{\nu}-S_{\mu\lambda}{}^{\nu}v^{\lambda}u^{\mu}dt-\Gamma^{\nu}_{(\mu\lambda)}v^{\lambda}u^{\mu}dt = v^{\nu}-\Gamma^{\nu}_{\mu\lambda}v^{\lambda}u^{\mu}dt.$$

The above geometric interpretation of the torsion tensor provided Graiff [122] with a basis for certain geometric studies which played an important role in papers by Einstein [24] and Hlavatý [125] on field theory.

77. The Interpretation of the Curvature Tensor. Now assume that the connection object Γ is symmetric, so that the torsion is zero, i.e. $S_{\mu\lambda}{}^{\nu} = 0$; we denote the aforementioned parallelogram by R and consider at the point $\underset{0}{p}$ an arbitrary contravariant vector w^{ν}. Next we translate it along the boundary of R and calculate the increment Dw^ν in the vector w, following cyclical parallel displacement back to $\underset{0}{p}$.

Let $\underset{3}{p}$ denote the fourth vertex of the parallelogram R, and let $\underset{1}{w}$, $\underset{12}{w}$, $\underset{2}{w}$, $\underset{21}{w}$, respectively, denote the vectors generated by the parallel displacements of the vector w from $\underset{0}{p}$ to $\underset{1}{p}$, of the vector $\underset{1}{w}$ from $\underset{1}{p}$ to $\underset{3}{p}$, of the vector $\underset{12}{w}$ from $\underset{0}{p}$ to $\underset{2}{p}$, and of the vector $\underset{2}{w}$ from $\underset{2}{p}$ to $\underset{3}{p}$. Letting by $\Gamma^\lambda_{\mu\nu}$ the values of the coordinates of the object Γ at the point $\underset{0}{p}$, we obtain for the coordinates of the object Γ at the points $\underset{1}{p}$ and $\underset{2}{p}$, respectively, the values

$$\Gamma^\lambda_{\mu\nu} + (\partial_\varrho \Gamma^\lambda_{\mu\nu})\underset{1}{v}^\varrho dt, \qquad \Gamma^\lambda_{\mu\nu} + (\partial_\varrho \Gamma^\lambda_{\mu\nu})\underset{2}{v}^\varrho dt.$$

Hence, using the formulae above we obtain, successively,

$$\underset{1}{w}^\nu = w^\nu - \Gamma^\nu_{\mu\lambda} w^\lambda \underset{1}{v}^\mu dt,$$

$$\underset{12}{w}^\nu = \underset{1}{w}^\nu - (\Gamma^\nu_{\mu\lambda} + \underset{1}{v}^\varrho \partial_\varrho \Gamma^\nu_{\mu\lambda} dt)\underset{1}{w}^\lambda \underset{2}{\bar{v}}^\mu dt,$$

$$\underset{2}{w}^\nu = w^\nu - \Gamma^\nu_{\mu\lambda} w^\lambda \underset{2}{v}^\mu dt,$$

$$\underset{21}{w}^\nu = \underset{2}{w}^\nu - (\Gamma^\nu_{\mu\lambda} + \underset{2}{v}^\varrho \partial_\varrho \Gamma^\nu_{\mu\lambda} dt)\underset{2}{w}^\lambda \underset{1}{\bar{v}}^\mu dt.$$

Substitution yields

$$\underset{12}{w}^\nu = w^\nu - \Gamma^\nu_{\mu\lambda} w^\lambda \underset{1}{v}^\mu dt - (\Gamma^\nu_{\mu\lambda} + \underset{1}{v}^\varrho \partial_\varrho \Gamma^\nu_{\mu\lambda} dt)(w^\lambda - \Gamma^\lambda_{\omega\sigma} w^\sigma \underset{1}{v}^\omega dt)(\underset{2}{v}^\mu - \Gamma^\mu_{\tau\varkappa}\underset{2}{v}^\varkappa \underset{1}{v}^\tau dt)dt.$$

Similarly,

$$\underset{21}{w}^\nu = w^\nu - \Gamma^\nu_{\mu\lambda} w^\lambda \underset{2}{v}^\mu dt - (\Gamma^\nu_{\mu\lambda} + \underset{2}{v}^\varrho \partial_\varrho \Gamma^\nu_{\mu\lambda} dt)(w^\lambda - \Gamma^\lambda_{\omega\sigma} w^\sigma \underset{2}{v}^\omega dt)(\underset{1}{v}^\mu - \Gamma^\mu_{\tau\varkappa}\underset{1}{v}^\varkappa \underset{2}{v}^\tau dt)dt.$$

The increment Dw^ν that we are looking for is therefore given by

$$Dw^\nu = \underset{12}{w}^\nu - \underset{21}{w}^\nu.$$

If we ignore terms containing $(dt)^3$ and $(dt)^4$, we obtain

$$Dw^\nu = w^\nu - \Gamma^\nu_{\mu\lambda} w^\lambda \underset{1}{v}^\mu dt - \Gamma^\nu_{\mu\lambda} w^\lambda \underset{2}{v}^\mu dt +$$

$$+ \big(\Gamma^\nu_{\mu\lambda} w^\lambda \Gamma^\mu_{\tau\varkappa} \underset{2}{v}^\varkappa \underset{1}{v}^\tau + \Gamma^\nu_{\mu\lambda}\Gamma^\lambda_{\omega\sigma} w^\sigma \underset{1}{v}^\omega \underset{2}{v}^\mu - (\partial_\varrho \Gamma^\nu_{\mu\lambda}) w^\lambda \underset{1}{v}^\varrho \underset{2}{v}^\mu\big)(dt)^2 -$$

$$-w^{\nu}+\Gamma^{\nu}_{\mu\lambda}w^{\lambda}\underset{2}{v}^{\mu}dt+\Gamma^{\nu}_{\mu\lambda}w^{\lambda}\underset{1}{v}^{\mu}dt-$$

$$-\left(\Gamma^{\nu}_{\mu\lambda}w^{\lambda}\Gamma^{\mu}_{\tau\varkappa}\underset{1}{v}^{\varkappa}\underset{2}{v}^{\tau}+\Gamma^{\nu}_{\mu\lambda}\Gamma^{\lambda}_{\omega\sigma}w^{\sigma}\underset{2}{v}^{\omega}\underset{1}{v}^{\mu}-(\partial_{\varrho}\Gamma^{\nu}_{\mu\lambda})w^{\lambda}\underset{1}{v}^{\mu}\underset{2}{v}^{\varrho}\right)(dt)^2,$$

which, after simplification (w^{ν} and terms containing dt cancel out), yields

$$Dw^{\nu} = \{(\partial_{\varrho}\Gamma^{\nu}_{\mu\lambda})w^{\lambda}\underset{1}{v}^{\mu}\underset{2}{v}^{\varrho}-(\partial_{\varrho}\Gamma^{\nu}_{\mu\lambda})w^{\lambda}\underset{1}{v}^{\varrho}\underset{2}{v}^{\mu}+\Gamma^{\nu}_{\mu\lambda}\Gamma^{\mu}_{\tau\varkappa}w^{\lambda}\underset{1}{v}^{\tau}\underset{2}{v}^{\varkappa}-$$

$$-\Gamma^{\nu}_{\mu\lambda}\Gamma^{\mu}_{\tau\varkappa}w^{\lambda}\underset{1}{v}^{\varkappa}\underset{2}{v}^{\tau}+\Gamma^{\nu}_{\mu\lambda}\Gamma^{\lambda}_{\sigma\omega}w^{\sigma}\underset{1}{v}^{\omega}\underset{2}{v}^{\mu}-\Gamma^{\nu}_{\mu\lambda}\Gamma^{\lambda}_{\omega\sigma}w^{\sigma}\underset{2}{v}^{\omega}\underset{1}{v}^{\mu}\}(dt)^2.$$

Making appropriate changes in the dummy indices, we have

$$Dw^{\nu} = \{(\partial_{\varrho}\Gamma^{\nu}_{\mu\lambda}-\partial_{\mu}\Gamma^{\nu}_{\varrho\lambda})w^{\lambda}\underset{1}{v}^{\mu}\underset{2}{v}^{\varrho}+(\Gamma^{\nu}_{\omega\lambda}\Gamma^{\omega}_{\mu\varrho}-\Gamma^{\nu}_{\omega\lambda}\Gamma^{\omega}_{\varrho\mu})w^{\lambda}\underset{1}{v}^{\mu}\underset{2}{v}^{\varrho}+$$

$$+(\Gamma^{\nu}_{\varrho\omega}\Gamma^{\omega}_{\mu\lambda}-\Gamma^{\nu}_{\mu\omega}\Gamma^{\omega}_{\varrho\lambda})w^{\lambda}\underset{1}{v}^{\mu}\underset{2}{v}^{\varrho}\}(dt)^2.$$

Moreover, because of the displacement symmetry which we assumed, we have

$$\Gamma^{\omega}_{\mu\varrho} = \Gamma^{\omega}_{\varrho\mu}$$

and we finally arrive at

$$Dw^{\nu} = R_{\varrho\mu\lambda}{}^{\nu}w^{\lambda}\underset{1}{v}^{\mu}\underset{2}{v}^{\varrho}(dt)^2 = R_{\varrho\mu\lambda}{}^{\nu}w^{\lambda}\underset{1}{v}^{[\mu}\underset{2}{v}^{\varrho]}(dt)^2$$

$$= -R_{\varrho\mu\lambda}{}^{\nu}w^{\lambda}\underset{1}{v}^{[\varrho}\underset{2}{v}^{\mu]}(dt)^2. \tag{6.26}$$

In other words, the increment in the vector w^{ν} following cyclical parallel displacement along the edges of the parallelogram R is obtained by transvecting the negatively taken curvature tensor up onto the product of the vector w and the bivector formed from the vectors constituting the parallelogram R.

Formula (6.26) should be interpreted more precisely as follows. Let $\underset{1}{p}$ and $\underset{2}{p}$ denote the terminal points of the vectors $\underset{1}{v}\,dt$ and $\underset{2}{v}\,dt$, respectively, and $\underset{3}{p}$ and $\underset{4}{p}$ denote the ends of the vectors $\underset{2}{\bar{v}}\,dt$ and $\underset{1}{\bar{v}}\,dt$. Now consider the geodesics associated with the five pairs of points $\underset{0}{p}$ and $\underset{1}{p}$, $\underset{1}{p}$ and $\underset{3}{p}$, $\underset{3}{p}$ and $\underset{4}{p}$, $\underset{4}{p}$ and $\underset{2}{p}$, and $\underset{2}{p}$ and $\underset{0}{p}$, which together constitute a "geodesic pentagon". We now displace the vector w parallelly along the edges of this pentagon in the direction indicated by the order of vertices $(\underset{0}{p}, \underset{1}{p}, \underset{3}{p}, \underset{4}{p}, \underset{2}{p}, \underset{0}{p})$ back to the starting point $\underset{0}{p}$. If $\bar{w}$ denotes the vector obtained by this parallel displacement, we have the relation

$$\lim_{dt\to 0} \frac{\overline{w}^\nu - w^\nu}{(dt)^2} = -R_{\varrho\mu\lambda}{}^\nu w^\lambda \underset{1}{v}{}^{[\varrho}\underset{2}{v}{}^{\mu]}, \tag{6.27}$$

and we thus have one of the geometric interpretations of the curvature tensor.

78. Concomitants of the Curvature Tensor. By contraction we can form from the curvature tensor $R_{\lambda\mu\nu}{}^\varrho$ three tensors of valence $(0, 2)$, viz. the tensor

$$V_{\lambda\mu} \overset{\text{def}}{=} R_{\lambda\mu\nu}{}^\nu = 2\partial_{[\lambda}\Gamma_{\mu]} \tag{6.28}$$

already introduced earlier (equation (6.17)), and the tensors

$$R_{\nu\mu\lambda}{}^\nu, \qquad R_{\lambda\nu\mu}{}^\nu,$$

which differ only in sign since $R_{\lambda\mu\nu}{}^\varrho$ is antisymmetric with respect to the first two indices. It follows that we have just two essentially different tensors, the *concomitants of the curvature tensor*

$$V_{\lambda\mu} \overset{\text{def}}{=} R_{\lambda\mu\nu}{}^\nu \quad \text{and} \quad R_{\lambda\mu} \overset{\text{def}}{=} R_{\nu\lambda\mu}{}^\nu. \tag{6.29}$$

The first tensor $V_{\lambda\mu}$ is antisymmetric.

If we introduce the tensor

$$W_{\lambda\mu} \overset{\text{def}}{=} R_{\mu\lambda} - R_{\lambda\mu}, \tag{6.30}$$

then in addition to $V_{\lambda\mu}$ we obtain a second skew-symmetric tensor. The two tensors $V_{\lambda\mu}$ and $W_{\lambda\mu}$ are related in a fairly simple manner in that we have

$$V_{\lambda\mu} - W_{\lambda\mu} = 2\nabla_\nu S_{\lambda\mu}{}^\nu + 4\nabla_{[\lambda} S_{\mu]\nu}{}^\nu + 4S_{\lambda\mu}{}^\varrho S_{\varrho\nu}{}^\nu. \tag{6.31}$$

It follows that if the object Γ is symmetric ($S = 0$), we then have

$$W_{\lambda\mu} = V_{\lambda\mu}. \tag{6.32}$$

If Γ is semisymmetric (5.37), i.e. if

$$S_{\lambda\mu}{}^\nu = S_{[\lambda} A^\nu_{\mu]} = \tfrac{1}{2}(S_\lambda A^\nu_\mu - S_\mu A^\nu_\lambda), \tag{6.33}$$

then we have successively, assuming further that $n \geqslant 2$,

$$S_{\mu\nu}{}^\nu = \frac{n-1}{2} S_\mu,$$

$$4S_{\lambda\mu}{}^\varrho S_{\varrho\nu}{}^\nu = (n-1)(S_\lambda S_\mu - S_\mu S_\lambda) = 0,$$

$$2\nabla_\nu S_{\mu\lambda}{}^\nu = \partial_\mu S_\lambda - \partial_\lambda S_\mu = -(\partial_\lambda S_\mu - \partial_\mu S_\lambda),$$

$$4\nabla_{[\lambda} S_{\mu]\nu}{}^\nu = (n-1)(\partial_\lambda S_\mu - \partial_\mu S_\lambda)$$

and, consequently,

$$V_{\lambda\mu} - W_{\lambda\mu} = (n-2)\,(\partial_\lambda S_\mu - \partial_\mu S_\lambda). \tag{6.34}$$

We thus have the following two corollaries.

COROLLARY 1. *If the object Γ is semisymmetric and the vector field S_λ is a gradient, relation* (6.32) *holds.*

COROLLARY 2. *If $n = 2$ and the object Γ is semisymmetric, relation* (6.32) *holds.*

Note, however, that for $n = 2$ each object Γ is semisymmetric: it suffices to set $S_1 = \Gamma^2_{12} - \Gamma^2_{22}$, $S_2 = \Gamma^1_{21} - \Gamma^1_{11}$. Hence, by virtue of Corollary 2 we have the following

THEOREM. *For $n = 2$, the two bivectors $W_{\lambda\mu}$ and $V_{\lambda\mu}$ are the same.*

79. A Geometric Interpretation of the Tensors $V_{\lambda\mu}$ and $R_{\lambda\mu}$. We shall now give a geometric interpretation of the tensors $V_{\lambda\mu}$ and $R_{\lambda\mu}$.

For holonomic systems formula (6.20) yields

$$\nabla_{[\mu}\nabla_{\lambda]}\mathfrak{v} = \tfrac{1}{2} r\mathfrak{v}V_{\mu\lambda}, \tag{6.35}$$

whence the vanishing of the tensor $V_{\mu\lambda}$ leads to the relation

$$\nabla_{[\mu}\nabla_{\lambda]}\mathfrak{v} = 0, \tag{6.36}$$

which holds for every field of densities $\mathfrak{v}$. We say that in this case there exist covariantly constant density fields, i.e. fields for which

$$\nabla_\lambda \mathfrak{v} \equiv 0 \quad (\lambda = 1, 2, \ldots, n). \tag{6.37}$$

Accordingly, at some point p of space we take an arbitrary density $\mathfrak{v}$ (of weight $-r$). Let q be some other arbitrary point of space. We now take two curves C_1 and C_2 joining p and q, and subject the density $\mathfrak{v}$ to parallel displacement to the point q, first along C_1, and then along C_2. In order to show that the value of the density displaced to q is independent of whether it is moved along path C_1 or path C_2, it suffices to demonstrate that parallel displacement of the density along a circuit consisting of C_1 from p to q and $-C_2$ from q to p brings us back to the initial value of the density. If the point q is "infinitesimally close" to the point p, the cycle pC_1qC_2p can be replaced by an "infinitesimal" bivector $b^{\lambda\mu}d\sigma$. Then, by reasoning similar to that which led to formula (6.26), we can calculate the increment $D\mathfrak{v}$ from the formula

$$D\mathfrak{v} = -r\mathfrak{v}R_{\lambda\mu\nu}{}^{\nu}b^{\lambda\mu}d\sigma = -r\mathfrak{v}V_{\lambda\mu}b^{\lambda\mu}d\sigma,$$

which shows that Dv is zero if $V_{\lambda\mu} = 0$. This reasoning cannot be employed when the point q is not inifinitesimally close to p. In that case, however, the condition for total integrability [1] of the system (6.37) is given by relations (6.36) which are satisfied by virtue of (6.35) under the assumption $V_{\lambda\mu} = 0$.

If the tensor $V_{\lambda\mu}$ is zero throughout the space, the object $\Gamma^{\nu}_{\lambda\mu}$ of displacement is called a *volume-preserving* object, after Schouten, for reasons which we now discuss.

Take at some point p a non-zero n-vector $\underset{0}{w}^{\lambda_1 \ldots \lambda_n}$. Since each non-zero coordinate of this n-vector is a density, as we know, a field of n-vector $w^{\lambda_1 \ldots \lambda_n}$ for which

$$\nabla_\mu w^{\lambda_1 \ldots \lambda_n} \equiv 0 \quad \text{and} \quad (w^{\lambda_1 \ldots \lambda_n})_p = \underset{0}{w}^{\lambda_1 \ldots \lambda_n}$$

exists throughout space. If in our space we set up a field of densities $\mathfrak{g}$ of weight -1 and of the same sign as $\mathfrak{v} \overset{\text{def}}{=} w^{1 \ldots n}$, then we know that $V \overset{\text{def}}{=} v/\mathfrak{g} > 0$ is an invariant and can be taken to be a volume measure of the n-vector $w^{\lambda_1 \ldots \lambda_n}$.

We now show that this volume does not change when the n-vector w is displaced parallelly to any other point in space.

We first establish the existence of coordinate system such that

$$\Gamma_{\mu'} \overset{*}{=} 0 \quad (\mu' = 1', 2', \ldots, n').$$

To this end, we have to show that because of the relation

$$\Gamma_{\mu'} = \Gamma_\mu A^\mu_{\mu'} - \partial_{\mu'} \ln |J|$$

there exists a transformation $(\lambda) \to (\lambda')$ for which

$$\partial_{\mu'} \ln |J| = A^\mu_{\mu'} \Gamma_\mu$$

or

$$\partial_\mu \ln J = \Gamma_\mu \quad (\mu = 1, 2, \ldots, n). \tag{6.38}$$

Let us look for a special transformation $(\lambda) \to (\lambda')$ of the form

$$\xi^{\nu'} = \xi^\nu \quad (\nu = 1, 2, \ldots, n-1), \qquad \xi^{n'} = \varphi(\xi^1, \xi^2, \ldots, \xi^n).$$

For such a transformation we have

$$A^{\nu'}_\lambda = \delta^{\nu'}_\lambda \quad (\nu = 1, 2, \ldots, n=1;\ \lambda = 1, 2, \ldots, n)$$

and

$$A^{n'}_\lambda = \partial_\lambda \varphi,$$

---

[1] That is, a condition for the existence of solutions for all possible initial values.

so that

$$J = \partial_n \varphi = \psi .$$

Relations (6.38) thus assume the form

$$\frac{\partial_\mu \psi}{\psi} = \Gamma_\mu, \tag{6.39}$$

where ψ is an unknown function which satisfies the system (6.39). The integrability condition for the system (6.39) is

$$\partial_{[\lambda}\Gamma_{\mu]} = 0,$$

which is satisfied in our case because of equation $V_{\lambda\mu} = 0$ and the theorem is therefore proved.

Since

$$\nabla_\mu \mathfrak{v} = \partial_\mu \mathfrak{v} + \mathfrak{v}\Gamma_\mu, \quad \nabla_\mu \mathfrak{g} = \partial_\mu \mathfrak{g} + \mathfrak{g}\Gamma_\mu,$$

it follows in particular that

$$\nabla_{\mu'} \mathfrak{v}' \stackrel{*}{=} \partial_{\mu'} \mathfrak{v}', \quad \nabla_{\mu'} \mathfrak{g}' \stackrel{*}{=} \partial_{\mu'} \mathfrak{g}'. \tag{6.40}$$

But on the other hand

$$\begin{aligned} \nabla_{\mu'} \mathfrak{v}' &= \nabla_{\mu'}(V\mathfrak{g}') = V\nabla_{\mu'}\mathfrak{g}' + \mathfrak{g}'\nabla_{\mu'}V \\ &= V\nabla_{\mu'}\mathfrak{g} + \mathfrak{g}'\partial_{\mu'}V, \end{aligned}$$

since V is a scalar, and for every scalar σ in every coordinate system we have

$$\nabla_\mu \sigma = \partial_\mu \sigma .$$

Using (6.40) together with the equality $\nabla_\mu \mathfrak{v} = 0$ ($\nabla_\mu w^{1\ldots n} = 0$), we accordingly have

$$0 = V\partial_{\mu'}\mathfrak{g}' + \mathfrak{g}'\partial_{\mu'}V. \tag{6.41}$$

If, in particular, we write

$$\mathfrak{g}' = \overset{(\lambda')}{\mathfrak{g}} = \eta \underset{1'}{e}^{[1'} \ldots \underset{n'}{e}^{n']},$$

where the sign of η is chosen so that $\overset{(\lambda')}{\mathfrak{v}}\ \overset{(\lambda')}{\mathfrak{g}} > 0$. Then

$$\partial_{\mu'}\mathfrak{g}' \stackrel{*}{=} 0$$

and equality (6.41) now implies that

$$\mathfrak{g}'\partial_{\mu'}V = 0,$$

and, so, since $\mathfrak{g}' \neq 0$, we conclude that

$$V = \text{const.}$$

This justifies the term "volume-preserving" object.

Another geometric interpretation of the tensor $V_{\lambda\mu}$ has been given in a paper by Gołąb [106].

Let us now give a geometric interpretation of the tensor $R_{\lambda\mu}$, called the *Ricci tensor* [106]. The interpretation we give is a generalization of an idea due to Bompiani [93], who gave it in the special case of a Riemannian metric space.

Consider an infinitesimal vector v^ν at the point. p. We set $\underset{1}{v} = v$ and choose at p a further $(n-1)$ infinitesimal vectors $\underset{j}{v}^\lambda$ $(j = 2, 3, \ldots, n)$ in such a way that the ordered sequence of vectors

$$\underset{i}{v}^\lambda \qquad (i = 1, 2, \ldots, n) \tag{6.42}$$

is a set of linearly independent vectors.

Then, as we know, there exists a set of n covariant vectors

$$\overset{k}{v}_\lambda \qquad (k = 1, 2, \ldots, n) \tag{6.43}$$

inverse to the set (6.42), which are defined uniquely by the relations

$$\underset{i}{v}^\lambda \overset{i}{v}_\mu = A_\mu^\lambda .$$

Each vector $\underset{i}{v}$ of the set (6.42) is uniquely assigned a bivector

$$\underset{i}{B} = \underset{i}{v}^{[\lambda} \underset{1}{v}^{\mu]} .$$

$\underset{1}{B}$, the first of these n bivectors, is a null bivector.

We now take any one of the covariant vectors of the set (6.43) $\overset{k}{v}$ and displace it parallelly along the edge of the bivector $\underset{k}{B}$. The corresponding increment is denoted by $D\overset{k}{v}$. Accordingly (1)

$$D\overset{k}{v}_\lambda = R_{\varrho\mu\lambda}{}^\nu \overset{k}{v}_\nu \underset{k}{B}^{\varrho\mu} = R_{\varrho\mu\lambda}{}^\nu \overset{k}{v}_\nu \underset{k}{v}^{[\varrho} \underset{1}{v}^{\mu]} = R_{\varrho\mu\lambda}{}^\nu \overset{k}{v}_\nu \underset{k}{v}^{\varrho} \underset{1}{v}^{\mu} .$$

We now take the sum over R, of all the increments. We then have

$$\sum_k D\overset{k}{v}_\lambda = \sum_k R_{\varrho\mu\lambda}{}^\nu \overset{k}{v}_\nu \underset{k}{v}^\varrho v^\mu = R_{\varrho\mu\lambda}{}^\nu A_\nu^\varrho v^\mu = R_{\nu\mu\lambda}{}^\nu v^\mu = R_{\mu\lambda} v^\mu . \tag{6.44}$$

Since the geometric meaning of the left-hand side is known, the formula above provides a geometric meaning for $R_{\mu\lambda} v^\mu$ and thus indirectly, for the tensor $R_{\mu\lambda}$ as well.

---

(1) This formula is derived analogously to formula (6.26). Note that in it, we must not sum over the index k.

80. The Relations Between the Coordinates of the Curvature Tensor. On p. 217 we found that the curvature tensor R is antisymmetric with respect to the first two indices, i.e. that

$$R_{(\omega\mu)\lambda}{}^{\nu} = 0.$$

Let us now calculate $R_{[\omega\mu\nu]}{}^{\lambda}$. Accordingly, we have

$$R_{[\omega\mu\nu]}{}^{\lambda} = 2\partial_{[[\omega}\Gamma^{\lambda}_{\mu]\nu]} + 2\Gamma^{\lambda}_{[[\mu|\tau|}\Gamma^{\tau}_{\mu]\nu]}.$$

In view of the outer brackets [ ], the inner ones can be omitted and we consequently obtain

$$R_{[\omega\mu\nu]}{}^{\lambda} = 2\partial_{[\omega}\Gamma^{\lambda}_{\mu\nu]} + 2\Gamma^{\lambda}_{[\omega|\tau|}\Gamma^{\tau}_{\mu\nu]} = 2\partial_{[\omega}(\Gamma^{\lambda}_{(\mu\nu)]} + S_{\mu\nu]}{}^{\lambda}) + 2\Gamma^{\lambda}_{[\omega|\tau|}(\Gamma^{\tau}_{(\mu\nu)]} + S_{\mu\nu}{}^{\tau}).$$

Moreover, if there are parentheses inside the square brackets, the entire expression must be zero. We thus have

$$R_{[\omega\mu\nu]}{}^{\lambda} = 2\partial_{[\omega}S_{\mu\nu]}{}^{\lambda} + 2\Gamma^{\lambda}_{[\omega|\tau|}S_{\mu\nu]}{}^{\tau}.$$

We now transform the right-hand side, bearing in mind that

$$\begin{aligned}\nabla_{\omega}S_{\mu\nu}{}^{\lambda} &= \partial_{\omega}S_{\mu\nu}{}^{\lambda} + \Gamma^{\lambda}_{\omega\varrho}S_{\mu\nu}{}^{\varrho} - \Gamma^{\varrho}_{\omega\mu}S_{\varrho\nu}{}^{\lambda} - \Gamma^{\varrho}_{\omega\nu}S_{\mu\varrho}{}^{\lambda}\\ &= \partial_{\omega}S_{\mu\nu}{}^{\lambda'} + (\Gamma^{\lambda}_{(\omega\varrho)} + S_{\omega\varrho}{}^{\lambda})S_{\mu\nu}{}^{\varrho} - (\Gamma^{\varrho}_{(\omega\mu)} + S_{\omega\mu}{}^{\varrho})S_{\varrho\nu}{}^{\lambda} - (\Gamma^{\varrho}_{(\omega\nu)} + S_{\omega\nu}{}^{\varrho})S_{\mu\varrho}{}^{\lambda}.\end{aligned}$$

After antisymmetrization with respect to ω, μ, λ some terms disappear and we have, remaining,

$$\begin{aligned}\nabla_{[\omega}S_{\mu\nu]}{}^{\lambda} &= \partial_{[\omega}S_{\mu\nu]}{}^{\lambda} + S_{[\omega|\varrho|}{}^{\lambda}S_{\mu\nu]}{}^{\varrho} - S_{[\omega\mu}{}^{\varrho}S_{|\varrho|\nu]}{}^{\lambda} - S_{[\omega\nu}{}^{\varrho}S_{\mu]\varrho}{}^{\lambda} + \Gamma^{\lambda}_{[(\omega|\varrho|)}S_{\mu\nu]}{}^{\varrho}\\ &= \partial_{[\omega}S_{\mu\nu]}{}^{\lambda} - S_{\varrho[\omega}{}^{\lambda}S_{\mu\nu]}{}^{\varrho} + S_{[\omega\mu}{}^{\varrho}S_{\nu]\varrho}{}^{\lambda} - S_{[\omega\nu}{}^{\varrho}S_{\mu]\varrho}{}^{\lambda} + \Gamma^{\lambda}_{(\varrho[\omega)}S_{\mu\nu]}{}^{\varrho}\\ &= \partial_{[\omega}S_{\mu\nu]}{}^{\lambda} + S_{[\mu\nu}{}^{\varrho}S_{\omega]\varrho}{}^{\lambda} + S_{[\omega\mu}{}^{\varrho}S_{\nu]\varrho}{}^{\lambda} + S_{[\omega\mu}{}^{\varrho}S_{\nu]\varrho}{}^{\lambda} + \Gamma^{\lambda}_{(\varrho[\omega)}S_{\mu\nu]}{}^{\varrho}\\ &= \partial_{[\omega}S_{\mu\nu]}{}^{\lambda} + 3S_{[\omega\mu}{}^{\varrho}S_{\nu]\varrho}{}^{\lambda} + \Gamma^{\lambda}_{(\varrho[\omega)}S_{\mu\nu]}{}^{\varrho}.\end{aligned}$$

On the other hand

$$\Gamma^{\lambda}_{[\omega|\tau|}S_{\mu\nu]}{}^{\tau} = (\Gamma^{\lambda}_{(\omega|\tau|)} + S_{[\omega|\tau|}{}^{\lambda})S_{\mu\nu]}{}^{\tau} = \Gamma^{\lambda}_{(\tau[\omega)}S_{\mu\nu]}{}^{\tau} - S_{\tau[\omega}{}^{\lambda}S_{\mu\nu]}{}^{\tau}.$$

Accordingly,

$$\begin{aligned}\partial_{[\omega}S_{\mu\nu]}{}^{\lambda} + \Gamma^{\lambda}_{[\omega|\tau|}S_{\mu\nu]}{}^{\tau} &= \nabla_{[\omega}S_{\mu\nu]}{}^{\lambda} - 3S_{[\omega\mu}{}^{\varrho}S_{\nu]\varrho}{}^{\lambda} - S_{\tau[\omega}{}^{\lambda}S_{\mu\nu]}{}^{\tau}\\ &= \nabla_{[\omega}S_{\mu\nu]}{}^{\lambda} - 3S_{[\omega\mu}{}^{\varrho}S_{\nu]\varrho}{}^{\lambda} + S_{[\mu\nu}{}^{\varrho}S_{\omega]\varrho}{}^{\lambda}\\ &= \nabla_{[\omega}S_{\mu\nu]}{}^{\lambda} - 2S_{[\omega\mu}{}^{\varrho}S_{\nu]\varrho}{}^{\lambda}.\end{aligned}$$

Finally, therefore, we have

$$R_{[\omega\mu\nu]}{}^{\lambda} = 2\nabla_{[\omega}S_{\mu\nu]}{}^{\lambda} - 4S_{[\omega\mu}{}^{\varrho}S_{\nu]\varrho}{}^{\lambda} \qquad (6.45)$$

where the right-hand side depends only on the torsion tensor and its covariant derivative.

In the special case when the connection object is symmetric, i.e. when

$$S_{\mu\nu}{}^{\lambda} = 0 \tag{6.46}$$

identity (6.45) reduces to

$$R_{[\omega\mu\nu]}{}^{\lambda} = 0. \tag{6.47}$$

Identity (6.47) is called the *Ricci identity* by Rashevskiĭ.

Further identities will be derived for the curvature tensor of Riemannian space.

In this case, in addition to the tensor $R_{\omega\mu\nu}{}^{\lambda}$ we can introduce the tensor $R_{\omega\mu\nu\lambda}$ generated by transvecting the metric tensor $g_{\lambda\mu}$ onto the curvature tensor

$$R_{\omega\mu\nu\lambda} \stackrel{\text{def}}{=} R_{\omega\mu\nu}{}^{\varrho} g_{\varrho\lambda}. \tag{6.48}$$

Just as the tensor $R_{\omega\mu\nu}{}^{\lambda}$, the tensor $R_{\omega\mu\nu\lambda}$ is also skewsymmetric with respect to the first two indices. We assert that it is also skew-symmetric with respect to the other two indices, i.e.

$$R_{\omega\mu(\nu\lambda)} = 0. \tag{6.49}$$

The proof of this consists of carrying out some lengthy, but elementary, calculations. We introduce the *Christoffel symbols of the first kind* (cf. Section 82)

$$C_{\lambda\mu\nu} \stackrel{\text{def}}{=} \left\{ {\varrho \atop \lambda\mu} \right\} g_{\varrho\nu}. \tag{6.50}$$

REMARK. In contrast to Christoffel symbols of the second kind $\left\{ {\varrho \atop \lambda\mu} \right\}$ (7.11) which do constitute a geometric object, Christoffel symbols of the first kind $C_{\lambda\mu\nu}$ do not.

We then have the formula

$$R_{\omega\mu\nu\lambda} = \partial_{[\omega|\lambda|} g_{\mu]\nu} + \partial_{[\mu|\nu|} g_{\omega]\lambda} + 2g^{\alpha\beta} C_{[\mu|\nu\alpha|} C_{\omega]\lambda\beta}. \tag{6.51}$$

From this formula, since $\partial_{\omega\lambda} = \partial_{\lambda\omega}$ and since the symbol $C_{\lambda\mu\nu}$ is symmetric with respect to the indices λ, μ, we obtain both relations (6.49) and the further relations

$$R_{\omega\mu\nu\lambda} = R_{\nu\lambda\omega\mu}. \tag{6.52}$$

We assert that the Ricci tensor $R_{\lambda\mu}$ is symmetric in Riemannian spaces.

Indeed, we have

$$R_{\lambda\mu} = R_{\omega\lambda\mu}{}^{\omega} = R_{\omega\lambda\mu\varrho}g^{\varrho\omega} = R_{\mu\varrho\omega\lambda}g^{\varrho\omega} = -R_{\mu\varrho\lambda\omega}g^{\varrho\omega}$$
$$= -R_{\mu\varrho\lambda}{}^{\varrho} = R_{\varrho\nu\lambda}{}^{\varrho} = R_{\mu\lambda}.$$

Thus, in Riemannian spaces

$$R_{\mu\lambda} = R_{\lambda\mu}. \tag{6.53}$$

Furthermore, we assert that in Riemannian spaces, we have

$$V_{\lambda\mu} = 0. \tag{6.54}$$

Indeed, for symmetric objects Γ—and the object $\left\{ {\nu \atop \lambda\mu} \right\}$ is symmetric by virtue of formula (6.32) we have

$$V_{\lambda\mu} = W_{\lambda\mu} = R_{\mu\lambda} - R_{\lambda\mu},$$

in view of which relation (6.53) implies equation (6.54).

Transvecting the metric tensor $g^{\mu\lambda}$ onto the Ricci tensor $R_{\lambda\mu}$, we obtain the scalar

$$R \overset{\text{def}}{=} R_{\lambda\mu}g^{\lambda\mu}. \tag{6.55}$$

When we divide this scalar by $n(n-1)$, i.e. when we set

$$\varkappa \overset{\text{def}}{=} \frac{R}{n(n-1)}, \tag{6.56}$$

we obtain the *scalar curvature* of the space V_n.

81. The Bianchi Identity. We now proceed to derive what is known as the Bianchi relation [1]. We first recall formula (6.14), viz.

$$\nabla_{[\mu}\nabla_{\nu]}w_\lambda = -\tfrac{1}{2}R_{\mu\nu\lambda}{}^{\varrho}w_\varrho - S_{\mu\nu}{}^{\varrho}\nabla_\varrho w_\lambda. \tag{6.57}$$

Note that the operator $\nabla_{[\mu}\nabla_{\nu]}$ applied to the product of any two quantities U and W possesses the property of the Leibniz rule, i.e.

$$\nabla_{[\mu}\nabla_{\nu]}(UW) = U\nabla_{[\mu}\nabla_{\nu]}W + W\nabla_{[\mu}\nabla_{\nu]}U. \tag{6.58}$$

Indeed,

$$\nabla_\nu(UW) = U\nabla_\nu W + W\nabla_\nu U,$$

[1] Strictly speaking, this was first obtained by G. Ricci (and published in 1889 by E. Padova). In 1902 L. Bianchi arrived at it independently and it now bears his name. This identity has been the subject of many generalizations by various authors, and has played an important role in the theory of relativity. An interesting geometric interpretation of this identity was given in 1923 by E. Cartan.

and consequently

$$\nabla_\mu \nabla_\nu (UW) = \nabla_\mu [U\nabla_\nu W + W\nabla_\nu U]$$
$$= (\nabla_\mu U)\nabla_\nu W + U\nabla_\mu \nabla_\nu W + (\nabla_\mu W)\nabla_\nu U + W\nabla_\mu \nabla_\nu U.$$

From this we get

$$\nabla_{[\mu}\nabla_{\nu]}(UW) = U\nabla_{[\mu}\nabla_{\nu]}W + W\nabla_{[\mu}\nabla_{\nu]}U + (\nabla_{[\mu}U)\nabla_{\nu]}W + (\nabla_{[\mu}W)\nabla_{\nu]}U.$$

The last two terms cancel, however, and formula (6.58) now follows.

Formula (6.58) and formulae (6.14) together imply that, for instance

$$\nabla_{[\mu}\nabla_{\nu]}u_\omega w_\lambda = -\tfrac{1}{2}R_{\mu\nu\omega}{}^{\varrho}u_\varrho w_\lambda - R_{\mu\nu\lambda}{}^{\varrho}u_\omega w_\varrho - S_{\mu\nu}{}^{\varrho}\nabla_\varrho u_\omega w_\lambda. \qquad (6.59)$$

Formula (6.59) can be generalized, viz.

$$\nabla_{[\mu}\nabla_{\nu]}a_{\omega\lambda} = -\tfrac{1}{2}R_{\mu\nu\omega}{}^{\varrho}a_{\varrho\lambda} - \tfrac{1}{2}R_{\mu\nu\omega}{}^{\varrho}a_{\omega\varrho} - S_{\mu\nu}{}^{\varrho}\nabla_\varrho a_{\lambda\omega}$$

or, in its full generality

$$\nabla_{[\mu}\nabla_{\nu]}T^{\lambda_1\ldots\lambda_p}{}_{\mu_1\ldots\mu_q} = +\tfrac{1}{2}\sum_{i=1}^{p} R_{\mu\nu\varrho}{}^{\lambda i}T^{\lambda_1\ldots\lambda_{i-1}\varrho\lambda_{i+}\ldots\lambda_p}{}_{\mu_1\ldots\mu_q} -$$
$$-\tfrac{1}{2}\sum_{j=1}^{q} R_{\mu\nu\mu_j}{}^{\varrho}T^{\lambda_1\ldots\lambda_p}{}_{\mu_1\ldots\mu_{j-1}\varrho\mu_{j+1}\ldots\mu_q} - S_{\mu\nu}{}^{\varrho}\nabla_\varrho T^{\lambda_1\ldots\lambda_p}{}_{\mu_1\ldots\mu_p}. \qquad (6.60)$$

This formula can still be generalized further to embrace any arbitrary tensor densities $\mathfrak{T}$.

In particular we apply the operator $\nabla_{[\mu}\nabla_{\nu]}$ to the field of quantities $\nabla_\omega w_\lambda$

$$\nabla_{[\mu}\nabla_{\nu]}\nabla_\omega w_\lambda = -\tfrac{1}{2}R_{\mu\nu\omega}{}^{\varrho}\nabla_\varrho w_\lambda - \tfrac{1}{2}R_{\mu\nu\lambda}{}^{\varrho}\nabla_\omega w_\varrho - S_{\mu\nu}{}^{\varrho}\nabla_\varrho\nabla_\omega w_\lambda.$$

In the foregoing identity, we now antisymmetrize with respect to the indices μ, ν, ω, to obtain

$$2\nabla_{[\mu}\nabla_\nu\nabla_{\omega]}w_\lambda = -R_{[\mu\nu\omega]}{}^{\varrho}\nabla_\varrho w_\lambda - R_{[\mu\nu|\lambda|}{}^{\varrho}\nabla_{\omega]}w_\varrho - 2S_{[\mu\nu}{}^{\varrho}\nabla_{|\varrho|}\nabla_{\omega]}w_\lambda.$$

Using relation (6.45), we can write

$$2\nabla_{[\mu}\nabla_\nu\nabla_{\omega]}w_\lambda = -2(\nabla_\varrho w_\lambda)\nabla_{[\mu}S_{\nu\omega]}{}^{\varrho} + 4(\nabla_\varrho w_\lambda)S_{[\mu\nu}{}^{\sigma}S_{\omega]\sigma}{}^{\varrho} -$$
$$-R_{[\mu\nu|\lambda|}{}^{\varrho}\nabla_{\omega]}w_\varrho - 2S_{[\mu\nu}{}^{\varrho}\nabla_{|\varrho|}\nabla_{|\varrho|}\nabla_{\omega]}w_\lambda. \qquad (6.61)$$

On the other hand, when we apply the operator ∇_μ to the equality

$$\nabla_{[\nu}\nabla_{\omega]}w_\lambda = -\tfrac{1}{2}R_{\nu\omega\lambda}{}^{\varrho}w_\varrho - S_{\nu\omega}{}^{\varrho}\nabla_\varrho w_\lambda$$

we find

$$2\nabla_\mu\nabla_{[\nu}\nabla_{\omega]}w_\lambda = -(\nabla_\mu R_{\nu\omega\lambda}{}^{\varrho})w_\varrho - R_{\nu\omega\lambda}{}^{\varrho}\nabla_\mu w_\varrho - 2(\nabla_\mu S_{\nu\omega}{}^{\varrho})\nabla_\varrho w_\varrho - 2S_{\nu\omega}{}^{\varrho}\nabla_\mu\nabla_\varrho w_\lambda.$$

After antisymmetrization with respect to the indices μ, ν, ω, in the formula above, we have

$$2\nabla_{[\mu}\nabla_{\nu}\nabla_{\omega]}w_{\lambda} = -(\nabla_{[\mu}R_{\nu\omega]\lambda}{}^{\varrho})w_{\varrho} - R_{[\nu\omega|\lambda|}{}^{\varrho}\nabla_{\mu]}w_{\varrho} - \\ -2(\nabla_{[\mu}S_{\nu\omega]}{}^{\varrho})\nabla_{\varrho}w_{\lambda} - 2S_{[\nu\omega}{}^{\varrho}\nabla_{\mu]}\nabla_{\varrho}w_{\lambda}. \qquad (6.62)$$

Comparing identities (6.61) and (6.62) we deduce that

$$-2\nabla_{\varrho}w_{\lambda}\nabla_{[\mu}S_{\nu\omega]}{}^{\varrho} + 4\nabla_{\varrho}w_{\lambda}S_{[\mu\nu}{}^{\sigma}S_{\omega]\sigma}{}^{\varrho} - R_{[\mu\nu|\lambda|}{}^{\varrho}\nabla_{\omega]}w_{\varrho} - \\ -2S_{[\mu\nu}{}^{\varrho}\nabla_{|\varrho|}\nabla_{\omega]}w_{\lambda} + w_{\varrho}\nabla_{[\mu}R_{\nu\omega]\lambda}{}^{\varrho} + R_{[\nu\omega|\lambda|}{}^{\varrho}\nabla_{\mu]}w_{\varrho} + \\ +2(\nabla_{[\mu}S_{\nu\omega]}{}^{\varrho})\nabla_{\varrho}w_{\lambda} + 2S_{[\nu\omega}{}^{\varrho}\nabla_{\mu]}\nabla_{\varrho}w_{\lambda} = 0. \qquad (6.63)$$

In this expression, the first and seventh terms cancel, and so do the third and sixth. Now consider the fourth and eighth:

$$-2S_{[\mu\nu}{}^{\varrho}\nabla_{|\varrho|}\nabla_{\omega]}w_{\lambda} + 2S_{[\nu\omega}{}^{\varrho}\nabla_{\mu]}\nabla_{\varrho}w_{\lambda}$$
$$= -2S_{[\nu\omega}{}^{\varrho}\nabla_{|\varrho|}\nabla_{\mu]}w_{\lambda} + 2S_{[\nu\omega}{}^{\varrho}\nabla_{\mu]}\nabla_{\varrho}w_{\lambda} = 2S_{[\nu\omega}{}^{\varrho}(\nabla_{\mu]}\nabla_{\varrho}w_{\lambda} - \nabla_{|\varrho|}\nabla_{\mu]}w_{\lambda}).$$

But

$$\nabla_{\mu}\nabla_{\varrho}w_{\lambda} - \nabla_{\varrho}\nabla_{\mu}w_{\lambda} = -R_{\mu\varrho\lambda}{}^{\sigma}w_{\sigma} - 2S_{\mu\varrho}{}^{\sigma}\nabla_{\sigma}w_{\lambda},$$

and so the sum of the fourth and eighth terms is

$$-2S_{[\nu\omega}{}^{\varrho}R_{\mu]\varrho\lambda}{}^{\sigma}w_{\sigma} - 4S_{[\nu\omega}{}^{\varrho}S_{\mu]\varrho}{}^{\sigma}\nabla_{\sigma}w_{\lambda}.$$

Following the simplification above, equality (6.63) can be written as

$$4S_{[\mu\nu}{}^{\sigma}S_{\omega]\sigma}{}^{\varrho}\nabla_{\varrho}w_{\lambda} + w_{\varrho}\nabla_{[\mu}R_{\nu\omega]\lambda}{}^{\varrho} - 2S_{[\nu\omega}{}^{\varrho}R_{\mu]\varrho\lambda}{}^{\sigma}w_{\sigma} - 4S_{[\nu\omega}{}^{\varrho}S_{\mu]\varrho}{}^{\sigma}\nabla_{\sigma}w_{\lambda} = 0.$$

In the foregoing sum the first and fourth terms cancel (replace the permutation $\nu\omega\mu$ by $\mu\nu\omega$ and interchange the dummy indices ϱ and σ) and we finally have the identity

$$w_{\varrho}\nabla_{[\mu}R_{\nu\omega]\lambda}{}^{\varrho} = 2S_{[\nu\omega}{}^{\varrho}R_{\mu]\varrho\lambda}{}^{\varrho}w_{\sigma}$$

or, on interchanging the dummy indices and replacing the permutation $\nu\omega\mu$ by $\mu\nu\omega$ on the right-hand side,

$$\omega_{\varrho}\nabla_{[\mu}R_{\nu\omega]\lambda}{}^{\varrho} = 2S_{[\mu\nu}{}^{\sigma}R_{\omega]\sigma\lambda}{}^{\varrho}w_{\varrho}.$$

Since this identity is to hold for all w_{ϱ}'s, it follows that

$$\nabla_{[\mu}R_{\nu\omega]\lambda}{}^{\varrho} = 2S_{[\mu\nu}{}^{\sigma}R_{\omega]\sigma\lambda}{}^{\varrho}. \qquad (6.64)$$

This precisely is the Bianchi identity (generalized) which we were seeking.

In the special case when the connection is symmetric ($S_{\mu\nu}{}^{\sigma} = 0$), we have

$$\nabla_{[\mu}R_{\nu\omega]\lambda}{}^{\varrho} = 0. \qquad (6.65)$$

For a symmetric displacement a shorter proof of relations (6.65) is available: we start with the identity

$$R_{\nu\omega\lambda}{}^{\varrho} = 2\partial_{[\nu}\Gamma^{\varrho}_{\omega]\lambda}+2\Gamma^{\varrho}_{[\nu|\sigma|}\Gamma^{\sigma}_{\omega]\lambda}$$

and apply the operator ∇_μ to it. We thus obtain

$$\nabla_\mu R_{\nu\omega\lambda}{}^{\varrho} = 2\partial_{\mu[\nu}\Gamma^{\varrho}_{\omega]\lambda}+2(\partial_\mu\Gamma^{\varrho}_{[\nu|\sigma|})\Gamma^{\sigma}_{\omega]\lambda}+2\Gamma^{\varrho}_{[\nu|\sigma|}\partial_{|\mu|}\Gamma^{\sigma}_{\omega]\lambda}+$$
$$+\Gamma^{\varrho}_{\mu\sigma}R_{\nu\omega\lambda}{}^{\sigma}-\Gamma^{\sigma}_{\nu\mu}R_{\sigma\omega\lambda}{}^{\varrho}-\Gamma^{\sigma}_{\mu\omega}R_{\nu\sigma\lambda}{}^{\varrho}-\Gamma^{\sigma}_{\mu\lambda}R_{\nu\omega\sigma}{}^{\varrho}.$$

Skewing the indices μ, ν, ω on both sides yields

$$\nabla_{[\mu}R_{\nu\omega]\lambda}{}^{\varrho} = 2\partial_{[\mu\nu}\Gamma^{\varrho}_{\omega]\lambda}+2(\partial_{[\mu}\Gamma^{\varrho}_{\nu|\sigma|})\Gamma^{\sigma}_{\omega]\lambda}+2\Gamma^{\varrho}_{[\nu|\sigma|}\partial_{\mu}\Gamma^{\sigma}_{\omega]\lambda}+$$
$$+\Gamma^{\varrho}_{\sigma[\mu}R_{\nu\omega]\lambda}{}^{\sigma}+\Gamma^{\sigma}_{[\mu\nu}R_{\omega]\sigma\lambda}{}^{\varrho}-\Gamma^{\sigma}_{[\mu\omega}R_{\nu]\sigma\lambda}{}^{\varrho}-\Gamma^{\sigma}_{\lambda[\mu}R_{\nu\omega]\sigma}{}^{\varrho}.$$

In effecting the foregoing transformation we have made use of the symmetry of the object Γ and the skew symmetry of the curvature tensor with respect to the first two indices. By virtue of the symmetry of Γ, the fifth and sixth terms on the right-hand side are zero. Since $\partial_{\mu\nu} = \partial_{\nu\mu}$, the first term on the right-hand side also vanishes. We are therefore left with the identity

$$\nabla_{[\mu}R_{\nu\omega]\lambda} = 2(\partial_{[\mu}\Gamma^{\varrho}_{\nu|\sigma|})\Gamma^{\sigma}_{\omega]\lambda}+2\Gamma^{\varrho}_{[\nu|\sigma|}\partial_{\mu}\Gamma^{\sigma}_{\omega]\lambda}+\Gamma^{\varrho}_{\sigma[\mu}R_{\nu\omega]\lambda}{}^{\sigma}-\Gamma^{\sigma}_{\lambda[\mu}R_{\nu\omega]\sigma}{}^{\varrho}. \quad (6.66)$$

Now take normal coordinates at the point p, at which we wish to establish identity (6.66); then $\Gamma^{\nu}_{\mu\lambda} \overset{*}{=} 0$ at the point under consideration and the right-hand side of equation (6.66) is thus zero, q.e.d.

We now carry out contraction with respect to ϱ and λ in (6.64), to obtain the formula

$$\nabla_{[\mu}V_{\nu\omega]} = 2S_{[\mu\nu}{}^{\sigma}V_{\omega]\sigma}, \quad (6.67)$$

which for a symmetric object reduces to

$$\nabla_{[\mu}V_{\nu\omega]} = 0.$$

On the other hand, contraction with respect to ϱ and ν in the Bianchi identity leads to the identity

$$\nabla_{[\mu}R_{\varrho\omega]\lambda}{}^{\varrho} = 2S_{[\mu\varrho}{}^{\sigma}R_{\omega]\sigma\lambda}{}^{\varrho}.$$

We rearrange the last identity. Since the tensor $R_{\varrho\omega\lambda}{}^{\sigma}$ is antisymmetric in ϱ, ω and $S_{\mu\varrho}{}^{\sigma}$ in μ, ϱ, the identity above can be rewritten as

$$\nabla_\mu R_{\varrho\omega\lambda}{}^{\varrho}-\nabla_\omega R_{\varrho\mu\lambda}{}^{\varrho}+\nabla_\varrho R_{\omega\mu\lambda}{}^{\varrho} = 2S_{\mu\varrho}{}^{\sigma}R_{\omega\sigma\lambda}{}^{\varrho}+2S_{\varrho\omega}{}^{\sigma}R_{\mu\sigma\lambda}{}^{\varrho}+2S_{\omega\mu}{}^{\sigma}R_{\sigma\lambda},$$

or equivalently

$$2\nabla_{[\mu}R_{\omega]\lambda}-\nabla_\varrho R_{\mu\omega\lambda}{}^{\varrho} = -4S_{\varrho[\mu}{}^{\sigma}R_{\omega]\sigma\lambda}{}^{\varrho}-2S_{\mu\omega}{}^{\sigma}R_{\sigma\lambda}. \quad (6.68)$$

For symmetric connections, the latter simplifies to the identity

$$2\nabla_{[\mu} R_{\omega]\lambda} - \nabla_{\varrho} R_{\mu\omega\lambda}{}^{\varrho} = 0. \tag{6.69}$$

Now transvect the tensor $g^{\omega\lambda}$ onto the left-hand side of this equation. Since $\nabla^{\mu} g^{\omega\lambda} = 0$, the factor $g^{\omega\lambda}$ can be placed under the symbol of covariant differentiation and we thus obtain

$$\nabla_{\mu} R_{\omega\lambda} g^{\omega\lambda} - \nabla_{\omega} R_{\mu\lambda} g^{\omega\lambda} - \nabla_{\varrho} R_{\omega\mu\lambda}{}^{\varrho} g^{\omega\lambda} = 0.$$

We next introduce the short-hand notations

$$R_{\mu}^{\omega} \stackrel{\text{def}}{=} R_{\mu\lambda} g^{\lambda\omega},$$

$$\nabla_{\mu} \stackrel{\text{def}}{=} g^{\mu\nu} \nabla_{\nu},$$

and recall the fact that

$$R_{\mu\omega\lambda}{}^{\varrho} = R_{\mu\omega\lambda\alpha} g^{\varrho\alpha}.$$

Then, since

$$R_{\omega\lambda} g^{\omega\lambda} = R$$

and

$$R_{\lambda\mu} = R_{\nu\lambda\mu\alpha} g^{\alpha\nu},$$

it follows that

$$\nabla_{\mu} R - \nabla_{\omega} R_{\mu}^{\omega} - \nabla_{\varrho} R_{\alpha\mu} g^{\varrho\alpha} = 0$$

or

$$\nabla_{\mu} R - \nabla_{\omega} R_{\mu}^{\omega} - \nabla_{\varrho} R_{\mu}^{\varrho} = 0,$$

i.e. that

$$\nabla_{\mu} R - 2\nabla_{\omega} R_{\mu}^{\omega} = 0$$

or, further,

$$\nabla^{\alpha} R g_{\alpha\mu} - 2\nabla^{\alpha} R_{\alpha\mu} = 0,$$

or

$$\nabla^{\alpha}(R g_{\alpha\mu} - 2R_{\alpha\mu}) = 0,$$

or, finally, that

$$\nabla^{\alpha}(R_{\alpha\mu} - \tfrac{1}{2} R g_{\alpha\mu}) = 0. \tag{6.70}$$

For $n = 4$ this relation amounts to *Einstein's* famous *field equation* of the general theory of relativity *in vacuo*.

In Riemannian spaces V_n, for a given simple bivector defined by a pair of independent vectors $\underset{1}{v}^{\nu}$, $\underset{2}{v}^{\nu}$, we can define the curvature $R(\underset{1}{v}, \underset{2}{v})$ of the two-dimensional direction by means of the formula

$$R(\underset{1}{v}, \underset{2}{v}) \stackrel{\text{def}}{=} \frac{R_{\omega\mu\lambda\nu} \underset{1}{v}^{\omega} \underset{2}{v}^{\mu} \underset{1}{v}^{\lambda} \underset{2}{v}^{\nu}}{(g_{\omega\lambda} g_{\mu\nu} - g_{\omega\nu} g_{\lambda\mu}) \underset{1}{v}^{\omega} \underset{2}{v}^{\mu} \underset{1}{v}^{\lambda} \underset{2}{v}^{\nu}}. \tag{6.71}$$

A point in space is said to be *isotropic* if at this point $R(\underset{1}{v}, \underset{2}{v})$ is constant for all pairs of independent vectors $\underset{1}{v}, \underset{2}{v}$. It was shown by F. Schur that if every point in a Riemannian space V_n $(n \geqslant 3)$ is isotropic, then the space has a constant curvature. W. Wrona [157] worked on generalizing this theorem of Schur.

CHAPTER VII

THE METRIC OF RIEMANNIAN SPACE

82. Christoffel Symbols. Uniquely Distinguishing an Object of Linear Connection for Riemannian Space. We have seen how Levi-Civita arrived at a concept of parallel displacement of vectors along a curve in metric (Riemannian) spaces, a concept which is unambiguous, and which when the metric is fixed (by means of a fundamental tensor) does not necessitate the introduction of any further independent geometric objects. It follows that the metric tensor $g_{\lambda\mu}$ allows us to make in a certain unique manner an invariant choice of an object of parallel displacement from infinitely many possible objects Γ. This distinguished object consists of the *Christoffel symbols of the second kind*, which, however, their discoverer reached by a different route. For our part, we shall proceed in yet another way.

Recall that an object of parallel displacement has three indices. To obtain a three-index object from the metric tensor $g_{\lambda\mu}$, we start from the ordinary partial derivatives

$$\partial_\nu g_{\lambda\mu}. \tag{7.1}$$

The fact that all three indices are subscripts here, whereas in the object $\Gamma^{\nu}_{\mu\lambda}$ two are subscripts and one is a superscript, should not disconcert us for the time being since we can always raise one of the indices by means of the tensor $g^{\nu\omega}$.

We ask whether expression (7.1) is a geometric object. To answer this, we form the expression $\partial_{\nu'} g_{\lambda'\mu'}$ and we attempt to write it in terms of $\partial_\nu g_{\lambda\mu}$. Accordingly, we write

$$\begin{aligned}\partial_{\nu'} g_{\lambda'\mu'} &= A^{\nu}_{\nu'}\,\partial_\nu (g_{\lambda\mu} A^{\lambda}_{\lambda'} A^{\mu}_{\mu'}) \\ &= A^{\nu}_{\nu'}\{A^{\lambda\mu}_{\lambda'\mu'}\,\partial_\nu g_{\lambda\mu} + A^{\lambda}_{\lambda'} g_{\lambda\mu}\,\partial_\nu A^{\mu}_{\mu'} + A^{\mu}_{\mu'} g_{\lambda\mu}\,\partial_\nu A^{\lambda}_{\lambda'}\}. \end{aligned} \tag{7.2}$$

It is thus evident that $\partial_{\nu'} g_{\lambda'\mu'}$ is not expressed solely in terms of $\partial_\nu g_{\lambda\mu}$ by $A^{\lambda}_{\lambda'}$, $A^{\lambda'}_{\lambda}$, and their derivatives, but also in terms of the tensor $g_{\lambda\mu}$ itself.

Hence, $\partial_\nu g_{\lambda\mu}$ is not a geometric object. We now form permutations from $\partial_\nu g_{\lambda\mu}$ by rearranging indices just as when constructing isomers from a given quantity. Because of the symmetry of the fundamental tensor, only three out of six possible permutations can give different values, and these are $\partial_\nu g_{\lambda\mu}$, $\partial_\lambda g_{\mu\nu}$, $\partial_\mu g_{\nu\lambda}$; now in short-hand notation we write

$$\Gamma_{\lambda\mu\nu} \stackrel{\text{def}}{=} \alpha \partial_\nu g_{\mu\lambda} + \beta \partial_\lambda g_{\mu\nu} + \gamma \partial_\mu g_{\nu\lambda}, \tag{7.3}$$

where α, β, γ are three constant scalar factors. We now calculate $\Gamma_{\lambda'\mu'\nu'}$. By virtue of equation (7.2) derived above, we have

$$\begin{aligned}\Gamma_{\lambda'\mu'\nu'} = {} & A^{\nu}_{\nu'}\{A^{\lambda}_{\lambda'}{}^{\mu}_{\mu'}\,\alpha\partial_\nu g_{\lambda\mu} + A^{\lambda}_{\lambda'}\,\alpha g_{\lambda\mu}\,\partial_\nu A^{\mu}_{\mu'} + A^{\mu}_{\mu'}\,\alpha g_{\lambda\mu}\,\partial_\nu A^{\lambda}_{\lambda'}\} + \\ & + A^{\lambda}_{\lambda'}\{A^{\mu}_{\mu'}{}^{\nu}_{\nu'}\,\beta\partial_\lambda g_{\mu\nu} + A^{\mu}_{\mu'}\,\beta g_{\mu\nu}\,\partial_\lambda A^{\nu}_{\nu'} + A^{\nu}_{\nu'}\,\beta g_{\mu\nu}\,\partial_\lambda A^{\mu}_{\mu'}\} + \\ & + A^{\mu}_{\mu'}\{A^{\nu}_{\nu'}{}^{\lambda}_{\lambda'}\,\gamma\partial_\mu g_{\nu\lambda} + A^{\nu}_{\nu'}\,\gamma g_{\nu\lambda}\,\partial_\mu A^{\lambda}_{\lambda'} + A^{\lambda}_{\lambda'}\,\gamma g_{\nu\lambda}\,\partial_\mu A^{\nu}_{\nu'}\}\end{aligned}$$

or

$$\begin{aligned}\Gamma_{\lambda'\mu'\nu'} = {} & A^{\lambda}_{\lambda'}{}^{\mu}_{\mu'}{}^{\nu}_{\nu'}\{\alpha\partial_\nu g_{\mu\nu} + \beta\partial_\lambda g_{\mu\nu} + \gamma\partial_\mu g_{\nu\lambda}\} + A^{\nu}_{\nu'}{}^{\lambda}_{\lambda'}\{\alpha g_{\lambda\mu}\,\partial_\nu A^{\mu}_{\mu'} + \beta g_{\mu\nu}\,\partial_\lambda A^{\mu}_{\mu'}\} + \\ & + A^{\nu}_{\nu'}{}^{\mu}_{\mu'}\{\alpha g_{\lambda\mu}\,\partial_\nu A^{\lambda}_{\lambda'} + \gamma g_{\nu\lambda}\,\partial_\mu A^{\lambda}_{\lambda'}\} + A^{\lambda}_{\lambda'}{}^{\mu}_{\mu'}\{\beta g_{\mu\nu}\,\partial_\lambda A^{\nu}_{\nu'} + \gamma g_{\nu\lambda}\partial_\mu A^{\nu}_{\nu'}\},\end{aligned}$$

or, finally,

$$\begin{aligned}\Gamma_{\lambda'\mu'\nu'} = {} & A^{\lambda}_{\lambda'}{}^{\mu}_{\mu'}{}^{\nu}_{\nu'}\Gamma_{\lambda\mu\nu} + A^{\lambda}_{\lambda'}{}^{\mu}_{\mu'}[\beta g_{\mu\nu}\,\partial_\lambda A^{\nu}_{\nu'} + \gamma g_{\nu\lambda}\,\partial_\mu A^{\nu}_{\nu'}] + \\ & + A^{\mu}_{\mu'}{}^{\nu}_{\nu'}[\alpha g_{\lambda\mu}\,\partial_\nu A^{\lambda}_{\lambda'} + \gamma g_{\nu\lambda}\,\partial_\mu A^{\lambda}_{\lambda'}] + A^{\nu}_{\nu'}{}^{\lambda}_{\lambda'}[\alpha g_{\lambda\mu}\,\partial_\nu A^{\mu}_{\mu'} + \beta g_{\mu\nu}\,\partial_\lambda A^{\mu}_{\mu'}].\end{aligned} \tag{7.4}$$

Next, we introduce the expression

$$\Gamma^{\varrho}_{\cdot\mu\nu} \stackrel{\text{def}}{=} g^{\lambda\varrho}\Gamma_{\lambda\mu\nu}. \tag{7.5}$$

By the formula above we have

$$\begin{aligned}\Gamma^{\varrho'}_{\cdot\mu'\nu'} & = g^{\lambda'\varrho'}\Gamma_{\lambda'\mu'\nu'} \\ & = A^{\lambda'}_{\lambda}{}^{\varrho'}_{\varrho}g^{\lambda\varrho}\{A^{\sigma}_{\lambda'}{}^{\mu}_{\mu'}{}^{\nu}_{\nu'}\Gamma_{\sigma\mu\nu} + A^{\sigma}_{\lambda'}{}^{\mu}_{\mu'}[\beta g_{\mu\nu}\,\partial_\sigma A^{\nu}_{\nu'} + \gamma g_{\nu\sigma}\,\partial_\mu A^{\nu}_{\nu'}] + \\ & \quad + A^{\mu}_{\mu'}{}^{\nu}_{\nu'}[\alpha g_{\sigma\mu}\,\partial_\nu A^{\sigma}_{\lambda'} + \gamma g_{\nu\sigma}\,\partial_\mu A^{\sigma}_{\lambda'}] + A^{\nu}_{\nu'}{}^{\sigma}_{\lambda'}[\alpha g_{\sigma\mu}\,\partial_\nu A^{\mu}_{\mu'} + \beta g_{\mu\nu}\,\partial_\sigma A^{\mu}_{\mu'}]\} \\ & = A^{\varrho'}_{\varrho}{}^{\mu}_{\mu'}{}^{\nu}_{\nu'}g^{\varrho\sigma}\Gamma_{\sigma\mu\nu} + A^{\varrho'}_{\varrho}{}^{\mu}_{\mu'}[\beta g_{\mu\nu}g^{\lambda\varrho}\,\partial_\lambda A^{\nu}_{\nu'} + \gamma g_{\nu\lambda}g^{\lambda\varrho}\partial_\mu A^{\nu}_{\nu'}] + \\ & \quad + A^{\lambda'}_{\lambda}{}^{\varrho'}_{\varrho}{}^{\mu}_{\mu'}{}^{\nu}_{\nu'}[\alpha g^{\lambda\varrho}g_{\mu\sigma}\,\partial_\nu A^{\sigma}_{\lambda'} + \gamma g^{\lambda\varrho}g_{\nu\sigma}\,\partial_\mu A^{\sigma}_{\lambda'}] + \\ & \quad + A^{\varrho'}_{\varrho}{}^{\nu}_{\nu'}[\alpha g_{\lambda\mu}g^{\lambda\varrho}\partial_\nu A^{\mu}_{\mu'} + \beta g_{\mu\nu}g^{\lambda\varrho}\partial_\lambda A^{\mu}_{\mu'}] \\ & = A^{\varrho'}_{\varrho}{}^{\mu}_{\mu'}{}^{\nu}_{\nu'}\Gamma^{\varrho}_{\cdot\mu\nu} + (\alpha+\gamma)A^{\varrho'}_{\nu}\,\partial_{\mu'}A^{\nu}_{\nu'} + \\ & \quad + (\beta+\gamma)A^{\lambda'}_{\lambda}{}^{\varrho'}_{\varrho}{}^{\nu}_{\nu'}g^{\lambda\varrho}g_{\nu\sigma}\,\partial_{\mu'}A^{\sigma}_{\lambda'} + (\alpha+\beta)A^{\varrho'}_{\varrho}{}^{\mu}_{\mu'}{}^{\lambda'}_{\lambda}g^{\lambda\varrho}g_{\mu\nu}\,\partial_{\lambda'}A^{\nu}_{\nu'}.\end{aligned}$$

Setting $\beta+\gamma = 0$ and $\alpha+\beta = 0$, that is [1]

$$\beta = -\alpha, \qquad \gamma = \alpha \tag{7.6}$$

we obtain a particularly simple transformation law

$$\Gamma^{\varrho'}_{\cdot\mu'\nu'} = \Gamma^{\varrho}_{\cdot\mu\nu} A^{\varrho'\mu\;\nu}_{\varrho\;\mu'\nu'} + 2\alpha A^{\varrho'}_{\nu} \partial_{\mu'} A^{\nu}_{\nu'}, \tag{7.7}$$

where $\Gamma^{\varrho}_{\cdot\mu\nu}$ now becomes a geometric object, since the transformation law does not contain the metric tensor $g_{\lambda\mu}$. Finally, if we set $2\alpha = 1$, that is

$$\alpha = \tfrac{1}{2} \tag{7.8}$$

we obtain the transformation law

$$\Gamma^{\varrho'}_{\cdot\mu'\nu'} = \Gamma^{\varrho}_{\cdot\mu\nu} A^{\varrho'\mu\;\nu}_{\varrho\;\mu'\nu'} + A^{\varrho'}_{\nu} \partial_{\mu'} A^{\nu}_{\nu'} \tag{7.9}$$

which is exactly the same as the transformation law for an object of parallel displacement.

Therefore, in equation (7.3) we take

$$\alpha = \tfrac{1}{2}, \qquad \beta = -\tfrac{1}{2}, \qquad \gamma = \tfrac{1}{2} \tag{7.10}$$

and we then arrive at

$$\Gamma^{\varrho}_{\cdot\mu\nu} = \tfrac{1}{2} g^{\varrho\lambda} \{\partial_{\mu} g_{\nu\lambda} + \partial_{\nu} g_{\lambda\mu} - \partial_{\lambda} g_{\mu\nu}\}. \tag{7.11}$$

Here, $\Gamma^{\varrho}_{\mu\nu}$ are *Christoffel symbols of the second kind* ($\Gamma_{\lambda\mu\nu}$ satisfying condition (7.10), i.e. the terms in the braces on the right-hand side of (7.11) are known as Christoffel symbols of the first kind) which nowadays are denoted by the symbol

$$\left\{ {\varrho \atop \mu\nu} \right\}, \tag{7.12}$$

which is seen to be a modification of the original symbol $\left\{ {\mu\nu \atop \varrho} \right\}$. As we see, Christoffel symbols are uniquely defined when the metric tensor $g_{\lambda\mu}$ is given.

The metric tensor thus provides a basis for uniquely distinguishing an object of parallel displacement (specifically, the one which Levi-Civita discovered) and for defining the concept of covariant derivative.

83. The Properties of Christoffel Symbols as an Object of Parallel Displacement. The covariant derivative defined in terms of the Christoffel

[1] We set these coefficients equal to zero because they appear next to terms containing coordinates of the tensor g, and since it is our intention to obtain a geometric object, we want to eliminate terms containing coordinates of the tensor g.

symbols, i.e.

$$\nabla_\mu v^\lambda = \partial_\mu v^\lambda + \left\{ {\lambda \atop \mu\nu} \right\} v^\nu \tag{7.13}$$

(and analogously for other kinds of quantities) is called the *congenital covariant derivative.*

Note that the congenital derivative (7.13) can be defined only when a field of fundamental tensors with the property that the matrix $[g_{\lambda\mu}]$ has maximum rank is given in space. In the space V_n, where an object (field) of parallel displacement is defined independently of the field of fundamental tensors, the covariant derivative

$$\nabla_\mu v^\lambda = \partial_\mu v^\lambda + \Gamma^\lambda_{\mu\nu} v^\nu$$

will not in general coincide with the inborn derivative (7.13). Two different symbols ∇_μ and $\overset{*}{\nabla}_\mu$ then have to be introduced. When there is no risk of ambiguity, we will continue to use ∇_μ to denote the inborn derivative in the space V_n.

Note that the Christoffel symbols satisfy the symmetry condition

$$\left\{ {\lambda \atop \mu\nu} \right\} = \left\{ {\lambda \atop \nu\mu} \right\}, \tag{7.14}$$

so that the torsion tensor is identically zero.

Let us now calculate the covariant derivative of the metric tensor $g_{\lambda\mu}$ in terms of the Christoffel symbols. We have

$$\begin{aligned}
\nabla_\nu g_{\lambda\mu} &= \partial_\nu g_{\lambda\mu} - \left\{ {\varrho \atop \lambda\nu} \right\} g_{\varrho\mu} - \left\{ {\varrho \atop \mu\nu} \right\} g_{\lambda\varrho} \\
&= \partial_\nu g_{\lambda\mu} - \tfrac{1}{2}\{ g^{\varrho\sigma}[\partial_\lambda g_{\sigma\nu} + \partial_\nu g_{\lambda\sigma} - \partial_\sigma g_{\lambda\nu}] g_{\varrho\mu} + \\
&\quad + g^{\varrho\sigma}[\partial_\mu g_{\sigma\nu} + \partial_\nu g_{\mu\sigma} - \partial_\sigma g_{\mu\nu}] g_{\lambda\varrho}\} \\
&= \partial_\nu g_{\lambda\mu} - \tfrac{1}{2}\{ A^\sigma_\mu [\partial_\lambda g_{\sigma\nu} + \partial_\nu g_{\lambda\sigma} - \partial_\sigma g_{\lambda\nu}] + \\
&\quad + A^\sigma_\lambda [\partial_\mu g_{\sigma\nu} + \partial_\nu g_{\mu\sigma} - \partial_\sigma g_{\mu\nu}]\} \qquad (7.15) \\
&= \partial_\nu g_{\lambda\mu} - \tfrac{1}{2}\{ \partial_\lambda g_{\mu\nu} + \partial_\nu g_{\lambda\mu} - \partial_\mu g_{\lambda\nu} + \\
&\quad + \partial_\mu g_{\lambda\nu} + \partial_\nu g_{\mu\lambda} - \partial_\lambda g_{\mu\nu}\} \\
&= \partial_\nu g_{\lambda\mu} - \partial_\nu g_{\lambda\mu} = 0 .
\end{aligned}$$

Accordingly, we have the following result:

The covariant derivative of a fundamental tensor calculated by means of the Christoffel symbols is identically zero.

Similarly, it can be shown that the covariant inborn derivative of the Ricci n-vectors e is also identically zero. To this end, we first calculate

$$\begin{Bmatrix} \lambda \\ \lambda\mu \end{Bmatrix}. \tag{7.16}$$

Hence

$$\begin{Bmatrix} \lambda \\ \lambda\mu \end{Bmatrix} = \tfrac{1}{2} g^{\lambda\varrho}[\partial_\lambda g_{\varrho\mu} + \partial_\mu g_{\lambda\varrho} - \partial_\varrho g_{\lambda\mu}]$$

$$= \tfrac{1}{2}\{g^{\lambda\varrho}\partial_\lambda g_{\varrho\mu} + g^{\lambda\varrho}\partial_\mu g_{\lambda\varrho} - g^{\varrho\lambda}\partial_\lambda g_{\varrho\mu}\} = \tfrac{1}{2} g^{\lambda\varrho}\partial_\mu g_{\lambda\varrho}.$$

By virtue of the well-known theorem concerning the differentiation of determinants, we have

$$\partial_\sigma \mathfrak{g} = \sum_{\lambda,\mu} \partial_\sigma g_{\lambda\mu} \cdot \operatorname{minor} g_{\lambda\mu}, \qquad \text{where} \qquad \mathfrak{g} = |g_{\lambda\mu}|.$$

Dividing both sides by $\mathfrak{g}$, we get

$$\partial_\sigma \ln|\mathfrak{g}| = \sum_{\lambda,\mu} (\partial_\sigma g_{\lambda\mu}) g^{\mu\lambda} = g^{\mu\lambda}\partial_\sigma g_{\lambda\mu}.$$

Consequently

$$\begin{Bmatrix} \lambda \\ \lambda\mu \end{Bmatrix} = \tfrac{1}{2} g^{\lambda\varrho}\partial_\mu g_{\lambda\varrho} = \tfrac{1}{2}\partial_\mu \ln|\mathfrak{g}| = \partial_\mu \ln|\mathfrak{g}|^{1/2}. \tag{7.17}$$

Next, let us calculate the covariant derivative of the density $|\mathfrak{g}|^{1/2}$ which, as we know, is a W-density of weight $+1$. Thus

$$\nabla_\mu |\mathfrak{g}|^{1/2} = \partial_\mu |\mathfrak{g}|^{1/2} - \begin{Bmatrix} \lambda \\ \lambda\mu \end{Bmatrix} |\mathfrak{g}|^{1/2} = \partial_\mu |\mathfrak{g}|^{1/2} - |\mathfrak{g}|^{1/2}\partial_\mu \ln|\mathfrak{g}|^{1/2} = 0.$$

In the space V_n, where the Christoffel symbols constitute an object of parallel displacement, this formula enables the covariant derivative of a p-vector, contracted with respect to the differentiation index, to be written in a form involving only ordinary differentiation and not containing the parameters $\Gamma^\nu_{\mu\lambda}$.

Suppose we have the field

$$v^{\lambda_1\cdots\lambda_p} = v^{[\lambda_1\cdots\lambda_p]}.$$

Let us calculate the value of the expression $\nabla_\mu v^{\mu\lambda_2\cdots\lambda_p}$.

By equation (5.80) we have

$$\nabla_\mu v^{\lambda_1\cdots\lambda_p} = \partial_\mu v^{\lambda_1\cdots\lambda_p} + \sum_{i=1}^{p} \Gamma^{\lambda_i}_{\mu\varrho} v^{\lambda_1\cdots\lambda_{i-1}\varrho\lambda_{i+1}\cdots\lambda_p}.$$

From this, by contraction we arrive at

$$\nabla_\mu v^{\mu\lambda_2\cdots\lambda_p} = \partial_\mu v^{\mu\lambda_2\cdots\lambda_p} + \Gamma^{\mu}_{\mu\varrho} v^{\varrho\lambda_2\cdots\lambda_p} + \sum_{i=2}^{p} \Gamma^{\lambda_i}_{\mu\varrho} v^{\mu\lambda_2\cdots\lambda_{i-1}\varrho\lambda_{i+1}\cdots\lambda_p}.$$

Now since the $\Gamma^{\lambda_i}_{\mu\varrho}$'s are symmetric in the lower indices μ, ϱ, whereas the $v^{\mu\cdots\varrho\cdots}$'s are antisymmetric, each of the sums

$$\Gamma^{\lambda_i}_{\mu\varrho} v^{\mu\lambda_2\cdots\lambda_{i-1}\varrho\lambda_{i+1}\cdots\lambda_p}$$

is zero and we therefore have

$$\nabla_\mu v^{\mu\lambda_2\cdots\lambda_p} = \partial_\mu v^{\mu\lambda_2\cdots\lambda_p} + \Gamma_\varrho v^{\varrho\lambda_2\cdots\lambda_p}.$$

On the other hand, we can also write

$$\begin{aligned}\nabla_\mu(|\mathfrak{g}|^{1/2} v^{\mu\lambda_2\cdots\lambda_p}) &= (\nabla_\mu |\mathfrak{g}|^{1/2}) v^{\mu\lambda_2\cdots\lambda_p} + |\mathfrak{g}|^{1/2} \nabla_\mu v^{\mu\lambda_2\cdots\lambda_p} = |\mathfrak{g}|^{1/2} \nabla_\mu v^{\mu\lambda_2\cdots\lambda_p} \\ &= |\mathfrak{g}|^{1/2}\{\partial_\mu v^{\mu\lambda_2\cdots\lambda_p} + \Gamma_\varrho v^{\varrho\lambda_2\cdots\lambda_p}\} = \partial_\mu\{|\mathfrak{g}|^{1/2} v^{\mu\lambda_2\cdots\lambda_p}\} + \\ &\quad + |\mathfrak{g}|^{1/2}\Gamma_\varrho v^{\varrho\lambda_2\cdots\lambda_p} - v^{\mu\lambda_2\cdots\lambda_p}\partial_\mu |\mathfrak{g}|^{1/2} = \partial_\mu\{|\mathfrak{g}|^{1/2} v^{\mu\lambda_2\cdots\lambda_p}\} \\ &\quad - v^{\mu\lambda_2\cdots\lambda_p}\{\partial_{\nu\mu}|\mathfrak{g}|^{1/2} - \Gamma_\mu |\mathfrak{g}|^{1/2}\} = \partial_\mu\{|\mathfrak{g}|^{1/2} v^{\mu\lambda_2\cdots\lambda_p}\},\end{aligned}$$

since $\partial_\mu |\mathfrak{g}|^{1/2} - \Gamma_\mu |\mathfrak{g}|^{1/2} = 0$, from which we get

$$\nabla_\mu v^{\mu\lambda_2\cdots\lambda_p} = \frac{1}{|\mathfrak{g}|^{1/2}} \partial_\mu\{|\mathfrak{g}|^{1/2} v^{\mu\lambda_2\cdots\lambda_p}\}. \tag{7.18}$$

Next, we take the n-vector $e^{\lambda_1\cdots\lambda_n}$ and calculate its covariant derivative in terms of the Christoffel symbols. Thus, we have

$$e^{\lambda_1\cdots\lambda_n} = |\mathfrak{g}|^{-1/2} \varepsilon^{\lambda_1\cdots\lambda_n}.$$

But the covariant derivative of an n-vector density ε is zero, regardless of what kind of parallel displacement object we take (formulae (5.95) and (5.96)). Therefore

$$\nabla_\mu e^{\lambda_1\cdots\lambda_n} = (\nabla_\mu |\mathfrak{g}|^{-1/2}) \varepsilon^{\lambda_1\cdots\lambda_n} + |\mathfrak{g}|^{-1/2} \nabla_\mu \varepsilon^{\lambda_1\cdots\lambda_n} = 0. \tag{7.19}$$

Similarly, we have

$$\nabla_\mu e_{\lambda_1\cdots\lambda_n} = 0. \tag{7.20}$$

84. Tensors Constant Under Covariant Differentiation. Let us consider the case of the two-dimensional Riemannian space V_2. We have seen that if the covariant derivative is defined in terms of Christoffel symbols, then both

$$\nabla_\mu g_{\lambda\nu} = 0,$$

and

$$\nabla_\mu e_{\lambda\nu} = 0,$$

where $e_{\lambda\nu}$ is the antisymmetric Ricci tensor. The question is whether there exist any other covariant two-index tensors whose covariant derivative is identically zero. The answer to this question is to be found in papers by Dubnov [100] and Lopschitz [136] who, using different methods, gave the conditions for the solubility of the covariant equation

$$\nabla_\mu x_{\lambda\nu} = T_{\lambda\nu\mu},$$

where T is a given tensor field, whereas x is an unkown tensor field. The conditions for this equation to be integrable are fairly complicated: they are of one type when the curvature tensor R of the space V_2 is identically zero, and of another type otherwise. They are expressed in terms of the metric tensor g (of the given field T, of course), the Ricci tensor e, and the curvature $\varkappa$ (1) of the space. If these conditions are satisfied and if $\varkappa \neq 0$, the general solution is of the form

$$x_{\lambda\nu} = \frac{1}{2}\left\{\frac{1}{\varkappa}e^{\varrho\sigma}e^{\omega\tau}g_{\lambda\omega}\nabla_\sigma T_{\tau\nu\varrho} + g_{\lambda\nu}\int g^{\omega\tau}T_{\omega\tau\mu}d\xi^\mu + e_{\lambda\nu}\int e^{\omega\tau}T_{\omega\tau\mu}d\xi^\mu + \alpha g_{\lambda\nu} + \beta e_{\lambda\nu}\right\},$$

where α, β are arbitrary constants.

In the special case when $T_{\lambda\nu\mu} = 0$, that is, when the equation

$$\nabla_\mu x_{\lambda\nu} = 0,$$

is to be integrated, the conditions for total integrability are satisfied and then, as indicated by the formula above, the general solution reduces to

$$x_{\lambda\nu} = \alpha g_{\lambda\nu} + \beta e_{\lambda\nu}.$$

It follows from this that the V_2 contains no covariant tensors apart from $g_{\lambda\nu}$ and $e_{\lambda\nu}$ with identically zero covariant derivative.

It is extremely interesting that integration of covariant equations of the type

$$\nabla_\mu x_{\lambda_1 \dots \lambda_n} = T_{\lambda_1 \dots \lambda_n \mu},$$

where T is given, and x is the tensor field sought, is effected in different ways depending on whether n is even or odd. When n is even, as we have seen above that a certain integration (quadrature) must be performed to

(1) Note that the curvature $\varkappa$ in Riemannian space is defined by the formula $\varkappa = \frac{1}{n(n-1)} R_{\mu\lambda} g^{\mu\lambda}$ (cf. equation (6.56)).

find the general solution. On the other hand, when n is odd the general solution is obtained in a finite form without any quadrature. By way of an example let us consider a certain result obtained by Graiff [122] concerning the Riemannian space V_n $(n > 3)$ with constant curvature $\varkappa$. For the equation

$$\nabla_\mu x_\lambda = T_{\lambda\mu},$$

the conditions of integrability (for $\varkappa \neq 0$) are

$$\frac{1}{(n-1)!\varkappa} e^{\lambda_1 \ldots \lambda_{n-2}\varrho\sigma} e_{\lambda_1 \ldots \lambda_{n-2}\lambda\nu} g^{\nu\omega} \nabla_\mu \nabla_\sigma T_{\omega\varrho} = T_{\lambda\mu}.$$

The general solution (in fact the only one) is then of the form

$$x_\lambda = \frac{1}{(n-1)!\varkappa} e^{\lambda_1 \ldots \lambda_{n-2}\varrho\sigma} e_{\lambda_1 \ldots \lambda_{n-2}\lambda\nu} g^{\nu\omega} \nabla_\sigma T_{\omega\varrho},$$

which means we have obtained it without any further integration.

We have seen that the inborn covariant derivative of the fundamental tensor is identically zero. In addition to the fundamental tensor, however, we can independently give a parallel displacement object $\Gamma^\nu_{\mu\lambda}$ and calculate the covariant derivative $\nabla_\mu g_{\lambda\nu}$ in terms of the object $\Gamma^\nu_{\mu\lambda}$. The question is when will this derivative be identically zero? To answer this, let us prove the following theorem.

THEOREM [(1)]. *If the object $\Gamma^\nu_{\mu\lambda}$ is symmetric and if*

$$\nabla_\mu g_{\lambda\nu} = \partial_\mu g_{\lambda\nu} - \Gamma^\sigma_{\mu\lambda} g_{\varrho\nu} - \Gamma^\sigma_{\mu\nu} g_{\varrho\nu} = 0,$$

then

$$\Gamma^\nu_{\mu\lambda} = \begin{Bmatrix} \nu \\ \mu\lambda \end{Bmatrix}.$$

PROOF. We put

$$R_{\mu\lambda}{}^\nu \stackrel{\text{def}}{=} \Gamma^\nu_{\mu\lambda} - \begin{Bmatrix} \nu \\ \mu\lambda \end{Bmatrix}. \tag{7.21}$$

It is easily shown that $R_{\mu\lambda}{}^\nu$ is a tensor, plainly symmetric in the subscripts, since the object $\Gamma^\nu_{\mu\lambda}$ is symmetric by assumption and the Christoffel symbol $\begin{Bmatrix} \nu \\ \mu\lambda \end{Bmatrix}$ is symmetric by definition,

---

[(1)] This theorem is a special case of Schouten's more general theorem according to which an object Γ can be defined uniquely by means of the fundamental tensor $g_{\lambda\mu}$, the antisymmetric part of the connection object $\Gamma^\nu_{[\mu\lambda]}$, and the derivative $\nabla_\mu g_{\lambda\nu}$ (Schouten [67]).

Subtracting the first of the equations

$$\partial_\mu g_{\lambda\nu} - \Gamma^\varrho_{\mu\lambda} g_{\varrho\nu} - \Gamma^\varrho_{\mu\nu} g_{\lambda\varrho} = 0,$$

$$\partial_\mu g_{\lambda\nu} - \left\{ {\varrho \atop \mu\lambda} \right\} g_{\varrho\nu} - \left\{ {\varrho \atop \mu\nu} \right\} g_{\lambda\varrho} = 0$$

from the second and using the notation (7.21), we obtain

$$R_{\mu\lambda}{}^{\varrho} g_{\varrho\nu} + R_{\mu\nu}{}^{\varrho} g_{\lambda\varrho} = 0 \qquad (\lambda, \mu, \nu = 1, 2, \ldots, n). \tag{7.22}$$

We now interchange the indices λ and μ, and make use of the symmetry of the tensor R, to find that

$$R_{\mu\lambda}{}^{\varrho} g_{\varrho\nu} + R_{\lambda\nu}{}^{\varrho} g_{\mu\varrho} = 0. \tag{7.23}$$

If we subtract relation (7.23) from (7.22), we get

$$R_{\nu\mu}{}^{\varrho} g_{\varrho\lambda} - R_{\nu\lambda}{}^{\varrho} g_{\mu\varrho} = 0. \tag{7.24}$$

Cyclical permutation of the indices λ, μ, ν in relations (7.22) yields

$$R_{\nu\mu}{}^{\varrho} g_{\varrho\lambda} + R_{\nu\lambda}{}^{\varrho} g_{\mu\varrho} = 0. \tag{7.25}$$

Next we add relations (7.24) and (7.25) together, and by virtue of the symmetry of g and R we get

$$R_{\mu\nu}{}^{\varrho} g_{\lambda\varrho} = 0 \qquad (\lambda, \mu, \nu = 1, 2, \ldots, n). \tag{7.26}$$

If in these relations the indices μ, ν are regarded as fixed and if λ is allowed to vary from 1 to n, we obtain a system of n homogeneous linear equations in the n unknowns $R_{\mu\nu}{}^{\varrho}$ ($\varrho = 1, 2, \ldots, n$):

$$R_{\mu\nu}{}^{\varrho} g_{\lambda\varrho} = 0 \qquad (\lambda = 1, 2, \ldots, n),$$

whose determinant of coefficients $\mathfrak{g} \neq 0$; the system therefore has only a zero solution, i.e. the only solution is

$$R_{\mu\nu}{}^{\varrho} = 0 \qquad (\varrho = 1, 2, \ldots, n).$$

Moreover, the same reasoning can be applied for every pair of indices (μ, ν) and we have thus proved that

$$R_{\mu\nu}{}^{\varrho} = 0 \qquad (\mu, \nu, \varrho = 1, 2, \ldots, n),$$

which implies the assertion of our theorem.

85. Lengths and Angles under Parallel Displacement. We shall show that under the parallel displacement of vectors by means of Christoffel symbols the lengths of vectors and the angles between them remain unaltered.

Suppose that we are given a field of parallel vectors v^λ (parallel along some curve C), i.e. the field is such that

$$\frac{Dv^\lambda}{d\tau} = \frac{dv^\lambda}{d\tau} + \left\{ {\lambda \atop \mu\nu} \right\} v^\nu \frac{d\xi^\mu}{d\tau} = 0.$$

We set

$$\alpha^2 \overset{\text{def}}{=} |v^\lambda|^2 = g_{\lambda\mu} v^\lambda v^\mu,$$

and our aim is to show that α^2 is constant along C. We have

$$\frac{d\alpha^2}{d\tau} = \frac{D\alpha^2}{d\tau} = \frac{Dg_{\lambda\mu}}{d\tau} v^\lambda v^\mu + g_{\lambda\mu} \frac{Dv^\lambda}{d\tau} v^\mu + g_{\lambda\mu} v^\lambda \frac{Dv^\mu}{d\tau}.$$

But, as demonstrated above, we have

$$\frac{Dg_{\lambda\mu}}{d\tau} = \nabla_\nu g_{\lambda\mu} \frac{d\xi^\nu}{d\tau} = 0,$$

and since $\dfrac{Dv^\lambda}{d\tau} = \dfrac{Dv^\mu}{d\tau} = 0$, it follows that $\dfrac{d\alpha^2}{d\tau} = 0$, which is what was to be proved.

To show that angles are preserved under the parallel displacement of vectors, it suffices, since we have already shown above that the lengths of the vectors are invariant, to show that the scalar product

$$\sigma [\underset{1}{v}, \underset{2}{v}] = g_{\lambda\mu} \underset{1}{v}^\lambda \underset{2}{v}^\mu$$

does not change under parallel displacement. Accordingly we have

$$\frac{d\sigma}{d\tau} = \frac{D\sigma}{d\tau} = \frac{Dg_{\lambda\mu}}{d\tau} \underset{1}{v}^\lambda \underset{2}{v}^\mu + g_{\lambda\mu} \left[\frac{D\underset{1}{v}^\lambda}{d\tau} \underset{2}{v}^\mu + \underset{1}{v}^\lambda \frac{D\underset{2}{v}^\mu}{d\tau} \right] = 0,$$

which shows that σ is indeed invariant and there is no more to prove.

86. Metrization of a Space Equipped with a Linear Connection. The fact that the covariant derivative of a metric tensor vanishes is expressed analytically by the equations

$$\partial_\mu g_{\lambda\mu} = \left\{ {\varrho \atop \lambda\mu} \right\} g_{\varrho\nu} + \left\{ {\varrho \atop \nu\mu} \right\} g_{\lambda\varrho}.$$

Now suppose that we are given a field of symmetric objects of parallel displacement

$$\Gamma^\nu_{\mu\lambda}, \tag{7.27}$$

and that we seek a tensor $g_{\lambda\mu}$ which satisfies the relations

$$\partial_\mu g_{\lambda\nu} = \Gamma^{\varrho}_{\lambda\mu} g_{\varrho\nu} + \Gamma^{\varrho}_{\mu\nu} g_{\lambda\varrho} \qquad (\lambda, \mu, \nu = 1, 2, \ldots, n). \tag{7.28}$$

The system of equations (7.28) always has the trivial solution

$$g_{\lambda\nu} = 0 \qquad (\lambda, \nu = 1, 2, \ldots, n), \tag{7.29}$$

but we are interested only in solutions for which

$$\mathfrak{g} = |g_{\lambda\nu}| \neq 0, \tag{7.30}$$

and such solutions need not always exist. This was first pointed out by Veblen. Indeed, counter-example can be given even for $n = 2$. Suppose that

$$\begin{aligned} \Gamma^1_{11} &\overset{*}{=} \Gamma^1_{22} \overset{*}{=} \Gamma^2_{11} \overset{*}{=} \Gamma^2_{22} \overset{*}{=} 0, \\ \Gamma^1_{12} &\overset{*}{=} \Gamma^1_{21} \overset{*}{=} \Gamma^2_{12} \overset{*}{=} \Gamma^2_{21} \overset{*}{=} 1. \end{aligned} \tag{7.31}$$

We shall show that the system of equations (7.28) only has solutions for which $\mathfrak{g} = 0$. To this end, we note that a necessary condition for the system of equations (7.28) to be soluble is easily obtained if we differentiate both sides with respect to ξ^σ, eliminate the partial derivatives $\partial_\sigma g_{\varrho\nu}$ and $\partial_\sigma g_{\lambda\varrho}$ on the right-hand sides, and make use of the fact that we must have

$$\partial_\sigma \partial_\mu g_{\lambda\nu} = \partial_\mu \partial_\sigma g_{\lambda\nu}.$$

Fairly simple manipulations lead to the equations

$$R_{\sigma\mu\lambda}{}^{\varrho} g_{\varrho\nu} + R_{\sigma\mu\nu}{}^{\varrho} g_{\lambda\varrho} = 0 \qquad (\sigma, \mu, \lambda, \nu = 1, 2, \ldots, n), \tag{7.32}$$

which is a system of homogeneous linear equations in the unknowns $g_{\lambda\nu}$. These (independent) equations number $\binom{n}{2}\binom{n+1}{2}$, which in general exceeds the number of unknowns $g_{\lambda\nu}$, which is $\binom{n+1}{2}$.

For $n = 2$ we have three unknowns g_{11}, g_{12}, g_{22} and three equations.

For the object Γ, defined by equations (7.31), simple calculation shows that the coordinates of the curvature tensor have the values

$$R_{121}{}^1 \overset{*}{=} 1, \quad R_{121}{}^2 \overset{*}{=} 1, \quad R_{122}{}^1 \overset{*}{=} -1, \quad R_{122}{}^2 = -1. \tag{7.33}$$

The system of equations (7.32) reduces for $n = 2$ to three equations

$$\begin{aligned} R_{121}{}^1 g_{11} + \qquad & R_{121}{}^2 g_{12} && = 0, \\ & R_{122}{}^1 g_{12} + R_{122}{}^2 g_{22} && = 0, \\ R_{122}{}^1 g_{11} + (R_{121}{}^1 &+ R_{122}{}^2) g_{12} + R_{121}{}^2 g_{22} && = 0. \end{aligned} \tag{7.34}$$

The determinant of the coefficients is easily shown to be given by

$$W = (R_{121}{}^{1}+R_{122}{}^{2})(R_{122}{}^{1}R_{121}{}^{2}-R_{121}{}^{1}R_{122}{}^{2}),$$

which in our case gives

$$W = 0,$$

using equation (7.33).

Accordingly, the system of equations (7.34) only has solutions for which $\mathfrak{g} = \det g_{\lambda\mu} = 0$.

Now let q denote the maximum rank of the matrix $[g_{\lambda\mu}]$ for all possible solutions $g_{\lambda\mu}$ of the system (7.28) (in our example 7.31 we have $q = 1$). The number q satisfies the inequality $0 \leqslant q \leqslant n$.

Obviously, q is invariant under transformations of coordinate system, and the number q could appropriately be called the *rank of the object Γ*. If the $\Gamma^{\lambda}_{\mu\nu}$'s are the Christoffel symbols, then of course the rank $q = n$.

If the $g_{\lambda\mu}$'s constitute the solution of equations (7.28), then the values

$$\bar{g}_{\lambda\mu} = Cg_{\lambda\mu}, \tag{7.35}$$

where C is an arbitrary constant, also constitute a solution.

If $g_{\lambda\mu}$ and $\bar{g}_{\lambda\mu}$ are two non-trivial proportional solutions, i.e. if

$$\bar{g}_{\lambda\mu} = \tau g_{\lambda\mu}, \tag{7.36}$$

then the coefficient τ must be constant, i.e. it cannot depend on ξ^{ν}. Indeed, relation (7.36) implies

$$\partial_{\nu}\bar{g}_{\lambda\mu} = \tau\partial_{\nu}g_{\lambda\mu}+g_{\lambda\mu}\partial_{\nu}\tau,$$

which, together with the relations

$$\partial_{\nu}g_{\lambda\mu} = \Gamma^{\varrho}_{\nu\lambda}g_{\varrho\mu}+\Gamma^{\varrho}_{\nu\mu}g_{\lambda\varrho},$$

$$\partial_{\nu}\bar{g}_{\lambda\mu} = \Gamma^{\varrho}_{\nu\lambda}\bar{g}_{\varrho\mu}+\Gamma^{\varrho}_{\nu\mu}\bar{g}_{\lambda\varrho}$$

leads to the equation

$$g_{\lambda\mu}\partial_{\nu}\tau = 0,$$

that is,

$$\partial_{\nu}\tau = 0,$$

and so to the conclusion that

$$\tau = \text{const.}$$

Two solutions $g_{\lambda\mu}$ and $\bar{g}_{\lambda\mu}$ which satisfy relation (7.35) will be called *inessentially different solutions*. Thus, for $n = 2$ system (7.28) may have only one essentially different solution in the case that $R_{\mu\lambda\nu}{}^{\varrho} \neq 0$.

We conclude from this that for $n = 2$, system (7.34) is of the form

$$\begin{aligned} ag_{11}+ \quad bg_{12} \quad &= 0, \\ cg_{11}+(a+d)g_{12}+bg_{22} &= 0, \\ cg_{12}+dg_{22} &= 0, \end{aligned} \tag{7.37}$$

where a, b, c, d are the short-hand notations

$$a = R_{121}{}^{1}, \quad b = R_{121}{}^{2}, \quad c = R_{122}{}^{1}, \quad d = R_{122}{}^{2}. \tag{7.38}$$

It can be demonstrated that the matrix of coefficients for equations (7.37)

$$\begin{bmatrix} a & b & 0 \\ c & a+d & b \\ 0 & c & d \end{bmatrix} \tag{7.39}$$

can never be of rank one. If the matrix (7.39) is of rank three the system has only the trivial solution (7.29). If the matrix is of rank two, then all solutions of the system (7.37) are given by the formulae

$$g_{11} = bdp, \quad g_{12} = -adp, \quad g_{22} = acp,$$

where p is an arbitrary parameter; all solutions are thus proportional to each other, i.e. there is only one essential solution. When the rank of the matrix (7.39) is zero, then $R_{\lambda\mu\nu}{}^{\varrho} = 0$. In this case we know (cf. p. 217) that there exists a coordinate system (λ) in which $\Gamma^{\nu}_{\mu\lambda} \overset{*}{=} 0$, and the system of equations (7.28) then takes the form

$$\partial_{\nu} g_{\lambda\mu} \overset{*}{=} 0,$$

which yields

$$g_{\lambda\mu} \overset{*}{=} \underset{\lambda\mu}{C},$$

where the $\underset{\lambda\mu}{C}$'s are constant. As is easily demonstrated, there then exist just three essentially different tensors since any other tensor is a linear combination of these three. The problem of classifying objects Γ of parallel displacement by means of the rank q has been solved here for $n = 2$ and symmetric objects. This problem can, however, be posed in general for any n. It has been solved by A. Jakubowicz [129] in the case of symmetric objects. The necessary and sufficient conditions for q to be equal to n were somewhat earlier by V. Hlavatý [125]. The problem is quite difficult and with the methods presently available calls for certain algebraic comitants of the curvature tensor $R_{\omega\mu\lambda}{}^{\nu}$ to be introduced.

CHAPTER VIII

SOME SPECIAL SPACES

87. Affine-Euclidean Spaces. Teleparallelism. We know that in general (provided the curvature tensor R is not zero) the parallel displacement of a vector along a closed curve leads to another vector. On the other hand, if

$$R_{\omega\mu\nu}{}^{\lambda} = 0, \tag{8.1}$$

then after a cycle we always return to the initial vector; this implies that the parallel displacement of any vector v from a point p to another point q along an arbitrary curve C yields at q a vector which is independent of the curve C.

A space with affine connection, for which the curvature tensor $R_{\omega\lambda\mu}{}^{\nu}$ is identically zero, is called an *affine-Euclidean space*.

If we make the further assumption that the object of linear connection is symmetric, then the condition

$$R_{\omega\lambda\mu}{}^{\nu} = 0$$

is equivalent to the existence of a coordinate system $(\xi^{\lambda'})$, which can be called an *affine system*, for which

$$\Gamma^{\nu'}_{\lambda'\mu'} \stackrel{*}{=} 0. \tag{8.2}$$

Affine-Euclidean spaces, for which we introduce no special notation, should not be confused with metric-Euclidean spaces denoted by R_n for which there exist certain preferred coordinate systems in which

$$g_{\lambda'\mu'} \stackrel{*}{=} \text{const},$$

and which, of course, are automatically affine-Euclidean spaces since for them condition (8.1) is satisfied identically. The converse is not true, however; not every affine-Euclidean space is metric-Euclidean since it does not necessarily have a field of metric tensors.

Assuming now that relation (8.1) is satisfied, let us take an arbitrary point $\underset{0}{p}$ in space. At this point, we take n linearly independent contravariant vectors

$$\underset{J}{\overset{0}{u}} \quad (J = \mathrm{I}, \mathrm{II}, \ldots, N)\ ^{(1)}. \tag{8.3}$$

Next, we consider n vector fields

$$\underset{J}{u} \quad (J = \mathrm{I}, \mathrm{II}, \ldots, N) \tag{8.4}$$

generated by successively moving (along which curve is immaterial) the vectors of the sequence (8.3) to any point p of our space. The system so obtained will be holonomic.

Indeed, let (λ) be the coordinate system in which

$$\Gamma^{\nu}_{\lambda\mu} \overset{*}{=} 0 \quad (\lambda, \mu, \nu = 1, 2, \ldots, n),$$

and such that the origin of the system lies at the point $\underset{0}{p}$. In this case, we obtain $\underset{J}{u^{\nu}}$ by integrating the system

$$\nabla_{\mu}\underset{J}{u^{\nu}} = 0$$

with the initial conditions

$$(\underset{J}{u^{\nu}})_{\xi^{\nu}=0} = \underset{J}{\overset{0}{u^{\nu}}} \quad (\nu = 1, 2, \ldots, n; J = \mathrm{I}, \mathrm{II}, \ldots, N).$$

But

$$\nabla_{\mu}\underset{J}{u^{\nu}} = \partial_{\mu}\underset{J}{u^{\nu}} + \Gamma^{\nu}_{\mu\lambda}\underset{J}{u^{\lambda}} = \partial_{\mu}\underset{J}{u^{\nu}}$$

and we thus obtain

$$\frac{\partial \underset{J}{u^{\nu}}}{\partial \xi^{\mu}} = 0 \quad (\mu, \nu = 1, 2, \ldots, n; J = \mathrm{I}, \mathrm{II}, \ldots, N),$$

which yields

$$\underset{J}{u^{\nu}} \overset{*}{=} \underset{J}{\overset{0}{u^{\nu}}}.$$

Hence, in the system (λ), the coordinates of the vectors of the field (8.4) are constant. The system (K) is accordingly holonomic since

$$\Omega_{IJ}{}^{K} = A^{\lambda\mu}_{IJ}\partial_{[\lambda}A^{K}_{\mu]} = 0,$$

because the A^{K}_{μ}'s are also constant. However, it is not until we restrict

(1) Of course, $N = n$.

the group G_1 to the group G_a, that we can say that our space A_n with condition (8.1) becomes the space E_n.

Assume now that in the holonomic coordinate system (ξ^ν) we are given at every point of space (or of a certain region) a system of n independent contravariant vectors

$$\underset{K}{u}^{\lambda} \tag{8.5}$$

and let $\overset{K}{u}_\nu$ denote the inverse system of covariant vectors. Let us set

$$\Lambda^{\nu}_{\lambda\mu} \overset{\text{def}}{=} \underset{K}{u}^{\nu}\partial_\lambda \overset{K}{u}_\mu \tag{$*$}$$

and examine the transformation law for the coordinates of the object Λ under transition to any other system (λ'). Thus, by definition we have

$$\Lambda^{\nu'}_{\lambda'\mu'} = \underset{K}{u}^{\nu'}\partial_{\lambda'}\overset{K}{u}_{\mu'}.$$

But

$$\underset{K}{u}^{\nu'} = \underset{K}{u}^{\nu}A^{\nu'}_{\nu}, \qquad \overset{K}{u}_{\mu'} = \overset{K}{u}_{\mu}A^{\mu}_{\mu'}$$

and consequently

$$\partial_{\lambda'}\overset{K}{u}_{\mu'} = A^{\lambda}_{\lambda'}\partial_\lambda \overset{K}{u}_{\mu'} = A^{\lambda}_{\lambda'}\partial_\lambda(\overset{K}{u}_\mu A^{\mu}_{\mu'}) = A^{\lambda}_{\lambda'}A^{\mu}_{\mu'}\partial_\lambda \overset{K}{u}_\mu + \overset{K}{u}_\mu A^{\lambda}_{\lambda'}\partial_\lambda A^{\mu}_{\mu'}.$$

From this we have further

$$\Lambda^{\nu'}_{\lambda'\mu'} = \underset{K}{u}^{\nu}A^{\nu'}_{\nu}(A^{\lambda\,\mu}_{\lambda'\mu'}\partial_\lambda \overset{K}{u}_\mu + \overset{K}{u}_\mu \partial_{\lambda'}A^{\mu}_{\mu'}) = A^{\lambda\,\mu\,\nu'}_{\lambda'\mu'\nu}\underset{K}{u}^{\nu}\partial_\lambda \overset{K}{u}_\mu + \underset{K}{u}^{\nu}\overset{K}{u}_\mu A^{\nu'}_{\nu}\partial_{\lambda'}A^{\mu}_{\mu'},$$

which, because of the relation

$$\underset{K}{u}^{\nu}\overset{K}{u}_\mu = \delta^{\nu}_{\mu}$$

finally leads to the formula

$$\Lambda^{\nu'}_{\lambda'\mu'} = \Lambda^{\nu}_{\lambda\mu}A^{\nu'\lambda\,\mu}_{\nu\,\lambda'\mu'} + A^{\nu'}_{\mu}\partial_{\lambda'}A^{\mu}_{\mu'},$$

which we recognize as the very familiar transformation law for the coordinates of an object of parallel displacement [(1)].

The system of vectors (8.5) thus defines a particular object of parallel displacement which, following Einstein, we call *teleparallelism* or *distant parallelism* for a reason which we shall shortly make clear.

Take any arbitrary vector v^ν attached at some point p of the region

---

(1) This fact was first recognized by Weitzenböck [88].

under consideration and displace it parallelly along some curve C to a point q, using the object Λ.

We can then write

$$v^\nu = \overset{K}{\alpha} \underset{K}{u}^\nu,$$

where $\overset{K}{\alpha}$ are certain scalars. We now compute the covariant derivative $\nabla_\lambda \underset{K}{u}^\nu$, defined in terms of the object Λ.

We have

$$\nabla_\lambda \underset{K}{u}^\nu = \partial_\lambda \underset{K}{u}^\nu + \Lambda^\nu_{\lambda\mu} \underset{K}{u}^\mu = \partial_\lambda \underset{K}{u}^\nu + \underset{I}{u}^\nu (\partial_\lambda \overset{I}{u}_\mu) \underset{K}{u}^\mu,$$

but since

$$0 = \partial_\lambda \delta^I_K = \partial_\lambda (\overset{I}{u}_\mu \underset{K}{u}^\mu) = (\partial_\lambda \overset{I}{u}_\mu) \underset{K}{u}^\mu + \overset{I}{u}_\mu \partial_\lambda \underset{K}{u}^\mu,$$

it follows that

$$\nabla_\lambda \underset{K}{u}^\nu = \partial_\lambda \underset{K}{u}^\nu - \underset{I}{u}^\nu \overset{I}{u}_\mu \partial_\lambda \underset{K}{u}^\mu = \partial_\lambda \underset{K}{u}^\nu - \delta^\nu_\mu \partial_\lambda \underset{K}{u}^\mu = \partial_\lambda \underset{K}{u}^\nu - \partial_\lambda \underset{K}{u}^\nu = 0.$$

In other words, each vector remains "itself" after parallel displacement. It therefore follows that system (8.5) is a geodesic system moving along the curve C and furthermore, that the coefficients $\overset{K}{\alpha}$ do not change under parallel displacement, i.e. that they are constant. The invariance of these coefficients further implies a fact of great significance, viz. that the parallel displacement defined by the object Λ does not depend on the particular path chosen: hence the name teleparallelism [1].

For the symmetric object Γ we saw that the parallel displacement being independent of the path was equivalent to the curvature tensor vanishing identically.

In our case, the object Λ is not, in general, symmetric, since the condition for symmetry

$$\Lambda^\nu_{[\lambda\mu]} = 0$$

would lead to the conclusion

$$\underset{K}{u}^\nu \partial_{[\lambda} \overset{K}{u}_{\mu]} = 0,$$

[1] This term was introduced by Einstein (1928), although the concept had been previously discussed by G. Vitali in 1923.

which, because the vectors of system (8.5) are linearly independent, is equivalent to the relationships

$$\partial_{[\lambda} \overset{K}{u}_{\mu]} = 0,$$

i.e. to the holonomicity of system (8.5). If, on the other hand, system (8.5) is non-holonomic, then it is easy to see that the displacement object Λ will possess torsion.

However, let us calculate the curvature tensor associated with Λ. We have

$$R_{\omega\mu\lambda}{}^{\nu} = 2\partial_{[\omega} \Lambda^{\nu}_{\mu]\lambda} + 2\Lambda^{\nu}_{[\omega|\varrho|} \Lambda^{\varrho}_{\mu]\lambda} = 2\partial_{[\omega}(\underset{K}{u}^{\nu} \partial_{\mu]} \overset{K}{u}_{\lambda}) + 2\underset{K}{u}^{\nu}(\partial_{[\omega} \overset{K}{u}_{|\varrho|}) \underset{I}{u}^{\varrho} \partial_{\mu]} \overset{I}{u}_{\lambda}$$

$$= 2\underset{K}{u}^{\nu} \partial_{[\omega\mu]} \overset{K}{u}_{\lambda} + 2(\partial_{[\omega} \underset{K}{u}^{\nu}) \partial_{\mu[} \overset{K}{u}_{\lambda} + 2\underset{K}{u}^{\nu} \underset{I}{u}^{\varrho} (\partial_{[\omega} \overset{K}{u}_{|\varrho|}) \partial_{\mu]} \overset{I}{u}_{\lambda}.$$

But

$$\partial_{[\omega\mu]} \overset{K}{u}_{\lambda} = 0$$

and

$$\underset{K}{u}^{\nu} \partial_{\omega} \overset{K}{u}_{\varrho} = \partial_{\omega}(\underset{K}{u}^{\nu} \overset{K}{u}_{\varrho}) - \overset{K}{u}_{\varrho} \partial_{\omega} \underset{K}{u}^{\nu} = -\overset{K}{u}_{\varrho} \partial_{\omega} \underset{K}{u}^{\nu},$$

from which it follows that

$$R_{\omega\mu\lambda}{}^{\nu} = 2(\partial_{[\omega} \underset{K}{u}^{\nu}) \partial_{\mu]} \overset{K}{u}_{\lambda} - 2\overset{K}{u}_{\varrho} \underset{I}{u}^{\varrho} (\partial_{[\omega} \underset{K}{u}^{\nu}) \partial_{\mu]} \overset{I}{u}_{\lambda}$$

$$= 2(\partial_{[\omega} \underset{K}{u}^{\nu}) \partial_{\mu]} \overset{K}{u}_{\lambda} - 2(\partial_{[\omega} \underset{K}{u}^{\nu}) \partial_{\mu]} \overset{K}{u}_{\lambda} = 0.$$

It is thus see that the curvature tensor of teleparallelism is identically equal to zero, just as it is for Euclidean space with symmetric displacement without torsion.

88. Metric-Euclidean Spaces. Let us assume that the Riemannian space V_n has a curvature tensor identically equal to zero:

$$R_{\omega\mu\lambda}{}^{\nu} = 0.$$

Then we know that there exists a coordinate system (λ) in which

$$\Gamma^{\nu}_{\mu\lambda} \overset{*}{=} 0.$$

Consequently we have

$$\nabla_{\mu} g_{\lambda\nu} \overset{*}{=} \partial_{\mu} g_{\lambda\nu} \overset{*}{=} 0,$$

and thus it is sufficient to set

$$g_{\lambda\nu} \stackrel{*}{=} \underset{\lambda\nu}{C}.$$

If in particular we take

$$g_{\lambda\nu} \stackrel{*}{=} \delta_{\lambda\nu}$$

we obtain an orthogonal Cartesian system and the basis vectors are of length 1. The Riemannian space is then called a *metric-Euclidean space* and is denoted by R_n. If, in it, we restrict the pseudogroup G_1 to the orthogonal subgroup G_m, we get an ordinary n-dimensional Euclidean Cartesian space out of the space R_n. Rectangular Cartesian systems, with equal units on all axes, can be characterized as systems in which the distance between two points $p(\underset{1}{\xi^\nu})$ and $q(\underset{2}{\xi^\nu})$ is given by the formula

$$\varrho(p, q) = \Big[\sum_{\nu=1}^{n} (\underset{2}{\xi^\nu} - \underset{1}{\xi^\nu})^2\Big]^{1/2}.$$

89. Ricci Coefficients of Rotation. Ricci introduced (1895) *coefficients of rotation* when in an n-dimensional space R_n he considered n mutually orthogonal curved congruences. This concept was later generalized by H. Levy (1925, 1927).

Suppose that at each point of a domain X_n we are given a system of n independent contravariant vectors

$$\underset{K}{u^\lambda} \quad (\lambda = 1, \ldots, n;\ K = \mathrm{I}, \mathrm{II}, \ldots, N). \tag{8.6}$$

We assume that the $\underset{K}{u}$ are fields of class C^1 and accordingly we can write down their partial derivatives

$$\partial_\mu \underset{K}{u^\lambda}. \tag{8.7}$$

As we know, these derivatives do not represent any geometric object.

Since the system (8.6) is by assumption a system of independent vectors, there exists a uniquely defined inverse system of covariant vectors

$$\overset{K}{\overline{u}}_\lambda. \tag{8.8}$$

We now define the quantities

$$\tilde{\gamma}^K_{LM} \stackrel{\text{def}}{=} \underset{L}{u^\alpha} \underset{M}{u^\beta} \partial_\beta \overset{K}{\overline{u}}_\alpha. \tag{8.9}$$

It is difficult to tell *a priori* whether the $\tilde{\gamma}^K_{LM}$ are geometric objects. Let us see what the transformation law is for the quantities $\tilde{\gamma}^{K}_{LM}$ so defined. We set

$$\tilde{\gamma}'^{K}_{LM} \overset{\text{def}}{=} \underset{L}{u}^{\alpha'} \underset{M}{u}^{\beta'} \partial_{\beta'} \overset{K}{\bar{u}}_{\alpha'} \tag{8.10}$$

and transform the right-hand side. Accordingly, we have

$$\underset{L}{u}^{\alpha'} = \underset{L}{u}^{\alpha} A^{\alpha'}_{\alpha}, \qquad \overset{K}{\bar{u}}_{\alpha'} = \overset{K}{\bar{u}}_{\alpha} A^{\alpha}_{\alpha'},$$

$$\partial_{\beta'} \overset{K}{\bar{u}}_{\alpha'} = A^{\beta}_{\beta'} \partial_{\beta}(\overset{K}{\bar{u}}_{\alpha} A^{\alpha}_{\alpha'}) = A^{\beta}_{\beta'} A^{\alpha}_{\alpha'} \partial_{\beta} \overset{K}{\bar{u}}_{\alpha} + \overset{K}{\bar{u}}_{\alpha} \partial_{\beta'} A^{\alpha}_{\alpha'},$$

from which we get

$$\begin{aligned} \underset{L}{u}^{\alpha'} \underset{M}{u}^{\beta'} \partial_{\beta'} \overset{K}{\bar{u}}_{\alpha'} &= \underset{L}{u}^{\varrho} A^{\alpha'}_{\varrho} \underset{M}{u}^{\sigma} A^{\beta'}_{\sigma} (A^{\beta}_{\beta'} A^{\alpha}_{\alpha'} \partial_{\beta} \overset{K}{\bar{u}}_{\alpha} + \overset{K}{\bar{u}}_{\alpha} \partial_{\beta'} A^{\alpha}_{\alpha'}) \\ &= \underset{L}{u}^{\alpha} \underset{M}{u}^{\beta} \partial_{\beta} \overset{K}{\bar{u}}_{\alpha} + \overset{K}{\bar{u}}_{\alpha} \underset{L}{u}^{\varrho} \underset{M}{u}^{\sigma} A^{\alpha'}_{\varrho} A^{\beta'}_{\sigma} \partial_{\beta'} A^{\alpha}_{\alpha'} \\ &= \tilde{\gamma}^{K}_{LM} - \overset{K}{\bar{u}}_{\alpha} \underset{L}{u}^{\varrho} \underset{M}{u}^{\sigma} A^{\alpha}_{\alpha'} \partial_{\sigma} A^{\alpha'}_{\varrho}. \end{aligned} \tag{8.11}$$

The second term on the right-hand side presents the $\tilde{\gamma}^{K}_{LM}$'s being invariant under transformations of coordinate systems unless the group G_1 is restricted to the affine group for which $\partial_{\sigma} A^{\alpha'}_{\varrho} = 0$.

There are two possible ways of eliminating this perturbing term: we can either effect antisymmetrization in the original definition of the symbol $\tilde{\gamma}^{K}_{LM}$ by adopting the definition

$$\tilde{\gamma}^{K}_{LM} \overset{\text{def}}{=} \underset{[L}{u}^{\alpha} \underset{M]}{u}^{\beta} \partial_{\beta} \overset{K}{\bar{u}}_{\alpha}, \tag{8.12}$$

while not equipping the space with an object of parallel displacement, or we can furnish the space with such an object and replace the symbol for the ordinary derivative by the covariant derivative

$$\gamma^{K}_{LM} \overset{\text{def}}{=} \underset{L}{u}^{\alpha} \underset{M}{u}^{\beta} \nabla_{\beta} \overset{K}{\bar{u}}_{\alpha}. \tag{8.13}$$

In this second case we immediately find that the γ^{K}_{LM}'s are scalars. The scalars γ^{K}_{LM}, of course, depend not only on the system of fields $\underset{K}{u}^{\lambda}$, but also on the object $\Gamma^{\nu}_{\lambda\mu}$ of parallel displacement. Let us see now what the transformation law for $\tilde{\gamma}^{K}_{LM}$ is.

Using the formulae above, we can write

$$\tilde{\gamma}'^{K}_{LM} = \underset{[L}{u}^{\alpha}\underset{M]}{u}^{\beta}\partial_{\beta}\overset{K}{\bar{u}}_{\alpha} - A^{\alpha}_{\alpha'}\overset{K}{\bar{u}}_{\alpha}\underset{[L}{u}^{\varrho}\underset{M]}{u}^{\sigma}\partial_{\sigma}A^{\alpha'}_{\varrho}.$$

But we have

$$\underset{[L}{u}^{\varrho}\underset{M]}{u}^{\sigma}\partial_{\sigma}A^{\alpha'}_{\varrho} = \tfrac{1}{2}[\underset{L}{u}^{\varrho}\underset{M}{u}^{\sigma}\partial_{\sigma}A^{\alpha'}_{\varrho} - \underset{M}{u}^{\varrho}\underset{L}{u}^{\sigma}\partial_{\sigma}A^{\lambda'}_{\varrho}] = \tfrac{1}{2}[\underset{L}{u}^{\varrho}\underset{M}{u}^{\sigma}\partial_{\sigma}A^{\alpha'}_{\varrho} - \underset{M}{u}^{\sigma}\underset{L}{u}^{\varrho}\partial_{\varrho}A^{\alpha'}_{\mu}]$$

$$= \tfrac{1}{2}\underset{L}{u}^{\varrho}\underset{M}{u}^{\sigma}(\partial_{\sigma}A^{\alpha'}_{\varrho} - \partial_{\varrho}A^{\alpha'}_{\sigma}) = 0,$$

where $\partial_{\sigma}A^{\alpha'}_{\varrho} = \partial_{\partial}A^{\alpha'}_{\sigma}$, or finally,

$$\tilde{\gamma}'^{K}_{LM} = \tilde{\gamma}^{K}_{LM},$$

and so the $\tilde{\gamma}^{K}_{LM}$ are also scalars.

In the case when the space does have a metric tensor $g_{\lambda\mu}$, we may then take

$$\overset{*}{\gamma}_{KLM} \overset{\text{def}}{=} \underset{L}{u}^{\alpha}\underset{M}{u}^{\beta}\nabla_{\beta}\underset{K}{u}_{\alpha}, \tag{8.14}$$

because the covariant derivative is taken according to the Christoffel symbols $\left\{ {\nu \atop \lambda\mu} \right\}$ and where

$$\underset{K}{u}_{\alpha} \overset{\text{def}}{=} \underset{K}{u}^{\beta}g_{\alpha\beta}.$$

We then get for $\overset{*}{\gamma}_{KLM}$ the original *Ricci rotation coefficients*, which can then be shown to be skew-symmetric in the first pair of indices, i.e.

$$\overset{*}{\gamma}_{KLM} = -\overset{*}{\gamma}_{LKM}. \tag{8.15}$$

We now calculate the γ^{K}_{LM}'s for the special case when the covariant derivative ∇_{β} is taken with respect to the object $\Lambda^{\nu}_{\lambda\mu}$ which is uniquely defined by a system of independent vectors $\underset{K}{u}^{\lambda}$ and which is given by formula (*) of p. 252

$$\Lambda^{\nu}_{\lambda\mu} = \underset{K}{u}^{\nu}\partial_{\lambda}\overset{K}{\bar{u}}_{\mu}. \tag{8.16}$$

In this case we have

$$\nabla_{\beta}\overset{K}{\bar{u}}_{\alpha} = \partial_{\beta}\overset{K}{\bar{u}}_{\alpha} - \Lambda^{\varrho}_{\beta\alpha}\overset{K}{\bar{u}}_{\varrho},$$

and when these relations are inserted into equation (8.13), we obtain

$$\gamma^K_{LM} = \underset{L}{u^\alpha}\underset{M}{u^\beta}\,\partial_\beta \overset{K}{\bar{u}}_\alpha - \Lambda^\varrho_{\beta\alpha}\underset{L}{u^\alpha}\underset{M}{u^\beta}\overset{K}{\bar{u}}_\varrho = \underset{L}{u^\alpha}\underset{M}{u^\beta}\,\partial_\beta \overset{K}{\bar{u}}_\alpha - \underset{L}{u^\alpha}\underset{M}{u^\beta}\overset{K}{\bar{u}}_\varrho \underset{P}{u^\varrho}\,\partial_\beta \overset{P}{\bar{u}}_\alpha$$

$$= \underset{L}{u^\alpha}\underset{M}{u^\beta}\,\partial_\beta \overset{K}{\bar{u}}_\alpha - \underset{L}{u^\alpha}\underset{M}{u^\beta}\,\partial_\beta \overset{K}{\bar{u}}_\alpha = 0,$$

which means that in this special case, all the γ^K_{LM}'s are zero.

Makai [137] has shown that each first-order differential concomitant of a system of independent vectors $\underset{K}{u^\lambda}$ is a function of the coefficients $\tilde{\gamma}^K_{LM}$.

To conclude this section, let us explain why the quantities γ^K_{LM} have come to be called rotation coefficients. Assume that the γ^K_{LM}'s are defined in terms of the covariant derivative ∇_β originating from the Christoffel symbols associated with the metric tensor $g_{\lambda\mu}$. Assume further that each base $\underset{K}{u^\lambda}$ is orthonormal. Suppose that at the point ξ^ν we are given an "infinitesimal" vector with coordinates $d\xi^\nu$. We carry out the parallel displacement of the vector $\underset{K}{u^\lambda}$ from the point ξ^ν to the point $\xi^\nu + d\xi^\nu$, and denote the result of this parallel displacement by $\underset{K}{\bar{u}^\lambda}$.

The "infinitesimal" rotation which vector $\underset{K}{\bar{u}^\lambda}$ must execute in order to coincide with the vector $\underset{K}{u^\lambda}(\xi^\nu + d\xi^\nu)$ is given by means of the bivector dF_{KL} defined by the relationships

$$dF_{KL} = \overset{*}{\gamma}_{LKM}\, d\xi^M. \tag{8.17}$$

It is now clear why the rotation coefficients for teleparallelism vanish identically.

90. Weyl Spaces. The space L_n is called a *semi-metric space* if there exists a tensor $g_{\lambda\mu}$ whose matrix of coordinates is of order n, and a vector (vector field) Q_λ with the property that

$$\nabla_\mu g_{\lambda\nu} = -Q_\mu g_{\lambda\nu}. \tag{8.18}$$

If in the special case the object $\Gamma^\nu_{\mu\lambda}$ is symmetric ($L_n = A_n$) and if relationships (8.18) are satisfied, the space is called a *Weyl space* [153]; we denote it by W_n.

Neither the tensor $g_{\lambda\mu}$ nor the vector Q_μ is defined uniquely for a given object Γ. Indeed, suppose that there exists a second pair of objects ($'Q_\mu$,

$'g_{\lambda\nu})$ with the same property, i.e.

$$\nabla_\mu 'g_{\lambda\nu} = -'Q_\mu 'g_{\lambda\nu}.$$

It can then be shown (Schouten [67]) by writing down the integrability conditions for system (8.18), which are of the form

$$R_{\omega\mu(\lambda}{}^{\varrho} g_{\nu)\varrho} = \nabla_{[\omega} Q_{\mu]} g_{\lambda\nu},$$

that we must then have

$$'g_{\lambda\nu} = \sigma g_{\lambda\nu},$$

where σ is a scalar field. It further follows from this that

$$\nabla_\mu 'g_{\lambda\nu} = \nabla_\mu(\sigma g_{\lambda\nu}) = (\partial_\mu \sigma) g_{\lambda\nu} + \sigma \nabla_\mu g_{\lambda\nu}$$

$$= g_{\lambda\nu}\partial_\mu \sigma - \sigma Q_\mu g_{\lambda\nu} = g_{\lambda\nu}(\partial_\mu \sigma - \sigma Q_\mu) = 'g_{\lambda\nu}\left(\frac{1}{\sigma}\partial_\mu \sigma - Q_\mu\right),$$

i.e. that

$$-'Q_\mu = -Q_\mu + \partial_\mu \ln|\sigma|;$$

thus if a pair of objects $(Q_\mu, g_{\lambda\nu})$ satisfies equation (8.18), then for every other pair satisfying that equation we have

$$'Q_\mu = Q_\mu - \partial_\mu \ln|\sigma|, \quad 'g_{\lambda\nu} = \sigma g_{\lambda\nu} \quad (\sigma \neq 0). \tag{8.19}$$

Since the factor σ is arbitrary, it follows that although a distance metric cannot be defined uniquely in a Weyl space, it is possible to define an angular metric, since

$$\frac{g_{\lambda\nu}u^\lambda v^\nu}{[('g_{\alpha\beta}u^\alpha u^\beta)('g_{\varrho\tau}v^\varrho v^\tau)]^{1/2}} = \frac{g_{\lambda\nu}u^\lambda v^\nu}{[(g_{\alpha\beta}u^\alpha u^\beta)(g_{\varrho\tau}v^\varrho v^\tau)]^{1/2}} \quad \text{for } \sigma > 0.$$

It can also be shown that under the parallel displacement of vectors along a curve, the angles and the ratios of lengths of vectors are preserved.

We now consider when a space W_n will be a Riemannian space, i.e. when, if the field Q_μ is given *a priori*, will there exist a scalar field σ such that $\nabla_\mu 'g_{\lambda\nu} = 0$, i.e. such that $'Q_\mu = 0$.

For this to be the case it is necessary and sufficient that

$$\partial_\mu \ln|\sigma| = Q_\mu,$$

that is, Q_μ must be a potential field. If Q_μ is a gradient, then there exists (up to a constant factor) a unique potential $\log|\underset{0}{\sigma}|$ with the property such that if we set

$$'g_{\lambda\nu} = \underset{0}{\sigma} g_{\lambda\mu}$$

we have a Riemannian space for $'g$.

It can be proved that a Weyl space is Riemannian if and only if $V_{\omega\mu} = 0$, i.e. if and only if the object preserves volume.

The formula

$$\Gamma_\mu = \partial_\mu \ln|\mathfrak{g}|^{1/2} + \tfrac{1}{2}nQ_\mu, \qquad \mathfrak{g} = |g_{\lambda\nu}| \tag{8.20}$$

can also be established for a Weyl space.

91. Einstein Spaces. We have seen that in Riemannian space the Ricci tensor $R_{\lambda\mu}$ is symmetric as well as the metric tensor $g_{\lambda\mu}$.

A relation

$$R_{\mu\lambda} = \tau g_{\lambda\mu} \tag{8.21}$$

is invariant and τ is then a scalar field.

We shall show that relation (8.21) is always valid for $n = 2$. In other words, the matrix

$$[g_{\lambda\mu}, R_{\lambda\mu}]$$

is always of rank one. The direct proof of relation (8.21) is by no means simple, and we shall in fact prove this relation by taking a somewhat artificial, indirect route. In Section 27 we considered the four-index tensor

$$g_{\lambda\mu\varrho\sigma} \stackrel{\text{def}}{=} g_{[\lambda|\mu|}g_{\varrho]\sigma} \tag{8.22}$$

as an algebraic concomitant of the tensor $g_{\lambda\mu}$.

We now define the isomer of this tensor

$$g_{\lambda\mu\varrho\sigma} \stackrel{\text{def}}{=} g_{\lambda\varrho\mu\sigma}. \tag{8.23}$$

It follows directly from the definition that the tensor $g_{\lambda\mu\varrho\sigma}$ has the following properties of symmetry or antisymmetry

$$g_{(\lambda\mu)\varrho\sigma} = 0, \qquad g_{\lambda\mu(\varrho\sigma)} = 0, \qquad g_{\lambda\mu\varrho\sigma} = g_{\varrho\sigma\lambda\mu}, \tag{8.24}$$

just like the tensor $R_{\lambda\mu\varrho\sigma}$.

The tensor $g_{\lambda\mu\varrho\sigma}$, as we showed on p. 78, possesses for $n = 2$ just one essential coordinate, e.g.

$$g_{1212} = \tfrac{1}{2}\mathfrak{g} = \tfrac{1}{2}(g_{11}g_{22} - g_{12}{}^2). \tag{8.25}$$

It thus follows that the tensor $R_{\lambda\mu\varrho\sigma}$ also has a single essential coordinate R_{1212}. We write

$$\sigma \stackrel{\text{def}}{=} \frac{R_{1212}}{g_{1212}} \tag{8.26}$$

and we have in general

$$R_{\lambda\mu\varrho\omega} = \sigma g_{\lambda\mu\varrho\omega}, \tag{8.27}$$

where σ is be a scalar coefficient.

We now calculate

$$\begin{aligned} R_{\lambda\mu} &= R_{\nu\lambda\mu}{}^{\nu} = R_{\nu\lambda\mu\omega}g^{\nu\omega} = \sigma g_{\nu\lambda\mu\omega}g^{\nu\omega} \\ &= \sigma g^{\nu\omega}g_{[\nu|\mu|}g_{\lambda]\omega} = \tfrac{1}{2}\sigma g^{\nu\omega}[g_{\nu\mu}g_{\lambda\omega}-g_{\lambda\mu}g_{\nu\omega}] \\ &= \tfrac{1}{2}\sigma\{A^{\omega}_{\mu}g_{\lambda\omega}-g^{\nu\omega}g_{\lambda\mu}\} \\ &= \tfrac{1}{2}\sigma(g_{\lambda\mu}-ng_{\lambda\mu}) = \tfrac{1}{2}\sigma(g_{\lambda\mu}-2g_{\lambda\mu}) = -\tfrac{1}{2}\sigma g_{\lambda\mu}. \end{aligned} \tag{8.28}$$

Setting

$$\tau = -\frac{\sigma}{2} = -\frac{R_{1212}}{2g_{1212}} = -\frac{R_{1212}}{\mathfrak{g}} \tag{8.29}$$

we find that relation (8.21) does indeed hold, which is what we wished to show. We now adopt the following definition.

DEFINITION. A Riemannian space V_n is said to be an *Einstein space* for $n \geqslant 3$ if relation (8.21) holds, i.e. if

$$R_{\lambda\mu} = \tau g_{\lambda\mu}.$$

REMARK. The condition $n \geqslant 3$ which appears in the definition above is present because otherwise no two-dimensional Riemannian space would be an Einstein space.

THEOREM. *The scalar factor* τ *in equation* (8.21) *of Einstein space is a constant.*

PROOF. We begin with formula (6.70)

$$\nabla^{\alpha}(R_{\alpha\mu}-\tfrac{1}{2}Rg_{\alpha\mu}) = 0.$$

By assumption we have

$$R_{\alpha\mu} = \tau g_{\alpha\mu},$$

in view of which the formula above can be rewritten as

$$\nabla^{\alpha}[(\tau-\tfrac{1}{2}R)g_{\alpha\mu}] = 0.$$

But

$$\nabla^{\alpha}g_{\alpha\mu} = 0,$$

whereby

$$g^{\alpha\mu}\nabla_{\alpha}(\tau-\tfrac{1}{2}R) = 0,$$

that is

$$\nabla_{\mu}(\tau-\tfrac{1}{2}R) = 0 \quad \text{or} \quad \partial_{\mu}(\tau-\tfrac{1}{2}R) = 0,$$

whence

$$\tau-\tfrac{1}{2}R = \text{const.}$$

Now

$$R = R_{\lambda\mu} g^{\lambda\mu} = \tau g_{\lambda\mu} g^{\lambda\mu} = \tau n$$

and it follows from this that

$$\tau - \tfrac{1}{2}\tau n = \text{const},$$

or, in other words,

$$\tau(n-2) = \text{const}.$$

Since by hypothesis $n \geqslant 3$, we finally get $\tau = \text{const}$, and the theorem is proved.

Einstein spaces thus have a constant scalar curvature $\varkappa$. A special class of Einstein space are spaces for which

$$R_{\lambda\mu} = 0. \tag{8.30}$$

These spaces have no special name of their own. For them, the scalar curvature vanishes identically. Note, however, that Riemannian spaces with identically vanishing scalar curvature need not be Einstein spaces, for we may have

$$R = R_{\lambda\mu} g^{\lambda\mu} = 0,$$

although the Ricci tensor does not vanish identically. Note also that for $n = 3$ the fact that the Ricci tensor vanishes does not imply that the curvature tensor also vanishes [117].

The special case of Einstein space is obtained when we require a Riemannian space V_n to be such that geodesic subspaces V_{n-1} perpendicular to every direction exist through every point. A subspace V_{n-1} embedded in a space V_n is said to be *geodesic at a given point p* if every geodesic of the space V_n emerging from p tangent to V_{n-1} lies in V_{n-1}. In R_n only hyperplanes are geodesic hypersurfaces.

Einstein spaces play a great role in the general theory of relativity. However, apart from their applications in physics, the theory of these spaces has developed considerably and has been expounded in an elegant and full monograph in the book by Petrov [58].

92. Spaces of Constant Curvature. These constitute an important special subclass of the class of Einstein spaces. They are Riemannian spaces for which the curvature tensor $R_{\omega\mu\lambda\nu}$ is proportional to the tensor (8.23) $g_{\omega\mu\lambda\nu}$, that is, Riemannian spaces for which relation (8.27) holds, or in other words, such that

$$R_{\omega\mu\lambda\nu} = \sigma g_{[\omega|\lambda|} g_{\mu]\nu}.$$

Now assume that $n \geqslant 3$ and that the relation above is valid, i.e. that

$$R_{\omega\mu\lambda\nu} = \tfrac{1}{2}\sigma[g_{\omega\lambda}g_{\mu\nu} - g_{\mu\lambda}g_{\omega\nu}]. \tag{8.31}$$

The calculation on p. 261 for $n = 2$ can be applied here without change and yields the conclusion

$$R_{\lambda\mu} = \tfrac{1}{2}\sigma(1-n)g_{\lambda\mu}.$$

Since

$$R = R_{\lambda\mu}g^{\lambda\mu} = \tfrac{1}{2}\sigma(1-n)n,$$

while, on the other hand,

$$\varkappa = \frac{R}{n(n-1)},$$

it follows from this result that we in fact have

$$\sigma = -2\varkappa. \tag{8.32}$$

We thus find that (8.31) implies that the space is an Einstein space and hence by virtue of the theorem on p. 261 that its scalar curvature is constant. Conversely, the fact that the scalar curvature $\varkappa$ is constant does not in general imply that the space is a space of constant curvature in the sense defined above, i.e. that relation (8.31) is satisfied. The fact that for $n = 2$ this relation implies the constancy of the curvature $\varkappa$ is the essence of the famous theorem of F. Schur of 1886.

93. Spaces of Recurrent Curvature. A space with an affine connection $\Gamma^{\nu}_{\lambda\mu}$ is called a *space of recurrent curvature*, or briefly a *recurrent space*, if its curvature tensor satisfies the relation (system of differential equations)

$$\nabla_{\varrho} R_{\omega\mu\lambda}{}^{\nu} = k_{\varrho} R_{\omega\mu\lambda}{}^{\nu}, \tag{8.33}$$

where k_{ϱ} is a field of covariant vectors. In the special case when the relation

$$\nabla_{\varrho} R_{\omega\mu\lambda}{}^{\lambda} = 0 \tag{8.34}$$

is satisfied, the space is said to be *symmetric*. E. Cartan introduced symmetric spaces and made a partial investigation of their structure.

Similarly, one can introduce the concept of recurrent Ricci spaces as spaces for which the covariant derivative of the Ricci tensor $R_{\lambda\mu}$ is equal to a vector field multiplied by the same Ricci tensor, i.e.

$$\nabla_{\nu} R_{\lambda\mu} = k_{\nu} R_{\lambda\mu}. \tag{8.35}$$

Of course, every recurrent space is a recurrent Ricci space but the converse is not true. A special class of recurrent Ricci spaces consists of those spaces for which the relation

$$\nabla_\nu R_{\lambda\mu} = 0. \tag{8.36}$$

THEOREM. *If $n = 2$ and if relation* (8.36) *is satisfied, then the object Γ of the space preserves volume, i.e. $V_{\lambda\mu} = 0$* [106].

Spaces with affine connection $\Gamma^\nu_{\lambda\mu}$ for which two fields exist, a field of covariant vectors k_ν and a field of tensors $g_{\lambda\mu}$ such that

$$\nabla_\nu g_{\lambda\mu} = k_\nu g_{\lambda\mu}, \tag{8.37}$$

have been given the name of Weyl spaces. Their structure, except for the special case $k = 0$, has not yet been investigated in detail (cf. Section 90).

If the space is an Einstein space

$$R_{\lambda\mu} = \tau g_{\lambda\mu},$$

we then have

$$\nabla_\nu R_{\lambda\mu} = \nabla_\nu(\tau g_{\lambda\mu}) = \tau \nabla_\nu g_{\lambda\mu} + (\partial_\nu \tau) g_{\lambda\mu} = g_{\lambda\mu} \partial_\nu \tau,$$

and the space is, therefore, a recurrent Ricci space, with k_ν a gradient.

94. Projective-Euclidean Spaces. We have seen that the object Γ of parallel displacement uniquely defines a system of geodesics (called *autoparallel lines* by some authors) and we have given without proof the condition for two objects $\underset{1}{\Gamma}$ and $\underset{2}{\Gamma}$ to have the same system of geodesics (5.112). Weyl found that there must then exist a vector field p_λ with the property that

$$\underset{1}{\Gamma}{}^\nu_{(\lambda\mu)} - \underset{2}{\Gamma}{}^\nu_{(\lambda\mu)} = 2A^\nu_{(\lambda} p_{\mu)}. \tag{8.38}$$

If we restrict ourselves to symmetric objects Γ (A_n spaces), this condition can be written somewhat more simply, viz.

$$\underset{1}{\Gamma}{}^\nu_{\lambda\mu} - \underset{2}{\Gamma}{}^\nu_{\lambda\mu} = 2A^\nu_{(\lambda} p_{\mu)}. \tag{8.39}$$

DEFINITION. A space A_n is called a *projective-Euclidean space* if there exists a vector field p_λ such that the space with the connection object

$$\tilde{\Gamma}^\nu_{\lambda\mu} = \Gamma^\nu_{\lambda\mu} + 2A^\nu_{(\lambda} p_{\mu)} \tag{8.40}$$

is affinely Euclidean (cf. Section 87).

If we write down the condition for the curvature tensor belonging to the object $\tilde{\Gamma}^{\nu}_{\lambda\mu}$ to vanish identically, we obtain the condition

$$R_{\omega\mu\lambda}{}^{\nu} = 2p_{[\omega\mu]}A^{\nu}_{\lambda} - 2A^{\nu}_{[\omega}\, p_{\mu]\lambda}, \qquad (8.41)$$

where $p_{\omega\mu}$ is the shortand notation for

$$p_{\omega\mu} \stackrel{\text{def}}{=} -\nabla_{\omega}\, p_{\mu} + p_{\omega}\, p_{\mu}. \qquad (8.42)$$

Since the object $\Gamma^{\nu}_{\lambda\mu}$ is given, while the field p_{λ} is what we are looking for, relations (8.41) in general constitute a system of $\frac{1}{3}n^2(n^2-1)$ equations (for this is the number of independent coordinates of the tensor $R_{\omega\mu\lambda}{}^{\nu}$ in general) in the n^2 unknowns $p_{\omega\mu}$. For $n = 2$ these two numbers are the same and straightforward calculation shows that in this case the system of equations (linear in $p_{\omega\mu}$) can be solved uniquely for $p_{\omega\mu}$. Treating (8.42), in turn, as a system with given $p_{\omega\mu}$'s and unknown p_{λ}'s, we obtain a system of first order differential equations. The conditions for this system of equations to be integrable can be shown to be

$$R_{\omega\mu\lambda}{}^{\nu} p_{\nu} = 2\nabla_{[\omega}\, p_{\mu]\lambda} + 2p_{[\omega\mu]}\, p_{\lambda} + 2p_{[\mu}\, p_{\omega]\lambda}$$

or, on elimination of R by use of relation (8.41)

$$p_{[\omega\mu]}p_{\lambda} - p_{[\omega}p_{\mu]\lambda} = \nabla_{[\omega}p_{\mu]\lambda} + p_{[\omega\mu]}p_{\lambda} + p_{[\mu}p_{\omega]\lambda}, \qquad (8.43)$$

which, when simplified, reduces to the condition

$$\nabla_{[\omega}p_{\mu]\lambda} = 0. \qquad (8.44)$$

Thus, in order for a space A_n to be Euclidean-projective there must exist a tensor $p_{\omega\mu}$ satisfying conditions (8.41) and (8.44) These conditions can be put in another way.

Contracting the indices ω, ν in equation (8.41), we obtain

$$R_{\mu\lambda} = -np_{\mu\lambda} + p_{\mu\lambda}; \qquad (8.45)$$

whereas on contracting the indices λ, ν, we get

$$V_{\omega\mu} = 2(n+1)p_{[\omega\mu]}. \qquad (8.46)$$

Simple algebraic rearrangements of these relations give

$$p_{\mu\lambda} = \frac{1}{1-n^2}\{(n+1)R_{\mu\lambda} + V_{\mu\lambda}\}. \qquad (8.47)$$

On the other hand, if we covariantly differentiate relation (8.41) and apply the Bianchi identity to the left-hand side, we find upon contracting it in two ways and comparing the results that

$$(n-2)\nabla_{[\omega}p_{\mu]\lambda} = 0, \tag{8.48}$$

which shows that for $n > 2$ the integrability conditions (8.44) are satisfied automatically as consequence of (8.41).

We now introduce the short-hand notations

$$W_{\lambda\mu} \stackrel{\text{def}}{=} \frac{1}{1-n^2}\,[(n+1)R_{\lambda\mu}+V_{\nu\mu}], \tag{8.49}$$

and

$$W_{\omega\mu\lambda}{}^{\nu} \stackrel{\text{def}}{=} R_{\omega\mu\lambda}{}^{\nu}-2W_{[\omega\mu]}A_{\lambda}^{\nu}+2A_{[\omega}^{\nu}W_{\mu]\lambda}. \tag{8.50}$$

Following Weyl, we can then formulate the following theorem.

For $n > 2$ a space A_n is projective-Euclidean if and only if the tensor $W_{\omega\alpha\lambda}{}^{\nu}$ is identically zero. For $n = 2$ a space A_2 is projective-Euclidean if and only if the tensor $W_{\lambda\mu}$ satisfies the conditions

$$\nabla_{[\omega}W_{\mu]\lambda} = 0. \tag{8.51}$$

The tensor $W_{\omega\mu\lambda}{}^{\nu}$ is called the *projective curvature tensor* or the *Weyl curvature tensor*. For $n = 2$ this tensor always vanishes identically.

Note the fundamental difference in the above theorem between the case $n > 2$ and $n = 2$. For $n > 2$ the condition $W_{\omega\mu\lambda}{}^{\nu} = 0$ is an algebraic relationship between the curvature tensor $R_{\omega\mu\lambda}$ and its algebraic concomitants $R_{\mu\lambda}$ and $V_{\lambda\mu}$ (which are implicit in the definition of the tensor $W_{\lambda\mu}$) whereas for $n = 2$ condition (8.51) is a system of differential equations.

We now prove the following theorem.

THEOREM. *The contracted Weyl tensor vanishes identically in a Riemannian space, i.e.*

$$W_{\nu\mu\lambda}{}^{\nu} = 0. \tag{8.52}$$

PROOF. Relationships (6.53) and (6.54) hold in Riemannian spaces, i.e.

$$R_{\lambda\mu} = R_{\mu\lambda}, \qquad V_{\lambda\mu} = 0.$$

It follows from this that

$$W_{\lambda\mu} = -\frac{R_{\lambda\mu}}{n-1}$$

and the Weyl tensor is given by the formula

$$W_{\omega\mu\lambda}{}^{\nu} = R_{\omega\mu\lambda}{}^{\nu}+A_{\omega}^{\nu}W_{\mu\lambda}-A_{\mu}^{\nu}W_{\omega\lambda} = R_{\omega\mu\lambda}{}^{\nu}+\frac{1}{n-1}(A_{\mu}^{\nu}R_{\omega\lambda}-A_{\omega}^{\nu}R_{\mu\lambda}).$$

Contracting this, we get

$$W_{\nu\mu\lambda}{}^{\nu} = R_{\mu\lambda} + \frac{1}{n-1}(R_{\mu\lambda} - nR_{\mu\lambda}) = 0.$$

THEOREM. *If a Riemannian space is a space with constant curvature, then it is projective-Euclidean.*

PROOF. By assumption we have

$$R_{\omega\mu\lambda\nu} = -\varkappa[g_{\omega\lambda}g_{\mu\nu} - g_{\mu\lambda}g_{\omega\nu}], \tag{8.53}$$

whence

$$R_{\omega\mu\lambda}{}^{\nu} = R_{\omega\mu\lambda\varrho}g^{\nu\varrho} = \varkappa[g_{\mu\lambda}A^{\nu}_{\omega} - g_{\omega\lambda}A^{\nu}_{\mu}].$$

Further we have

$$V_{\lambda\mu} = 0,$$

and hence

$$W_{\lambda\mu} = \frac{n+1}{1-n^2}R_{\lambda\mu} = -\frac{R_{\lambda\mu}}{n-1}.$$

Moreover, formula (8.53) yields

$$R_{\mu\lambda} = \varkappa[ng_{\mu\lambda} - g_{\mu\lambda}] = \varkappa(n-1)g_{\mu\lambda}.$$

Consequently, we have

$$W_{\lambda\mu} = -\varkappa g_{\lambda\mu}.$$

We now calculate the Weyl tensor

$$\begin{aligned} W_{\omega\mu\lambda}{}^{\nu} &= R_{\omega\mu\lambda}{}^{\nu} + (A^{\nu}_{\omega}W_{\mu\lambda} - A^{\nu}_{\mu}W_{\omega\lambda}) \\ &= \varkappa[g_{\mu\lambda}A^{\nu}_{\omega} - g_{\omega\lambda}A^{\nu}_{\mu}] - \varkappa[A^{\nu}_{\omega}g_{\mu\lambda} - A^{\nu}_{\mu}g_{\omega\lambda}] = 0. \end{aligned}$$

It therefore follows that for $n > 2$ the space is projective-Euclidean. We suppose now that $n = 2$ and we calculate $\nabla_{[\omega}W_{\mu]\lambda}$. We have

$$\begin{aligned} \nabla_{[\omega}W_{\mu]\lambda} &= \tfrac{1}{2}(\nabla_{\omega}W_{\mu\lambda} - \nabla_{\mu}W_{\omega\lambda}) = \tfrac{1}{2}\{\nabla_{\omega}(-\varkappa g_{\mu\lambda}) - \nabla_{\mu}(-\varkappa g_{\omega\lambda})\} \\ &= \tfrac{1}{2}\{g_{\omega\lambda}\nabla_{\mu}\varkappa - g_{\mu\lambda}\nabla_{\omega}\varkappa\} = \tfrac{1}{2}\{g_{\omega\lambda}\partial_{\mu}\varkappa - g_{\mu\lambda}\partial_{\omega}\varkappa\}. \end{aligned}$$

However, the space has a constant curvature $\varkappa$ by assumption. Hence $\partial_{\mu}\varkappa = \partial_{\omega}\varkappa = 0$ and so the condition $\nabla_{[\omega}W_{\mu]\lambda} = 0$ is satisfied; this implies that the space is projective-Euclidean.

We give, without proof, several more theorems concerning projective transformations of the object Γ. By a projective transformation we mean one of the form (8.40), i.e. a transformation

$$\Gamma^{\nu}_{\lambda\mu} \to \Gamma^{\nu}_{\lambda\mu} + 2A^{\nu}_{(\lambda}p_{\mu)}.$$

It may be asked when a Riemannian space V_n with an object of connection $\Gamma^{\nu}_{\lambda\mu} = \left\{ {\nu \atop \lambda\mu} \right\}$ can be transformed into another Riemannian space V_n.

A necessary and sufficient condition is that there exist a vector field p_λ which is a gradient and a field of metric tensors $\tilde{g}_{\lambda\mu}$ with the property that

$$\nabla_\mu \tilde{g}_{\lambda\nu} = 2p_\mu \tilde{g}_{\lambda\mu} + p_\nu \tilde{g}_{\lambda\mu} + p_\lambda \tilde{g}_{\mu\nu}. \tag{8.54}$$

Another theorem, due to Beltrami (1868), states that a Riemannian space can be projectively transformed into a space with constant curvature if and only if it is itself a space of constant curvature.

T. Y. Thomas defined in spaces A_n an object of so-called *projective connection*

$$\Pi^{\nu}_{\lambda\mu} \overset{\text{def}}{=} \Gamma^{\nu}_{\lambda\mu} - \frac{2}{n+1} A^{\nu}_{(\lambda} \Gamma_{\mu)}, \qquad \Gamma_\mu = \Gamma^{\nu}_{\mu\nu}, \tag{8.55}$$

which is an algebraic concomitant of the object Γ. It can be shown that this is a geometric object which is of the second class since the object Γ itself of the second class. The transformation law for this object is

$$\Pi^{\nu'}_{\lambda'\mu'} = \Pi^{\nu}_{\lambda\mu} A^{\nu'\lambda\mu}_{\nu\lambda'\mu'} + A^{\nu'}_{\nu} \partial_{\lambda'} A^{\nu}_{\mu'} + \frac{2}{n+1} A^{\nu'}_{(\lambda'} \partial_{\mu')} \log|J| \tag{8.56}$$

and it can easily be verified that it possesses the group property.

For the subgroup of transformations

$$|J| = \text{const},$$

we see that this law becomes the transformation law for the object Γ.

The Thomas object can be shown to possess the property that if $\tilde{A}_n$ is generated from A_n by projective transformation, the objects of projective connection for the two spaces A_n and $\tilde{A}_n$ are the same.

95. Conformal Transformations of Riemannian Space. If we are given two Riemannian spaces V_n and $\tilde{V}_n$ referred to the same coordinate system (ξ^λ) and if the metric tensors $g_{\lambda\mu}$ and $\tilde{g}_{\lambda\mu}$ of the two spaces are related by

$$\tilde{g}_{\lambda\mu} = \varrho g_{\lambda\mu}, \qquad \varrho > 0, \tag{8.57}$$

where, of course, the coefficient ϱ is a scalar field, the angles between pairs of vectors $\underset{1}{v}$ and $\underset{2}{v}$ attached at the same points are equal. Indeed, if we denote these angles by φ and $\tilde{\varphi}$ respectively, then by virtue of the definition in Section 38 we have

$$\cos\varphi = \frac{g_{\lambda\mu}\underset{1}{v^{\lambda}}\underset{2}{v^{\mu}}}{|\underset{1}{v}|\cdot|\underset{2}{v}|} = \frac{g_{\lambda\mu}\underset{1}{v^{\lambda}}\underset{2}{v^{\mu}}}{\sqrt{g_{\alpha\beta}\underset{1}{v^{\alpha}}\underset{1}{v^{\beta}}}\sqrt{g_{\alpha\beta}\underset{2}{v^{\alpha}}\underset{2}{v^{\beta}}}},$$

$$\cos\tilde{\varphi} = \frac{\tilde{g}_{\lambda\mu}\underset{1}{v^{\lambda}}\underset{2}{v^{\mu}}}{\sqrt{\tilde{g}_{\alpha\beta}\underset{1}{v^{\alpha}}\underset{1}{v^{\beta}}}\sqrt{\tilde{g}_{\alpha\beta}\underset{2}{v^{\alpha}}\underset{2}{v^{\beta}}}} = \frac{\varrho g_{\lambda\mu}\underset{1}{v^{\lambda}}\underset{2}{v^{\mu}}}{\sqrt{\varrho g_{\alpha\beta}\underset{1}{v^{\alpha}}\underset{1}{v^{\beta}}}\sqrt{\varrho g_{\alpha\gamma}\underset{2}{v^{\alpha}}\underset{2}{v^{\beta}}}}$$

$$= \frac{g_{\lambda\mu}\underset{1}{v^{\lambda}}\underset{2}{v^{\mu}}}{\sqrt{g_{\alpha\beta}\underset{1}{v^{\alpha}}\underset{1}{v^{\beta}}}\sqrt{g^{\alpha\beta}\underset{2}{v^{\alpha}}\underset{2}{v^{\beta}}}} = \cos\varphi.$$

The mapping between the spaces V_n and $\tilde{V}_n$ can therefore be said to be *conformal.*

Equally, if in a given space V_n with a tensor field $g_{\lambda\mu}(\xi)$ we carry out a change of metric

$$g_{\lambda\mu} \to \varrho g^{\lambda\mu} \qquad (\varrho > 0),$$

such a change can be called a *conformal transformation.* It is easy to derive relationships between the covariant derivative $\nabla_\mu v^\lambda$ or $\nabla_\mu u_\lambda$, associated with the tensor $g_{\lambda\mu}$, and the covariant derivative $\tilde{\nabla}_\mu v^\lambda$ or $\tilde{\nabla}_\mu u_\lambda$, associated with the tensor $g_{\lambda\mu}$. Note, to begin with, that

$$\tilde{g}^{\lambda\mu} = \frac{1}{\varrho}g^{\lambda\mu}.$$

This enables us to derive the relationships between corresponding Christoffel symbols of the second kind. We have

$$\left\{\begin{matrix}\tilde{\nu}\\ \lambda\mu\end{matrix}\right\} = \tfrac{1}{2}\tilde{g}^{\nu\omega}[\partial_\lambda g_{\omega\mu}+\partial_\mu \tilde{g}_{\lambda\omega}-\partial_\omega \tilde{g}_{\lambda\mu}]$$

$$= \frac{g^{\nu\omega}}{2\varrho}\{\varrho[\partial_\lambda g_{\omega\mu}+\partial_\mu g_{\lambda\omega}-\partial_\omega g_{\lambda\mu}]+g_{\omega\mu}\partial_\lambda\varrho+g_{\lambda\omega}\partial_\mu\varrho-g_{\lambda\mu}\partial_\omega\varrho\}$$

$$= \left\{\begin{matrix}\nu\\ \lambda\mu\end{matrix}\right\}+\tfrac{1}{2}[A^\nu_\mu s_\lambda+A^\nu_\lambda s_\mu-g_{\lambda\mu}g^{\nu\omega}s_\omega], \tag{8.58}$$

where we have adopt the short-hand notation

$$s_\lambda \overset{\text{def}}{=} \partial_\lambda \log\varrho. \tag{8.59}$$

Note that if in particular

$$\varrho = \text{const},$$

then we have $s_\lambda = 0$ and, consequently,

$$\left\{ \begin{matrix} \tilde{\nu} \\ \lambda\mu \end{matrix} \right\} = \left\{ \begin{matrix} \nu \\ \lambda\mu \end{matrix} \right\},$$

i.e. in this case, the conformal transformation does not alter the object of parallel displacement.

Accordingly we have

$$\tilde{\nabla}_\mu v^\lambda = \partial_\mu v^\lambda + \left\{ \begin{matrix} \tilde{\lambda} \\ \omega\mu \end{matrix} \right\} v^\omega = \nabla_\mu v^\lambda + \tfrac{1}{2}[v^\lambda s_\mu - v_\mu s^\lambda + A^\lambda_\mu s_\omega v^\omega]. \tag{8.60}$$

Similarly, we get

$$\tilde{\nabla}_\mu u_\lambda = \nabla_\mu u_\lambda + \tfrac{1}{2}[g_{\lambda\mu} u_\omega s^\omega - u_\lambda s_\mu - u_\mu s_\lambda], \tag{8.61}$$

where

$$s^\lambda \overset{\text{def}}{=} g^{\lambda\omega} s_\omega. \tag{8.62}$$

We introduce the further abbreviation

$$s_{\lambda\mu} \overset{\text{def}}{=} 2\nabla_\lambda s_\mu - s_\lambda s_\mu + \tfrac{1}{2} g_{\lambda\mu} s_\omega s^\omega, \tag{8.63}$$

using which, we find first that

$$\nabla_\mu s_\lambda = \partial_\mu s_\lambda - \left\{ \begin{matrix} \omega \\ \mu\lambda \end{matrix} \right\} s_\omega = \partial_\lambda \sigma_\mu - \left\{ \begin{matrix} \omega \\ \lambda\mu \end{matrix} \right\} s_\omega = \nabla_\lambda s_\mu \tag{8.64}$$

($\partial_\mu s_\lambda = \partial_\lambda s_\mu$ so that s_λ is by definition a gradient) and consequently,

$$s_{\mu\lambda} = s_{\lambda\mu}. \tag{8.65}$$

Secondly, we derive the relationship between the curvature tensors $R_{\omega\mu\lambda}{}^\nu$ and $\tilde{R}_{\omega\mu\lambda}{}^\nu$ of the two spaces V_n and $\tilde{V}_n$. Somewhat tedious calculations, which are omitted here, yield the result

$$\tilde{R}_{\omega\mu\lambda}{}^\nu = R_{\mu\omega\lambda}{}^\nu + \tfrac{1}{4} g^{\alpha\nu}[g_{\omega\lambda} s_{\mu\alpha} + g_{\mu\alpha} s_{\omega\lambda} - g_{\omega\alpha} s_{\mu\lambda} - g_{\mu\lambda} s_{\omega\alpha}]. \tag{8.66}$$

This formula implies that if

$$s_{\lambda\mu} = 0,$$

then the conformal transformation does not alter the curvature tensor.

The question arises as to when a given Riemannian space V_n can be mapped conformally onto a space $\tilde{V}_n$ with a vanishing curvature tensor, i.e. onto a space R_n. A space such that this is possible is called *conformally-Euclidean*.

Treating this problem analytically, we ask for what $g_{\lambda\mu}$'s does there exist a scalar field ϱ with the property that $\tilde{R}_{\omega\mu\lambda}{}^{\nu} = 0$. Using formula (8.66) we are led to the condition

$$R_{\omega\mu\lambda}{}^{\nu} = -\tfrac{1}{4}g^{\alpha\nu}[g_{\omega\lambda}s_{\mu\alpha}+g_{\mu\alpha}s_{\omega\lambda}-g_{\omega\alpha}s_{\mu\lambda}-g_{\mu\lambda}s_{\omega\alpha}]. \tag{8.67}$$

If this condition is treated algebraically, as a system of linear equations with unknown $s_{\lambda\mu}$'s and given $g^{\alpha\nu}$'s and $R_{\omega\mu\lambda}{}^{\nu}$'s, then this system will not have any solutions in general.

If we multiply both sides by $g_{\nu\beta}$, take the sum over ν, and change indices we see that this system is equivalent to the system:

$$R_{\omega\mu\lambda\nu} = -\tfrac{1}{4}[g_{\omega\lambda}s_{\mu\nu}+g_{\mu\nu}s_{\omega\lambda}-g_{\omega\nu}s_{\mu\lambda}-g_{\mu\lambda}s_{\omega\nu}]. \tag{8.68}$$

These (independent) equations number $\frac{1}{12}n^2(n^2-1)$, whereas there are $\frac{1}{2}n(n+1)$ unknowns, because of the symmetry of $s_{\lambda\mu}$. For $n = 2$ we have one equation with 3 unknowns, for $n = 3$ we have 6 equations with 6 unknowns, and for $n > 3$ the equations outnumber the unknowns. Solutions should not therefore be expected to exist for $n > 3$ in general.

For $n = 2$ we have just one equation, obtained by setting, for example, $\omega = 1$, $\mu = 2$, $\lambda = 1$, $\nu = 2$

$$-\tfrac{1}{4}(g_{11}s_{22}+g_{22}s_{11}-2g_{12}s_{12}) = R_{1212} \tag{8.69}$$

and there are ∞^2 solutions for $s_{\lambda\mu}$.

For $n = 3$ we get a system of 6 equations in 6 unknowns, whose determinant of coefficients is $g^2 > 0$ and which is uniquely soluble for $s_{\lambda\mu}$.

However, the solution of the algebraic system (8.68) must also satisfy relations (8.63) which constitute a system of partial differential equations of the second order in the unknown function ϱ, since $s_\lambda = \partial_\lambda \log\varrho$.

If we write down the conditions for this system of equations to be integrable, then lengthy calculations which we omit here yield the result

$$R_{\omega\mu\lambda}{}^{\nu}s_\nu = -\nabla_{[\omega}s_{\mu]\lambda}-\tfrac{1}{4}s^{\alpha}[g_{\omega\lambda}s_{\mu\alpha}+g_{\mu\alpha}s_{\omega\lambda}-g_{\omega\alpha}s_{\mu\lambda}-g_{\mu\lambda}s_{\alpha\omega}] \tag{8.70}$$

and on eliminating $R_{\omega\mu\lambda}{}^{\nu}$ by using equation (8.67) we obtain the simple relationship

$$\nabla_{[\omega}s_{\mu]\lambda} = 0. \tag{8.71}$$

If the tensor $g_{\omega\nu}$ is transvected onto the tensor $R_{\omega\mu\lambda\nu}$, then making use relation (8.68), on the left-hand side we get the Ricci tensor $R_{\mu\lambda}$ while on the right-hand side we have

$$\tfrac{1}{4}\{(n-2)s_{\mu\lambda}+g_{\mu\lambda}s_{\alpha\beta}g^{\alpha\beta}\},$$

or, in other words, we have

$$R_{\mu\lambda} = \tfrac{1}{4}\{(n-2)s_{\mu\lambda}+g_{\mu\lambda}s_{\alpha\beta}g^{\alpha\beta}\}. \tag{8.72}$$

This is a system of equations linear in $s_{\mu\lambda}$, which for $n > 2$ can be solved uniquely for $s_{\lambda\mu}$. We obtain

$$s_{\lambda\mu} = \frac{4}{n-2}\left\{R_{\lambda\mu}-\frac{Rg_{\mu\lambda}}{2(n-1)}\right\}. \tag{8.73}$$

For $n = 2$ as we already know, we have

$$R_{\lambda\mu} = \tau g_{\lambda\mu},$$

where

$$\tau = \frac{R}{n} = \varkappa(n-1) = \varkappa.$$

When the equation above is used, relation (8.72) for $n = 2$ can be written in the form

$$\varkappa g_{\lambda\mu} = \tfrac{1}{4}g_{\lambda\mu}s_{\alpha\beta}g^{\alpha\beta}, \tag{8.74}$$

from which follows the equality

$$s_{\alpha\beta}g^{\alpha\beta} = 4\varkappa. \tag{8.75}$$

However, since

$$R_{1212} = -\varkappa \mathfrak{g} \quad \text{and} \quad g^{11} = \frac{g_{22}}{\mathfrak{g}}, \quad g^{12} = -\frac{g_{12}}{\mathfrak{g}}, \quad g^{22} = \frac{g_{11}}{\mathfrak{g}},$$

relation (8.75) is identical with relation (8.69). In this case it can be shown that the integrability conditions (8.71) are satisfied automatically, whence it follows that every two-dimensional Riemannian space can be mapped conformally onto a plane, which is a well-known classical result.

Assume now that $n > 2$. In place of the tensor $s_{\lambda\mu}$ we introduce the tensor L given by the definition

$$L_{\lambda\mu} \overset{\text{def}}{=} \tfrac{1}{4}(2-n)s_{\lambda\mu}, \tag{8.76}$$

which we can, then use to write condition (8.71) in the form

$$L_{\lambda\mu} = -R_{\lambda\mu}+\frac{Rg_{\lambda\mu}}{2(n-1)}. \tag{8.77}$$

Application of the Bianchi identity to relation (8.77) yields

$$g_{\lambda[\omega}\nabla_{\alpha}L_{\mu]\nu}-g_{\nu[\omega}\nabla_{\alpha}L_{\mu]\lambda} = 0.$$

By transvection of the tensor $g^{\omega\lambda}$ onto this relation, we obtain

$$(n-3)\nabla_{[\alpha}L_{\mu]\nu} = 0. \tag{8.78}$$

We thus find that for $n > 3$ we have the relation $\nabla_{[\alpha} L_{\mu]\nu} = 0$, which is, of course, equivalent to relation (8.71), as a consequence of condition (8.68). For $n = 3$ we can verify by elementary, although lengthy calculation that the curvature tensor is always of the form

$$R_{\omega\mu\lambda\nu} = \frac{1}{n-2}\,[g_{\omega\lambda} L_{\mu\nu} + g_{\mu\nu} L_{\omega\lambda} - g_{\mu\lambda} L_{\omega\nu} - g_{\omega\nu} L_{\mu\lambda}], \tag{8.79}$$

where $L_{\mu\nu}$ is given by formula (8.76).

The foregoing remarks are summed up in the following general *theorem of Schouten* (1921):

THEOREM 1. *Every space* V_2 *is conformally-Euclidean. The space* V_3 *is conformally-Euclidean if and only if the tensor* $L_{\lambda\mu}$, *defined by relation* (8.77), *satisfies the equation*

$$\nabla_{[\omega} L_{\mu]\nu} = 0. \tag{8.80}$$

For $n > 3$, *the space* V_n *is conformally-Euclidean if and only if the tensor* $L_{\lambda\mu}$, *defined by equation* (8.77), *satisfies the system of algebraic equations* (8.68).

We introduce the tensor

$$C_{\omega\mu\lambda\nu} \overset{\text{def}}{=} R_{\omega\mu\lambda\nu} - \frac{1}{n-2}\,[g_{\omega\lambda} L_{\mu\nu} + g_{\mu\nu} L_{\omega\lambda} - g_{\mu\lambda} L_{\omega\nu} - g_{\omega\nu} L_{\mu\lambda}], \tag{8.81}$$

which is called the *tensor of conformal curvature*.

It vanishes identically for $n = 3$, while for $n > 3$, the preceding theorem implies that its vanishing is a necessary and sufficient condition for a space to be conformally mappable onto Euclidean space.

Note that every space of constant curvature is conformally-Euclidean. Indeed, in this case we have

$$R_{\omega\mu\lambda\nu} = -\varkappa\{g_{\omega\lambda} g_{\mu\nu} - g_{\mu\lambda} g_{\omega\nu}\}.$$

From this we have

$$R_{\mu\lambda} = g_{\mu\lambda}(n-1), \qquad L_{\mu\lambda} = -\tfrac{1}{2}(n-2)\varkappa g_{\mu\lambda},$$

where $\varkappa = \text{const}$ and it follows that conditions (8.80) and (8.68) are both satisfied.

One could ask a question analogous to that posed in the preceding section, viz. what conformal transformations map an Euclidean space to another Euclidean space. This question leads us to the solution of the equation

$$s_{\lambda\mu} = 0.$$

Using notation (8.63), this equation can be written

$$2\nabla_\mu s_\lambda = s_\lambda s_\mu - \tfrac{1}{2} g_{\lambda\mu} s_\omega s^\omega. \tag{8.82}$$

Since the integrability conditions for this system, namely $\nabla_{[\omega} s_{\mu]\lambda} = 0$, are satisfied identically in this case, equation (8.82) is integrable unboundedly. An exhaustive discussion of the solutions of equation (8.82) subsequently leads to the theorems of Lie (1872) and Haantjes [69], which are generalizations of the classical Liouville theorem.

THEOREM 2. *Every conformal mapping of R_n onto R_n is for $n > 2$ either a similarity mapping or a combination of inversion with a similarity mapping.*

THEOREM 3. *Every real conformal mapping of R_n onto R_n is either a similarity mapping*

$$\eta^\lambda = \alpha \xi^\lambda \qquad (\alpha = \text{const}),$$

or a combination of inversion

$$\eta^\lambda = \frac{\xi^\lambda}{r^2} \qquad (r^2 = g_{\lambda\mu} x^\lambda x^\mu)$$

with a similarity mapping, or, lastly, a combination of a similarity mapping with a mapping of the form

$$\eta^\lambda = \frac{\xi^\lambda - \frac{1}{2} r^2 z^\lambda}{1 - z^\lambda \xi_\lambda}$$

(z^λ *is a constant null vector, i.e.* $z^\lambda z_\lambda = 0$).

It should be noted that whereas for $n = 2$ the conformal mappings of R_2 onto R_2 are given by arbitrary functions, the set of conformal mappings of R_n onto R_n for $n > 2$ constitutes a family with a finite number of parameters.

Thomas [69] defined in spaces V_n a first-order differential concomitant of the tensor $g_{\lambda\mu}$ (or, putting it another way, the first-order algebraic concomitant of the tensor $g_{\lambda\mu}$ and the object $\left\{ {\nu \atop \lambda\mu} \right\}$), which is invariant under the conformal transformations $g_{\lambda\mu} \to \tilde{g}_{\lambda\mu} = \varrho \cdot g_{\lambda\mu}$. This concomitant is known as the *conformal connection of Thomas* (cf. the projective connection of Thomas (8.55)) and is defined by

$$T^\nu_{\lambda\mu} \overset{\text{def}}{=} \left\{ {\nu \atop \lambda\mu} \right\} - \frac{1}{n} \left[\left\{ {\omega \atop \lambda\omega} \right\} A^\nu_\mu + \left\{ {\omega \atop \mu\omega} \right\} A^\nu_\lambda - g^{\nu\omega} g_{\lambda\mu} \left\{ {\tau \atop \omega\tau} \right\} \right]. \tag{8.83}$$

Similarly, it can be verified that the quantities

$$\mathfrak{g}_{\lambda\mu} \stackrel{\text{def}}{=} g_{\lambda\mu} \cdot \mathfrak{g}^{-1/n},$$

and

$$\mathfrak{g}^{\lambda\mu} \stackrel{\text{def}}{=} g^{\lambda\mu} \cdot \mathfrak{g}^{1/n},$$

which are tensor densities with weights $-2/n$ and $2/n$, respectively, are also invariants with respect to conformal transformations of the metric tensor. The connection $T^{\nu}_{\lambda\mu}$ is a geometric object of the second class obeying the transformation law

$$T^{\nu'}_{\lambda'\mu'} = T^{\nu}_{\lambda\mu} A^{\nu'\lambda\ \mu}_{\nu\ \lambda'\mu'} + A^{\nu'}_{\nu}\, \partial_{\lambda'} A^{\nu}_{\mu'} + \frac{1}{n}\{A^{\nu'}_{\lambda'}\varphi_{\mu'} + A^{\nu'}_{\mu'}\varphi_{\lambda'} - \mathfrak{g}_{\lambda'\mu'}\mathfrak{g}^{\nu'\omega'}\varphi_{\omega'}\}, \tag{8.84}$$

where

$$\varphi_{\lambda'} = \partial_{\lambda'} \log|J|. \tag{8.85}$$

It can be checked that if (and only if)

$$J = \text{const},$$

does the transformation law for the Thomas object $T^{\nu}_{\lambda\mu}$ reduce to the transformation law for the object $\left\{ {\nu \atop \lambda\mu} \right\}$.

96. Hermitian Spaces. So far we have generally assumed that both the space X_n and the geometric objects considered in it are real. Much of our discussion to this point, however, remains valid when the coordinates ξ^{λ} are complex numbers just as when the coordinates of the objects under consideration are complex numbers.

If ξ^{λ} denote complex numbers and the transformation $\xi^{\lambda} \to \xi^{\lambda'}$, or, to be precise,

$$\xi^{\lambda'} = \varphi^{\lambda'}(\xi^1, \ldots, \xi^n) \tag{8.86}$$

is given in terms of complex functions φ, then each of the transformations (8.86) can be uniquely assigned a transformation

$$\bar{\xi}^{\lambda'} = \bar{\varphi}^{\lambda'}(\xi^1, \ldots, \xi^n), \tag{8.87}$$

where a bar above a letter denotes its complex conjugate.

If as before we write

$$A^{\lambda'}_{\lambda} = \frac{\partial \varphi^{\lambda'}}{\partial \xi^{\lambda}} \tag{8.88}$$

the first-order partial derivatives of transformation (8.87), i.e.

$$\frac{\partial \bar{\varphi}^{\lambda'}}{\partial \xi^{\lambda}} \tag{8.89}$$

should be written as

$$\overline{A}_{\lambda}^{\lambda'} . \tag{8.90}$$

This notation should not lead to any confusion, especially as the analysis of complex functions shows that the conjugate value of a derivative is equal to the derivative of the conjugate function.

However, since in dealing with the transformation laws we shall have both parameters $A_{\lambda}^{\lambda'}$ and $\overline{A}_{\lambda}^{\lambda'}$, and since correspondingly quantities of two kinds (or mixture of both kinds) will occur, we are obliged to introduce two kinds of indices. Accordingly, we introduce the notation

$$\overline{A}_{\bar{\lambda}}^{\bar{\lambda}'} \stackrel{\text{def}}{=} \frac{\partial \bar{\varphi}^{\lambda'}}{\partial \xi^{\lambda}} . \tag{8.91}$$

Thus, we consider two kinds of contravariant vectors with the transformation laws

$$v^{\lambda'} = A_{\lambda}^{\lambda'} v^{\lambda} \qquad \text{(of first kind)},$$

and

$$v^{\bar{\lambda}'} = \overline{A}_{\bar{\lambda}}^{\bar{\lambda}'} v^{\bar{\lambda}} \qquad \text{(of second kind)} \tag{8.92}$$

and consequently we also consider all further quantities (tensors, tensor densities) and other objects of the first class of both kinds.

Mixed objects may also exist; for instance, a two-fold contravariant tensor with the transformation law

$$h_{\lambda'\bar{\mu}'} = h_{\lambda\bar{\mu}} A_{\lambda'}^{\lambda} \overline{A}_{\bar{\mu}'}^{\bar{\mu}} . \tag{8.93}$$

Of course, such transformation laws must also be shown to possess the group property; this amounts to proving the formulae

$$\overline{A}_{\bar{\lambda}'}^{\bar{\lambda}''} \overline{A}_{\bar{\lambda}}^{\bar{\lambda}'} = \overline{A}_{\bar{\lambda}}^{\bar{\lambda}''} . \tag{8.94}$$

It should be borne in mind, however, that indices of two different kinds cannot be contracted.

Two kinds of densities will also exist, depending on whether the Jacobian

$$J = \det A_{\lambda}^{\lambda'} , \tag{8.95}$$

or the Jacobian

$$J = \det \overline{A}_{\bar{\lambda}}^{\bar{\lambda}'} , \tag{8.96}$$

appears in the transformation law.

Objects for which we consider two kinds of transformation (8.86) and (8.87) and where two kinds of indices occur, are called *Hermitian objects.*

If in a complex space X_n we are given a field of a certain quantity, various assumptions can be made as to the regularity of the coordinates of his quantity, which involve the coordinates of the variable point.

If we restrict ourselves to analytic transformations of coordinates it might at first appear that we can consider objects whose coordinates are analytic functions of the coordinates of a variable point in this space. Unfortunately, it turns out that this property of analyticity of the variables ξ^λ or $\bar{\xi}^{\bar{\lambda}}$ is not one which is invariant under allowable transformations of coordinates. On the other hand, a property which is invariant is that of so-called *semi-analyticity*, which means that the coordinates of the object are analytic functions of the $2n$ real variables η^λ, ζ^λ:

$$\eta^\lambda = \tfrac{1}{2}(\xi^\lambda + \bar{\xi}^{\bar{\lambda}}), \qquad \zeta^\lambda = \tfrac{1}{2}i(\xi^\lambda - \bar{\xi}^{\bar{\lambda}}). \tag{8.97}$$

Let us consider whether the symmetry of the Hermitian tensor

$$h_{\lambda\bar{\mu}} = h_{\bar{\mu}\lambda} \tag{8.98}$$

is an invariant property. We have

$$h_{\lambda'\bar{\mu}'} = h_{\bar{\mu}\lambda} A^\lambda_{\lambda'} \bar{A}^{\bar{\mu}}_{\bar{\mu}'}, \qquad h_{\bar{\mu}'\lambda'} = h_{\lambda\bar{\mu}} \bar{A}^{\bar{\mu}}_{\bar{\mu}'} A^\lambda_{\lambda'}.$$

It follows from this, by virtue of assumption (8.98) and the commutativity of the product, that

$$h_{\lambda'\bar{\mu}'} = h_{\bar{\mu}'\lambda'}. \tag{8.99}$$

DEFINITION. A complex space X_n endowed with a semi-analytic field of Hermitian tensors $h_{\bar{\lambda}\mu}$ with property (8.98) and satisfying the auxiliary condition

$$\overline{h_{\bar{\mu}\lambda}} = h_{\lambda\bar{\mu}}, \tag{8.100}$$

is called a *unitary space* and is denoted by U_n.

The theory of unitary spaces was inaugurated in 1929 by Schouten [145] who later developed it further together with van Dantzig [147]. These spaces play a major role in theoretical physics. An object of linear connection can be defined in them and a theory of curvature can be developed.

97. Almost Complex Spaces. A space X_n whose points have complex coordinates ξ^λ can be regarded as a $2n$-dimensional real space when we put

$$\xi^\lambda = \eta^\lambda + i\zeta^\lambda \tag{8.101}$$

and treat η^λ, ζ^λ ($\lambda = 1, 2, \ldots, n$) as real.

If X_n is a manifold (cf. Section 116) and if there exists an atlas for it such that for the parts common to two neighbourhoods on every map the transformations of the transitions from ξ^λ to $\xi^{\lambda'}$ are given by

$$\eta^{\lambda'} = \varphi^{\lambda'}(\eta^\mu, \zeta^\mu), \qquad \zeta^{\lambda'} = \psi^{\lambda'}(\eta^\mu, \zeta^\mu), \tag{8.102}$$

where φ and ψ are analytic functions of $2n$ real variables, then the space X_n is called a *complex manifold.*

It can be shown quite easily that every complex manifold is always orientable.

It can be shown further that every complex manifold contains a field of mixed tensors

$$F_a^b \qquad (a, b = 1, 2, \ldots, 2n) \tag{8.103}$$

with the property

$$F_b^a F_b^c = -A_a^c. \tag{8.104}$$

Ehresmann [103] called a manifold for which there exists a field of tensors F_b^a with property (8.104) an *almost complex manifold.* These manifolds have been exhaustively investigated and classified.

An important role in almost complex manifolds is played by the tensor of Nijenhuis [139]. This is a first-order differential concomitant of the tensor F_a^b and is defined by

$$N_{ab}^c = 2F_{[a}^d\{\partial_{|d|} F_{b]}^c - \partial_{b]} F_d^c\}. \tag{8.105}$$

If the tensor F_a^b has the components

$$F_\lambda^\nu = i\delta_\lambda^\nu, \qquad F_{\bar\lambda}^\nu = F_\lambda^{\bar\nu} = 0, \qquad F_{\bar\lambda}^{\bar\nu} = -i\delta_\lambda^\nu,$$

where $\lambda, \nu = 1, \ldots, n$; $\bar\lambda, \bar\nu = n+1, \ldots, 2n$, then the space is complex.

A very important theorem (B. Eckmann, A. Frölicher [103]) states that a necessary condition for an almost complex space to be complex is that its Nijenhuis tensor N_{ab}^c vanish identically.

Lichnerowicz [135] has shown that a Hermitian metric can be introduced in every almost complex space, i.e. that such a space can always be made a unitary space U_n.

CHAPTER IX

DIFFERENTIAL OPERATORS AND INTEGRAL THEOREMS

98. Invariant Differential Operators. In the vector analysis of three-dimensional Euclidean space R_3 we introduce three operators: namely, an operator which, when applied to a scalar field σ yields a vector field called the *gradient* of the field σ; a second, which when applied to a vector field v, produces a scalar field called the *divergence* of the field v; and, finally, a third one, which when applied to a vector field v, yields a new vector field called the *rotation* of field v.

In orthogonal Cartesian coordinates ξ^i $(i = 1, 2, 3)$, we have the following formulae:

1. Writing

$$v = \operatorname{grad}\sigma, \tag{9.1}$$

we have

$$v^i = \frac{\partial\sigma}{\partial\xi^i} \quad (i = 1, 2, 3), \tag{9.2}$$

2.
$$\operatorname{div} v = \partial_i v^i. \tag{9.3}$$

3. If we denote the coordinates of the rotation of the field v by R^i, we have

$$R^1 = \partial_2 v^3 - \partial_3 v^2, \quad R^2 = \partial_3 v^1 - \partial_1 v^3, \quad R^3 = \partial_1 v^2 - \partial_2 v^1. \tag{9.4}$$

These operations can be generalized, both to spaces of higher dimensions and in respect of the kind of space or the kind of quantity to which we apply the relevant operator.

Since the application of an operator will lead to an appropriate quantity (in general, of a kind different from that on which the operator acts) it will not be necessary to specify the coordinate systems. Of course, in certain preferred coordinate systems the coordinates of the resulting quantities may be more simply expressed in terms of the coordinates of the quantity acted on by the operator than they are in other systems.

To define the gradient, the space X_n does not have to be provided with either a metric tensor or an object of linear displacement.

When a scalar field σ of class C^1 is given, then

$$w_\mu = \partial_\mu \sigma \tag{9.5}$$

is always a field of covariant vectors which we call the *gradient* of the field σ and we write

$$w = \operatorname{Grad} \sigma. \tag{9.6}$$

If, furthermore, the space X_n is furnished with a fundamental tensor $g_{\lambda\mu}$, so that $[g_{\lambda\mu}]$ is of order n, then each covariant vector w_μ can be uniquely assigned a contravariant vector

$$\overline{w}^\nu \stackrel{\text{def}}{=} g^{\nu\lambda} w_\lambda;$$

the vector $\overline{w}^\nu$ can also be called the *gradient* of the field σ. If $\tilde{\sigma}$ is a field of W-scalars

$$\tilde{\sigma}' = \operatorname{sgn} J \cdot \tilde{\sigma},$$

then

$$\tilde{w}_\mu \stackrel{\text{def}}{=} \partial_\mu \tilde{\sigma}$$

is a field of G-vectors, i.e.

$$\tilde{w}_{\mu'} = \operatorname{sgn} J \cdot A^\mu_{\mu'} \tilde{w}_\mu .$$

If we replace the scalar field σ with a field of density $\mathfrak{v}$, then—as we know—the coordinates $\partial_\mu \mathfrak{v}$ no longer represent any geometric object.

To obtain a quantity, we must first equip the space X_n with a field of parallel displacement objects Γ, so that it becomes a space L_n in which case $\nabla_\mu \mathfrak{v}$ represents a vector density. It may be called the *gradient* of the field $\mathfrak{v}$, although not all authors do this.

Suppose now that a field of covariant p-vectors

$$w_{\lambda_1 \dots \lambda_p} = w_{[\lambda_1 \dots \lambda_p]} . \tag{9.7}$$

is given in X_n. It is true that $\partial_\mu w_{\lambda_1 \dots \lambda_p}$ do not represent the coordinates of any $(p+1)$-indexed quantity; nevertheless

$$\partial_{[\mu} w_{\lambda_1 \dots \lambda_p]} \tag{9.8}$$

give a tensor of valence $(0, p+1)$, viz. a covariant $(p+1)$-vector which follows from the fact that

$$\partial_{[\mu'} w_{\lambda'_1 \ldots \lambda'_p]} = \partial_{[\mu'} w_{|\lambda_1 \ldots \lambda_p|} A^{\lambda_1 \ldots \lambda_p}_{\lambda'_1 \ldots \lambda'_p]}$$
$$= A^{\mu\lambda_1 \ldots p}_{\mu'\lambda'_1 \ldots \lambda'_p} \partial_{[\mu} w_{\lambda_1 \ldots \lambda_p]} + \partial_{[\mu'} A^{\lambda_1 \ldots \lambda_p}_{\lambda'_1 \ldots \lambda'_p]} w_{\lambda_1 \ldots \lambda_p},$$

while $\partial_{[\mu'} A^{\lambda_1 \ldots \lambda_p}_{\lambda'_1 \ldots \lambda'_p]} = 0$, since $\partial_{[\mu'} A^{\lambda_i}_{\lambda'_i]} = 0$, whence it follows that (9.8) is a tensor. Thus, the $(p+1)$-vector (9.8) is called the *rotation* of the p-vector (9.7) and we write

$$\partial_{[\mu} w_{\lambda_1 \ldots \lambda_p]} = \operatorname{Rot} w_{\lambda_1 \ldots \lambda_p} = \operatorname{Rot}_\mu w_{\lambda_1 \ldots \lambda_p} \text{ [1]}. \tag{9.9}$$

In the particular case of the vector field w_λ, the rotation is a field of bivectors $\partial_{[\mu} w_{\lambda]}$.

If we are dealing with a space A_n, the rotation can also be displayed in the equivalent form $\nabla_{[\mu} w_{\lambda_1 \ldots \lambda_p]}$.

Let us take the special case of the space V_3 and suppose that in it we are given a field of contravariant vectors

$$v^\nu. \tag{9.10}$$

We write

$$\bar{v}^{\cdot}_\nu = v_\nu{}^{\cdot} = g_{\nu\lambda} v^\nu.$$

Next, we form

$$\nabla_\mu \bar{v}_\nu = g_{\nu\lambda} \nabla_\mu v^\lambda$$

and we finally set

$$w^\nu = \operatorname{Rot} v \overset{\text{def}}{=} e^{\nu\mu\lambda} \nabla_\lambda \bar{v}_\mu = e^{\nu\lambda\mu} \nabla_{[\lambda} \bar{v}_{\mu]} = e^{\nu\lambda\mu} g_{\varrho[\mu} \nabla_{\lambda]} v^\varrho, \tag{9.11}$$

where $e^{\nu\lambda\mu}$ is the Ricci tensor. Thus for $n = 3$, we can define in metric space the rotation of a contravariant vector field as a field of contravariant G-vectors. In the special case when $V_3 = R_3$ this definition goes over into the classical definition.

Some authors give the name rotation to

$$\partial_{[\mu} w_{\lambda_1 \ldots \lambda_p]},$$

instead of to

$$(p+1)\, \partial_{[\mu} w_{\lambda_1 \ldots \lambda_p]} \text{ [2]}. \tag{9.12}$$

Rotation cannot be defined, however, for a field of contravariant vectors.

---

(1) After Weyssenhoff [155].

(2) This definition (9.12) is more convenient because the subsequent integral formulae turn out to be simpler. L. E. J. Brouwer refers to the $(p+1)$-vector as the Stokes tensor [88].

The definitions above immediately lead to the identities

$$\mathrm{Rot}[\mathrm{Grad}\,\sigma] \equiv 0, \qquad \mathrm{Rot}[\mathrm{Rot}\,w] = 0. \tag{9.13}$$

Indeed

$$\mathrm{Rot}\,[\mathrm{Grad}\,\sigma] = \mathrm{Rot}_\mu(\partial_\lambda \sigma) = 2\partial_{[\mu}\partial_{\lambda]}\sigma \equiv 0.$$

Similarly

$$\begin{aligned}\mathrm{Rot}_\nu\,[\mathrm{Rot}_\mu w_{\lambda_1\ldots\lambda_p}] &= \mathrm{Rot}_\nu[(p+1)\,\partial_{[\mu} w_{\lambda_1\ldots\lambda_p]}] \\ &= (p+2)\,(p+1)\,\partial_{[\nu}\partial_{[\mu} w_{\lambda_1\ldots\lambda_p]]} \\ &= (p+2)\,(p+1)\,\partial_{[\nu}\partial_{\mu} w_{\lambda_1\ldots\lambda_p]} = 0\end{aligned}$$

as

$$\partial_\nu\,\partial_\mu w_{\lambda_1\ldots\lambda_p} = \partial_\mu\,\partial_\nu\, w_{\lambda_1\ldots\lambda_p}.$$

We have stated that the concept of rotation introduced above is a generalization of the classical concept, i.e. that introduced for fields of contravariant vectors in a three-dimensional Euclidean space and, more especially, space referred to rectangular Cartesian coordinates. The perceptive reader has no doubt noticed a certain contradiction in the second formula of (9.13), for if we take the special case $p = 1$, we get

$$\mathrm{Rot}_\mu\, w_\lambda = 2\partial_{[\mu}\, w_{\lambda]}.$$

Moreover, on considering the special case of three-dimensional Euclidean space with a rectangular Cartesian coordinate system ($g_{ij} \overset{*}{=} \delta_{ij}$), we can uniquely assign the vector w_i a contravariant vector

$$\overline{w}^i = g^{ij}\, w_j = w_i$$

with identical coordinates and then

$$2\partial_{[j}\, w_{i]} = \partial_j\, w_i - \partial_i\, w_j = \partial_j\, w^i - \partial_i\, w^j,$$

i.e. these coordinates coincide with the coordinates of the classical rotation: rot w. If we now repeat this procedure treating rot w as a field of contravariant vectors, the newly-obtained field rot (rot w) will in general not be identically equal to zero, as we would expect. Straight forward calculation shows that we have

$$\mathrm{rot}(\mathrm{rot}\, w) = \mathrm{grad}(\mathrm{div}\, w) - \Delta w, \tag{9.14}$$

where the gradient and divergence operations are understood in the classical sense, and Δw denotes a vector whose coordinates are the Laplacians of the coordinates of the vector w.

REMARK. In a general Riemannian space V_n the operator Δ is defined for a vector field v^ν in the following manner:

$$\Delta v^\nu \overset{\text{def}}{=} g^{\mu\lambda}\nabla_\lambda \nabla_\mu v^\nu$$

(cf. the concept of the second parameter of Beltrami (p. 321), where the analogous operator is defined for a scalar field and not for a vector field).

The apparent contradiction between formula (9.14) and formula (9.13) stems from the fact that Rot is an operator which changes the valence of the quantity to which it is applied.

If we take a field of vector densities $\mathfrak{w}_\lambda$ (of non-zero weight) instead of a field of covariant vectors w_λ, then $\partial_{[\mu}\mathfrak{w}_{\lambda]}$ no longer represents a quantity. In a space A_n, however, it is possible to consider a quantity $\nabla_{[\mu}\mathfrak{w}_{\lambda]}$ which, however, we do not call the rotation of the field $\mathfrak{w}_\lambda$.

We have seen that the rotation operator increases the valence of the quantity to which it is applied.

We now introduce another operator which will decrease the valence of the quantity on which it operates. In fact, consider a field of contravariant p-vector densities

$$\mathfrak{v}^{\lambda_1 \ldots \lambda_p} \tag{9.15}$$

of weight $+1$, which therefore transform according to the law

$$\mathfrak{v}^{\lambda_1' \ldots \lambda_p'} = J^{-1} A^{\lambda_1' \ldots \lambda_p'}_{\lambda_1 \ldots \lambda_p} \mathfrak{v}^{\lambda_1 \ldots \lambda_p} \tag{9.16}$$

and write

$$\operatorname{Div} \mathfrak{v} \overset{\text{def}}{=} \partial_\mu \mathfrak{v}^{\mu\lambda_2 \ldots \lambda_p} = \operatorname{Div}_\mu \mathfrak{v}^{\mu\lambda_2 \ldots \lambda_p\,(1)}. \tag{9.17}$$

It can be shown that *divergence* so defined represents a field of $(p-1)$-vector densities of weight $+1$ (2),

We shall do this for $p = 1$. If

$$\mathfrak{v}^{\lambda'} = J^{-1}\mathfrak{v}^\lambda A^{\lambda'}_\lambda,$$

then

$$\begin{aligned}
\partial_{\mu'}\mathfrak{v}^{\lambda'} &= A^\mu_{\mu'}\partial_\mu \mathfrak{v}^{\lambda'} \\
&= A^\mu_{\mu'}\partial_\mu\{J^{-1}\mathfrak{v}^\lambda A^{\lambda'}_\lambda\} \\
&= A^\mu_{\mu'}\{-\mathfrak{v}^\lambda A^{\lambda'}_\lambda J^{-1}\partial_\mu \ln|J| + J^{-1}A^{\lambda'}_\lambda \partial_\mu \mathfrak{v}^\lambda + J^{-1}\mathfrak{v}^\lambda \partial_\mu A^{\lambda'}_\lambda\} \\
&= J^{-1}A^\mu_{\mu'}\{A^{\lambda'}_\lambda \partial_\mu \mathfrak{v}^\lambda + \mathfrak{v}^\lambda[\partial_\mu A^{\lambda'}_\lambda - A^{\lambda'}_\lambda \partial_\mu \ln|J|]\}.
\end{aligned}$$

---

(1) This notation is introduced following J. Weyssenhoff.

(2) The proof of this can be found for example, in the book by Schouten [68].

From this we have

$$\partial_{\lambda'} \mathfrak{v}^{\lambda'} = J^{-1}\partial_{\lambda}\mathfrak{v}^{\lambda} + J^{-1}\mathfrak{v}^{\lambda}[\partial_{\lambda'} A_{\lambda}^{\lambda'} - \partial_{\lambda}\ln|J|].$$

However, since the difference within the last brackets is zero, we finally have

$$\partial_{\lambda'} \mathfrak{v}^{\lambda'} = J^{-1}\partial_{\lambda}\mathfrak{v}^{\lambda}, \tag{9.18}$$

which shows that $\partial_{\lambda}\mathfrak{v}^{\lambda}$ is a density of weight $+1$.

It follows from the calculations above that for a density with a different weight the theorem no longer holds since the term $\partial_{\lambda}\ln|J|$ is then multiplied by a factor other than (-1) and so does not cancel with the term $\partial_{\lambda'} A_{\lambda}^{\lambda'}$.

On the other hand, if we confine ourselves to a pseudogroup of transformations for which

$$J = \text{const}, \tag{9.19}$$

then formula (9.18) yields

$$\partial_{\lambda'} \mathfrak{v}^{\lambda'} = \partial_{\lambda}\mathfrak{v}^{\lambda},$$

which shows that $\partial_{\lambda}\mathfrak{v}^{\lambda}$ is a scalar. However, if v^{ν} is a vector field we have

$$v^{\lambda'} = v^{\lambda} A_{\lambda}^{\lambda'}.$$

Differentiation yields

$$\begin{aligned}
\partial_{\lambda'} v^{\lambda'} &= \partial_{\lambda'}(v^{\lambda} A_{\lambda}^{\lambda'}) = A_{\lambda'}^{\mu}\partial_{\mu}(v^{\lambda} A_{\lambda}^{\lambda'}) \\
&= A_{\lambda'}^{\mu} A_{\lambda}^{\lambda'}\partial_{\mu} v^{\lambda} + v^{\lambda} A_{\lambda'}^{\mu}\partial_{\mu} A_{\lambda}^{\lambda'} \\
&= A_{\lambda}^{\mu}\partial_{\mu} v^{\lambda} + v^{\lambda}\partial_{\lambda'} A_{\lambda}^{\lambda'} \\
&= \partial_{\lambda} v^{\lambda} + v^{\lambda}\partial_{\lambda}\log|J|.
\end{aligned}$$

It is thus seen that with a general pseudogroup of transformations $\partial_{\lambda} v^{\lambda}$ is not a geometric object, although it does become one (and a scalar object at that) if the pseudogroup is restricted to (9.19), i.e. to the unimodular group. Accordingly, in affine space and, hence, in particular in Euclidean spaces, $\partial_{\lambda} v^{\mu}$ is a scalar and is the classical divergence of the vector field v^{λ}.

If the space X_n is a space L_n, then for a given field

$$\mathfrak{v}^{\lambda_1 \ldots \lambda_p}$$

we can consider the quantity

$$\nabla_{\mu}\mathfrak{v}^{\mu\lambda_2 \ldots \lambda_p}, \tag{9.20}$$

which, just as (9.17), has a valence of $(p-1, 0)$ and also represents a density field of $(p-1)$-vectors of weight r, if the density (9.15) has a weight of r. However, when $r = 1$, the quantity (9.20) does not always coincide with the density (9.17). The following relationship (Schouten [69], p. 171) can be established

$$\nabla_\mu \mathfrak{v}^{\mu\lambda_2\dots\lambda_p} = \partial_\mu \mathfrak{v}^{\mu\lambda_2\dots\lambda_p} + 2S_{\lambda\mu}{}^{\lambda} \mathfrak{v}^{\mu\lambda_2\dots\lambda_p} + (p-1)\mathfrak{v}^{\mu[\lambda_2\dots\lambda_{p-1}|\nu|} S_{\mu\nu}{}^{\lambda_p]}. \quad (9.21)$$

It should be noted, however, that the derivation of this formula for $p = 1$, as given by Schouten, fails since $\lambda_2 \dots \lambda_p$ do not exist for $p = 1$, and thus the term $\Gamma_{\varkappa\lambda}{}^{[\varkappa_p} \mathfrak{v}^{|\varkappa|\varkappa_2\dots\varkappa_{p-1}]\lambda}$ is seen to be meaningless. Nevertheless, for $p = 1$ we have

$$\nabla_\mu \mathfrak{v}^\nu = \partial_\mu \mathfrak{v}^\nu + \Gamma_{\mu\lambda}^{\nu} \mathfrak{v}_\lambda - \Gamma_{\mu\lambda}^{\lambda} \mathfrak{v}^\nu.$$

When the indices μ, ν are contracted, this yields

$$\begin{aligned} \nabla_\mu \mathfrak{v}^\mu &= \partial_\mu \mathfrak{v}^\mu + \Gamma_{\mu\lambda}^{\mu} \mathfrak{v}^\lambda - \Gamma_{\mu\lambda}^{\lambda} \mathfrak{v}^\mu \\ &= \partial_\mu \mathfrak{v}^\mu + \Gamma_{\mu\lambda}^{\mu} \mathfrak{v}^\lambda - \Gamma_{\lambda\mu}^{\mu} \mathfrak{v}^\lambda \\ &= \partial_\mu \mathfrak{v}^\mu + (\Gamma_{\mu\lambda}^{\mu} - \Gamma_{\lambda\mu}^{\mu}) \mathfrak{v}^\lambda \\ &= \partial_\mu \mathfrak{v}^\mu + 2S_{\mu\lambda}^{\mu} \mathfrak{v}^\lambda, \end{aligned}$$

and consequently, formula (9.21) remains valid for $p = 1$ as well. We find, therefore, that when the object Γ is symmetric we have $S_{\mu\lambda}{}^{\nu} = 0$ and then

$$\nabla_\mu \mathfrak{v}^{\mu\lambda_2\dots\lambda_p} = \partial_\mu \mathfrak{v}^{\mu\lambda_2\dots\lambda_p}, \quad (9.22)$$

i.e., the contracted covariant derivative of density reduces to the ordinary derivative of density.

If $r = 0$, that is, if the field $\mathfrak{v}$ is a field of p-vectors $v^{\lambda_1\dots\lambda_p}$, then

$$\nabla_\mu v^{\mu\lambda_2\dots\lambda_p}$$

is also a field of $(p-1)$-vectors (of valence $(p-1, 0)$). If in particular the space L_n is the space V_n, a differential operator can also be defined which produces to a field of covariant $(p-1)$-vectors from a field of covariant p-vectors $w_{\lambda_1\dots\lambda_p}$. We write

$$w^{\lambda_1}_{\cdot\;\lambda_2\dots\lambda_p} \overset{\text{def}}{=} g^{\lambda_1\varrho} w_{\varrho\lambda_2\dots\lambda_p}$$

and then form the expression

$$\nabla_\mu w^{\mu}_{\cdot\;\lambda_2\dots\lambda_p}, \quad (9.23)$$

which will be a field of covariant $(p-1)$-vectors. Weitzenböck [88] calls expression (9.23) the ***tensor of Brouwer*** who was the first (1906) to study this operator in Euclidean space. For $p = 1$ this operator gives the divergence of a vector field w^λ, viz. $g^{\lambda\nu}\nabla_\lambda w_\nu$.

If the space is based on the group G_m, the difference between vector densities and vectors disappears and the divergence then coincides with the divergence in the classical sense.

The divergence of a field of vector w-densities can also be defined in a similar fashion; it will again be a vector W-density but with a valence one unit less. The definition of divergence implies that

$$\mathrm{Div}(\mathrm{Div}\,\mathfrak{v}) \equiv 0 \quad (p \geqslant 2), \tag{9.24}$$

since $\partial_\mu \partial_\nu \mathfrak{v}^{\nu\mu\lambda_3\ldots\lambda_p} = 0$, by the antisymmetry of the quantity $\mathfrak{v}^{\lambda_1\ldots\lambda_p}$.

There is nothing in classical vector analysis (in three-dimensional Euclidean space) equivalent to formula (9.24) since $\mathrm{div}\, w$ for $p = 1$ is a scalar field σ, and thus the operator $\mathrm{div}\,\sigma$ is quite meaningless.

99. The Integrability Conditions for Differential Tensor Fields. Suppose that we are given a field of quantities $\mathfrak{T}$ along a curve C lying in L_n:

$$\xi^\nu = \xi^\nu(\tau). \tag{9.25}$$

The absolute derivative is then known to be

$$\frac{D\mathfrak{T}}{d\tau} = D_\tau \mathfrak{T} = U, \quad \text{say}. \tag{9.26}$$

Conversely, we may assume that

$$\int_C U\, d\tau = \mathfrak{T},$$

if relation (9.26) is valid. The problem of integrating a field of tensor quantities along a given curve C reduces to one of integrating a system of ordinary linear differential equations of the first order. Suppose, for example, that the field of quantities U is a field of contravariant vectors v^λ. Then the field $\mathfrak{T}$ that we are looking for will also be (along C) a field of contravariant vectors x^λ. By definition we must have

$$D_\tau x^\lambda = v^\lambda,$$

that is,

$$\frac{dx^\lambda}{d\tau} + \Gamma^\lambda_{\mu\nu} x^\nu \frac{d\xi^\mu}{d\tau} = v^\lambda \quad (\lambda = 1, 2, \ldots, n). \tag{9.27}$$

Here, both $\Gamma^\lambda_{\mu\nu}$ and $\dfrac{d\xi^\mu}{d\tau}$ are given functions of the parameter τ and the system of equations (9.27) is a system of n ordinary linear differential equations of the first order for the n unknown functions x^λ. This system will have a single solution if we state the initial conditions, i.e. if we set

$$\underset{0}{x^\lambda}(\tau) = \underset{0}{x^\lambda} \qquad (\lambda = 1, 2, \dots, n),$$

where the $\underset{0}{x^\lambda}$ are given numbers. This means that if at an arbitrarily chosen point on the curve we take any arbitrary contravariant vector, the field x^λ will then be uniquely determined.

It is a different matter when the field U is given throughout the space L_n or in some particular n-dimensional domain D of that space. The problem of integration, i.e. the operation inverse to covariant differentiation, then consists in solving equations of the type

$$\nabla_\mu X^{\lambda_1 \dots \lambda_p}{}_{\mu_1 \dots \mu_q} = T^{\lambda_1 \dots \lambda_p}{}_{\mu\mu_1 \dots \mu_q}, \tag{9.28}$$

where T is a given field of quantities having a valence $(p, q+1)$ in the domain D. This problem is difficult and has not yet been solved in its full generality.

Consider, for example, the equation

$$\nabla_{[\mu} x_{\lambda_1 \dots \lambda_p]} = T_{\mu\lambda_1 \dots \lambda_p}, \tag{9.29}$$

where T is a given field of covariant $(p+1)$-vectors.

The integrability conditions for equation (9.29) are (Schouten [67], p. 115):

$$\nabla_{[\lambda} T_{\mu\lambda_1 \dots \lambda_p]} = 0.$$

On the other hand, if a given field T, which depends on both ξ^ν and x, is not assumed to represent a field of antisymmetric quantities, the integrability conditions for the equation

$$\nabla_\mu x_{\lambda_1 \dots \lambda_p} = T_{\mu\lambda_1 \dots \lambda_p}$$

are

$$\nabla_{[\nu} T_{\mu]\lambda_1 \dots \lambda_p} = -\tfrac{1}{2} \sum_{i=1}^{p} R_{\nu\mu\lambda_i}{}^{\varrho} x_{\lambda_1 \dots \lambda_{i-1} \varrho \lambda_{i+1} \dots \lambda_p},$$

and, hence, are not in a form in which the unknown x's do not appear.

For the equations

$$\nabla_\mu x_{\lambda_1 \dots \lambda_n} = a_\mu x_{\lambda_1 \dots \lambda_n},$$

where a_μ is a given vector field, the integrability conditions are

$$\nabla_{[\omega} a_{\mu]} = -\tfrac{1}{2} R_{\omega\mu\lambda}{}^{\lambda} = -\tfrac{1}{2} V_{\omega\mu}.$$

Not much progress has been made on the problem of giving not only the integrability conditions but also the solutions. Apart from the papers of Dubnov, Lopschitz, and Graiff, cited on p. 243, 244 and that of Hlavatý, little has been done to date. It should be noted, however, that in many cases, only absolute calculus has made it possible to write the integrability conditions for systems of equations in a concise and lucid form.

100. The Formulae of Green, Stokes, and Gauss–Ostrogradski. There is a certain amount of confusion regarding the terminology employed for the formulae mentioned in the heading of this section and cited below. Different authors use different names. It may truly be said that these formula represent an elegant achievement of mathematical analysis. Their implications are profound because they are global rather than local in character and, moreover, they have a vast multitude of applications. Generally speaking, these theorems enable an integral of higher dimensions to be transformed into one of lower dimensions (or conversely). The essential nature of these theorems did not become clear until they were written in vector or tensor form, which revealed the invariant and, hence, geometric character of these formulae. The literature dealing with this cycle of theorems is extensive. Authors have struggled to make use of various methods of proof, to make these proofs more rigorous, and finally, to generalize them. The generalizations concerned either transposing the results to spaces of higher dimensions or weakening the assumptions of regularity either as regards the domains of integration or the integrands, or, finally, generalizing the quantities appearing in the integrand. These theorems are still waiting for a suitable monograph to be written presenting all aspects of this cycle of theorems in a way which is both up-to-date and of a satisfactory standard as regards mathematical rigour. As far as the relatively far-reaching generality as of to the assumptions of regularity and rigour of argumentation are concerned, the matter has been resolved by M. Krzyżański [(1)] for the Gauss–Ostrogradski theorem, but only in three-dimensional Euclidean space.

Multiple integrals appear in the aforementioned integral theorems. It turns out that the classical definition of a multiple integral is inappropriate for these theorems (as, indeed, it is for many other purposes) and it has therefore had to be modified. This modification and its consistent implementation are best effected in a paper by Weyssenhoff [155] although

(1) M. Krzyżański, *Partial Differential Equations of Second Order*, Part I and II, Warszawa 1971.

insufficient emphasis is placed there on the rigorous formulation of the assumptions concerning the regularity of the domain of integration.

In view of these complications and bearing in mind the scope of this textbook, we are obliged to omit the proofs of these integral theorems and confine ourselves merely to formulating them.

101. The Oriented Multiple Integral. The classical definition of a double integral

$$\iint_D f(x, y)\,dx\,dy \tag{9.30}$$

defines this integral to be the limit of the sum

$$\sum_{i,k} f(\xi_i, \eta_k)\,(x_{i+1}-x_i)\,(y_{k+1}-y_k),$$

where $x_{i+1}-x_i > 0$, and $y_{k+1}-y_k > 0$.

It is true that this definition implies that the value of the integral can be associated with the average value of the integrand and the measure (absolute) of the domain of integration. However, the theorem concerning change of variables no longer takes on as simple a form as for a single integral, since

$$\iint_D f(x, y)\,dx\,dy = \iint_{T(D)} f[\varphi(u, v), \psi(u, v)]\,|J|\,du\,dv, \tag{9.31}$$

where J denotes the Jacobian of transformation T

$$x = \varphi(u, v), \qquad y = \psi(u, v), \tag{9.32}$$

while $T(D)$ denotes the domain in the plane (u, v) which maps into the domain D under transformation (9.32).

Formula (9.31) brings to light a serious inconvenience in that the integrand on the right-hand side contains the absolute value of the Jacobian of the transformation, and not the Jacobian itself.

This inconvenience is eliminated by the following modification of the concept of multiple integral. Consider the space X_n containing the bounded and measurable n-dimensional domain D.

We next consider in the space X_n a coordinate system (λ) which, as we know, defines the orientation of the space X_n. In addition, we impart to the domain D a certain arbitrary orientation $(*)$ and denote the oriented domain D by D^*. If we are given the integrable function

$$f(\xi^1, \ldots, \xi^n) \tag{9.33}$$

in the domain D we define the orientable integral

$$\int_{D^*} f(\xi^1, \ldots, \xi^n) \, [d\xi^{\lambda_1} \ldots d\xi^{\lambda_n}] \,^{(1)}, \tag{9.34}$$

where $(\lambda_1, \ldots, \lambda_n)$ is a permutation of the sequence $(1, 2, \ldots, n)$ as follows

$$\int_{D^*} f(\xi^1, \ldots, \xi^n) \, [d\xi^{\lambda_1} \ldots d\xi^{\lambda_n}] \overset{\text{def}}{=} \varepsilon \int_{D} f(\xi^1, \ldots, \xi^n) d\xi^1 \ldots d\xi^n. \tag{9.35}$$

Here, ε is defined by the formula

$$\varepsilon = \varepsilon_1 \varepsilon_2, \tag{9.36}$$

where $\varepsilon_1 = \pm 1$ according as the orientation of the domain D^* is, or is not, in agreement with the orientation defined by the coordinate system (λ), while $\varepsilon_2 = \pm 1$ according as the permutation $(\lambda_1, \ldots, \lambda_n)$ is even or odd.

Hence, in particular (for any D^*) we have

$$\int_{D^*} f(x, y) \, [dx \, dy] = - \int_{D^*} f(x, y) \, [dy \, dx]. \tag{9.37}$$

It follows from such a definition of an oriented integral that the formula

$$\int_{D^*} f(\xi^1, \ldots, \xi^n) \, [d\xi^1 \ldots d\xi^n] = \int_{T(D^*)} f[\varphi^1, \ldots, \varphi^n] J^{-1} [d\xi^{1'} \ldots d\xi^{n'}] \tag{9.38}$$

holds for a change of variables if the transformation T is defined by the equations

$$\xi^\nu = \varphi^\nu(\xi^{1'}, \ldots, \xi^{n'}) \tag{9.39}$$

where, as usual, we have put

$$J = |A_\lambda^{\lambda'}| = \left| \frac{\partial \xi^{\nu'}}{\partial \xi^\nu} \right|. \tag{9.40}$$

The orientation $(*)$ of the domain of integration is not affected when the variables are changed, although for $J < 0$ the orientation of the space itself, which is defined by the coordinate system, does change.

A surface X_m embedded in the space X_n is said to be *orientable* or *two-sided* if, starting with any arbitrary point p on the surface and fixing a local orientation of the surface X_m at that point by means of a simple m-vector

---

(1) We prefer not to write multiple integrals by using either the symbol $\int \ldots \int$ or the notation of J. Hadamard in which an n-fold integral is denoted by $\int\int\int$, and an $(n-1)$-fold integral by $\int\int$.

$\underset{0}{w}^{\lambda_1 \ldots \lambda_m}$, we can construct for every regular closed curve C, lying in X_n and beginning and ending at p, a continuous field of non-zero m-vectors $w^{\lambda_1 \ldots \lambda_m}$ along C, for which

$$(w^{\lambda_1 \ldots \lambda_m})_p = \underset{0}{w}^{\lambda_1 \ldots \lambda_m}.$$

Every curve X_1 is orientable, whereas a surface X_2, even though embedded in X_3, may be non-orientable (e.g. the *Möbius band*).

Let an orientable hypersurface X_{n-1} be embedded in X_n. Now, let us take a continuous field of (contravariant) vectors v along X_{n-1} not lying in X_{n-1}. Since we have an oriented space X_n (e.g. by means of a coordinate system), we can orient the hypersurface X_{n-1} uniquely by requiring that the n-vector

$$v^{[\lambda_1} w^{\lambda_2 \ldots \lambda_n]},$$

where the $(n-1)$-vector w serves to give the orientation of X_{n-1}, give for X_n an orientation in agreement with that prescribed for X_n. The hypersurface X_{n-1} is then said to be oriented by means of the field of vectors v and such an orientation is called *outer*.

In particular, if X_{n-1} is a closed orientable hypersurface, then for a given orientation of the space X_n it is possible to orient X_{n-1} uniquely by means of a field of vectors v along X_{n-1} which is directed outward from the domain bounded by X_{n-1}. This latter concept can be made precise not only in spaces E_n, but also in spaces X_n based on the group G_1.

The concept of outer orientation can be generalized to all spaces X_m, where $m < n-1$, except that $n-m$ linearly independent vector fields $\underset{i}{v}$ $(i = 1, 2, \ldots, n-m)$ not lying in X_m must then be defined in X_m.

Now suppose that we are given an orientable, closed (and, of course, regular) surface X_m $(1 \leqslant m \leqslant n-1)$ embedded in X_n. Let the continuous functions

$$F_{\lambda_1 \ldots \lambda_m} = F_{[\lambda_1 \ldots \lambda_m]} \text{ (1)} \tag{9.41}$$

be defined on X_m, and let X_m be oriented, which property we indicate by adding an asterisk: X_m^*.

We now define the *hypersurface integral*

$$I = \int_{X_m^*} F_{\lambda_1 \ldots \lambda_m} \, d\tau^{\lambda_1 \ldots \lambda_m} \tag{9.42}$$

---

(1) It is immaterial whether or not the set of functions F represents some geometric object.

as follows. Let

$$\xi^\nu = \xi^\nu(\eta^1, \ldots, \eta^m) \tag{9.43}$$

be a parametric representation of the surface X_m and let D^* denote the range of the variable for η^a $(a = 1, 2, \ldots, m)$ oriented by the curvilinear coordinate system (η^a). We now consider the integral

$$\int_{D^*} F_{\lambda_1 \ldots \lambda_m}(\eta) \det[C_1^{\lambda_1} \ldots C_m^{\lambda_m}] \, [d\eta^1 \ldots d\eta^m], \tag{9.44}$$

where

$$C_k^{\lambda_j} = \frac{\partial \xi^{\lambda_j}}{\partial \eta^k} \qquad (k, j = 1, 2, \ldots, m). \tag{9.45}$$

We write

$$I \overset{\text{def}}{=} \varepsilon \int_{D^*} F_{\lambda_1 \ldots \lambda_m} \det[C_1^{\lambda_1} \ldots C_m^{\lambda_m}] \, [d\eta^1 \ldots d\eta^m], \tag{9.46}$$

where

$$\varepsilon^2 = 1 \tag{9.47}$$

with $\varepsilon = +1$, when the orientation of X_m defined by the m-vector

$$C_1^{[\lambda_1} \ldots C_m^{\lambda_m]} \tag{9.48}$$

is in agreement with the orientation D^*, and $\varepsilon = -1$ when it is not.

102. Integral Formulae. Having defined integral (9.42), we can now proceed to formulate the generalized Gauss–Stokes theorem.

THEOREM (Gauss–Stokes). *Given in X_n an orientable surface X_m $(2 \leqslant m < n)$ whose boundary X_{m-1} is a closed orientable closed surface of class C^1 with the property that there exists a coordinate system (λ) such that X_m is part of the plane $\bar{X}_m$ defined by the equations*

$$\xi^a = 0 \qquad (a = m+1, \ldots, n) \tag{9.49}$$

and, moreover, such that every parametric line $\xi^a = \tau$, $\xi^b = \text{const}$ for $b \neq a$ intersects the boundary of X_m, that is X_{m-1}, in at most two points. Let us orient X_m and X_{m-1} so that for a continuous field of vectors v along X_{m-1}, which lie in X_m but not in X_{m-1} and which are directed outward from X_{m-1}, the orientation by means of the m-vector

$$d\tau^{\lambda_1 \ldots \lambda_m} \overset{\text{def}}{=} m! \, C_1^{[\lambda_1} \ldots C_m^{\lambda_m]} d\eta^1 \ldots d\eta^m \tag{9.50}$$

coincides with the orientation defined by the m-vector

$$v^{[\lambda_1} d\bar{\eta}^{\lambda_1} \dots d\bar{\eta}^{\lambda_m]}, \tag{9.51}$$

if $\bar{\eta}^a$ *are curvilinear coordinates on* X_{m-1}.

With these assumptions as to X_m, X_{m-1}, *and their orientations, namely* X_m^*, X_{m-1}^*, *we have that if a field of class* C^1 $(m-1)$*-vectors* $w_{\lambda_1 \dots \lambda_{m-1}}$ *is given in the closed domain* X_m, *then*

$$\int_{X_m^*} \partial_{\lambda_1} w_{\lambda_2 \dots \lambda_m} d\tau^{\lambda_1 \dots \lambda_m} = \int_{X_{m-1}^*} w_{\lambda_1 \dots \lambda_{m-1}} d\tau^{\lambda_1 \dots \lambda_{m-1}}. \tag{9.52}$$

This formula is obviously invariant; it is valid in any allowable coordinate system, not just in the particular system for which we made our assumptions concerning the nature of the domains X_m and X_{m-1}.

We now draw some conclusions from formula (9.52).

Since by the definition

$$d\tau^{\lambda_1 \dots \lambda_m} = \det (C_1^{\lambda_1}, \dots, C_m^{\lambda_m}) d\eta^1 \dots d\eta^m$$

$d\tau^{\lambda_1 \dots \lambda_m}$ is antisymmetric with respect to all the indices $\lambda_1, \dots, \lambda_m$, it follows that

$$\partial_{\lambda_1} w_{\lambda_2 \dots \lambda_m} d\tau^{\lambda_1 \dots \lambda_m} = \partial_{[\lambda_1} w_{\lambda_2 \dots \lambda_m]} d\tau^{\lambda_1 \dots \lambda_m},$$

which, by virtue of formulae (9.9) and (9.12), is equal to

$$\frac{1}{m} \mathrm{Rot}_{\lambda_1} w_{\lambda_2 \dots \lambda_m} d\tau^{\lambda_1 \dots \lambda_m}.$$

Equation (9.52) can thus be rewritten as

$$\int_{X_m^*} \mathrm{Rot}_{\lambda_1} w_{\lambda_2 \dots \lambda_m} d\tau^{\lambda_1 \dots \lambda_m} = m \int_{X_{m-1}^*} w_{\lambda_1 \dots \lambda_{m-1}} d\tau^{\lambda_1 \dots \lambda_{m-1}}. \tag{9.53}$$

For $m = 2$ this yields in particular

$$\int_{X_2^*} \mathrm{Rot}_\lambda w_\mu d\tau^{\lambda\mu} = 2 \int_{X_1^*} w_\lambda d\tau^\lambda = 2 \int_{X_1^*} w_\lambda d\xi^\lambda$$

or

$$\int_{X_2^*} (\partial_\lambda w_\mu - \partial_\mu w_\lambda) d\tau^{\lambda\mu} = 2 \int_{X_1^*} w_\lambda d\xi^\lambda. \tag{9.54}$$

If we also have at the same time $n = 2$, the formula above reduces to the Green's theorem.

Indeed, for $n = 2$ we have

$$d\tau^{\lambda\mu} = \begin{vmatrix} \dfrac{\partial\xi^\lambda}{\partial\eta^1} & \dfrac{\partial\xi^\lambda}{\partial\eta^2} \\ \dfrac{\partial\xi^\mu}{d\eta^1} & \dfrac{\partial\xi^\mu}{\partial\eta^2} \end{vmatrix} \partial\eta^1 d\eta^2 = 2\frac{\partial\xi^{[\lambda}}{\partial\eta^1}\cdot\frac{\partial\xi^{\mu]}}{\partial\eta^2}d\eta^1 d\eta^2.$$

But in this case $(m = n)$

$$\xi^\lambda = \eta^\lambda, \qquad \frac{\partial\xi^\lambda}{\partial\eta^a} = \delta^\lambda_a,$$

and, accordingly, $d\tau^{12} = d\eta^1 d\eta^2 = d\xi^1 d\xi^2$.

Equation (9.53) is thus equivalent to the equation

$$\int\limits_{X_2^*} [\partial_1 w_2 - \partial_2 w_1][d\xi^1 d\xi^2] = \int\limits_{X_1^*} (w_1 d\xi^1 + w_2 d\xi^2). \tag{9.55}$$

If the domain X_2 is given an orientation X_2^*, for example, that defined by the coordinate system (λ), then the closed curve X_1^* must be given an orientation X_1 such that the bivector $v^{[\lambda}t^{\mu]}$, where v is an outward-directed vector along X_1, and t is a vector tangent to X_1 and directed in keeping

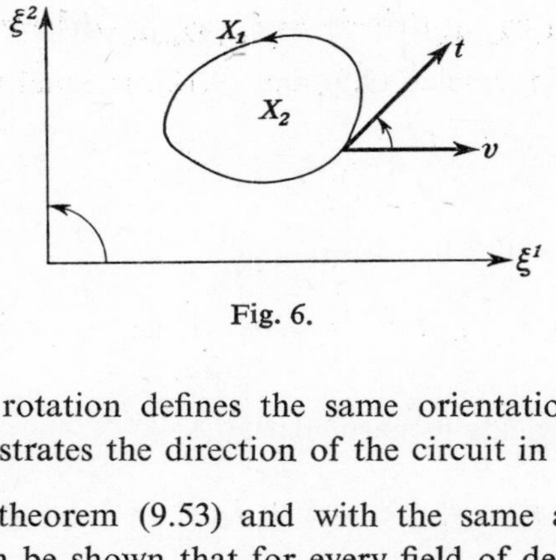

Fig. 6.

with the sense of rotation defines the same orientation as the bivector $\underset{1}{e}^{[\lambda}\underset{2}{e}^{\mu]}$. Figure 6 illustrates the direction of the circuit in X_1^*.

Analogously to theorem (9.53) and with the same assumptions as to X_m^* and X_{m-1}^* it can be shown that for every field of densities

$$\mathfrak{v}^{\lambda_1\ldots\lambda_p} \qquad (p = n - m + 1) \tag{9.56}$$

of class C^1 in the closed domain X_m the formula

$$\frac{p}{m}\int\limits_{X_m^*} \partial_\mu \mathfrak{v}^{[\lambda_1\ldots\lambda_{p-1}|\mu|} d\tau^{\lambda_p\ldots\lambda_n]} = \int\limits_{X_{m-1}^*} \mathfrak{v}^{[\lambda_1\ldots\lambda_p} d\tau^{\lambda_{p+1}\ldots\lambda_n]} \tag{9.57}$$

is valid. From this it follows in particular for $m = n$ ($p = 1$) that

$$\frac{1}{n}\int\limits_{\overset{*}{X}_n} \partial_\mu \mathfrak{v}^\mu d\tau^{\lambda_1\ldots\lambda_n} = \int\limits_{\overset{*}{X}_{n-1}} \mathfrak{v}^{[\lambda_1} d\tau^{\lambda_2\ldots\lambda_n]}. \tag{9.58}$$

For $n = 3$ and Euclidean space we obtain the familiar Gauss–Ostrogradski formula. In Euclidean space the density $\mathfrak{v}^\mu$ is a vector v^μ ($J^2 = 1$)

$$\partial_\mu \mathfrak{v}^\mu = \partial_\mu v^\mu = \partial_1 v^1 + \partial_2 v^2 + \partial_3 v^3 = \operatorname{div} v.$$

Hence

$$\begin{aligned} v^{[\lambda} d\tau^{\mu\nu]} &= \tfrac{1}{6}\{v^\lambda d\tau^{\mu\nu} + v^\mu d\tau^{\nu\lambda} + v^\nu d\tau^{\lambda\mu} - v^\lambda d\tau^{\nu\mu} - v^\mu d\tau^{\lambda\nu} - v^\nu d\tau^{\mu\lambda}\} \\ &= \tfrac{1}{3}[v^\lambda d\tau^{\mu\nu} + v^\mu d\tau^{\nu\lambda} + v^\nu d\tau^{\lambda\mu}]. \end{aligned}$$

Formula (9.58) thus reduces to the formula

$$\int\limits_{X_3^*} \operatorname{div} v d\tau^{123} = 3\int\limits_{X_2^*} v^{[1} d\tau^{23]} = \int\limits_{X_2^*} [v^1 d\tau^{23} + v^2 d\tau^{31} + v^3 d\tau^{12}]. \tag{9.59}$$

In concluding this section, we should note that this collection of theorems which express the relationships between m-fold and $(m-1)$-fold integrals is very extensive, and the monograph by Buhl [7] does not exhaust this part of the subject by any means.

The Dubnov problem [101] of finding an integral theorem of this type from which all other would follow as special cases has in our opinion not yet been solved.

Part III

The Applications of Tensor Calculus

CHAPTER X

THE APPLICATIONS OF TENSOR CALCULUS TO GEOMETRY

103. Geodesic Lines as the Extremals of a Certain Variational Problem. Jean Bernoulli considered the following problem for the first time in 1697. Given two points p_1 and p_2 on a surface S (two-dimensional and embedded in a three-dimensional Euclidean space), find the shortest of all rectifiable arcs joining p_1 and p_2 and lying in S. This is a classical problem in the calculus of variations. Its solution in the simplest case, when $S = V_2 \subset R_3$, is as follows.

Suppose that a surface $S = V_2$ embedded in a Euclidean space R_3 is given by means of the parametric equations

$$\xi^\lambda = \xi^\lambda(\eta^1, \eta^2) \quad (\lambda = 1, 2, 3). \tag{10.1}$$

Here, the ξ^λ's are rectangular Cartesian coordinates, and the right-hand sides are of class C^3 in a two-dimensional domain Ω, so that the matrix

$$\left[\frac{\partial \xi^\lambda}{\partial \eta^\alpha}\right] \tag{10.2}$$

is of the second order in Ω. Then, if curves on the surface with the property that the principal normal at every point of them coincides with the normal to the surface at that point are called *geodesics*, we have the following theorems.

THEOREM 1. *For each point p of a surface S and each direction d there exists exactly one geodesic curve passing through p and having the direction d at p.*

THEOREM 2. *For any two points p_1 and p_2 of a surafce S which are sufficiently close to each other there exists a single geodesic joining p_1 and p_2.*

If we suppress the condition of regularity of the surface S and only posit that the surface is continuous and that any two points p_1, p_2 can be joined by rectifiable arcs, then — as was first proved by D. Hilbert — at least one arc of minimum length can be drawn through every pair of points p_1, p_2.

The rigorous proof of these theorems can be found, for instance, in the book by Tonelli [81].

The problem can be generalized by considering a Riemannian space V_n with a metric tensor $g_{\lambda\mu}$, where the matrix of the coordinates of $g_{\lambda\mu}$ is of rank n, giving rise to a positive-definite quadratic form. Since the length of the arc of a curve

$$\xi^\nu = \xi^\nu(\tau) \qquad (\alpha \leqslant \tau \leqslant \beta), \tag{10.3}$$

in such a space is defined (p. 158) by means of the integral

$$\int_\alpha^\beta \left[g_{\lambda\mu}[\xi(\tau)] \frac{d\xi^\lambda}{d\tau} \cdot \frac{d\xi^\mu}{d\tau} \right]^{1/2} d\tau. \tag{10.4}$$

we are able to pose the problem of finding the arc of minimum length out of all rectifiable arcs (i.e. those arcs for which the integral above has a finite value) which join two given points p, q of the space V_n.

This variational problem comes within the scope of the Weierstrass theory, where the integrand F is of the form

$$F[\xi^1, \dots, \xi^n; p^1, \dots, p^n] \tag{10.5}$$

and is, moreover, a first-order homogeneous function (1) of the variables p^ν. In the case under consideration

$$F = [g_{\lambda\mu}[\xi] p^\lambda p^\mu]^{1/2} \tag{10.6}$$

and to find the extremal of this variational problem we need to solve the *Euler–Lagrange equations*

$$\frac{\partial F}{\partial \xi^\lambda} - \frac{d}{d\tau}\left(\frac{\partial F}{\partial p^\lambda}\right) = 0 \qquad (\lambda = 1, 2, \dots, n). \tag{10.7}$$

When the function F is given by the special formula (10.6), we have

$$\frac{\partial F}{\partial \xi^\lambda} = \frac{(\partial_\lambda g_{\mu\nu}) p^\mu p^\nu}{2F},$$

(1) That is, F satisfies the condition $F(hp_1, \dots, hp_n) \equiv hF(p_1, \dots, p_n)$ for every $h > 0$.

$$\begin{aligned}\frac{\partial F}{\partial p^\lambda} &= \frac{g_{\lambda\mu} p^\mu}{F}, \\ \frac{\partial^2 F}{\partial p^\lambda \partial p^\mu} &= \frac{g_{\lambda\mu}}{F} - \frac{g_{\lambda\nu} g_{\mu\varrho} p^\nu p^\varrho}{F^3}, \\ \frac{\partial^2 F}{\partial p^\lambda \partial \xi^\mu} &= \frac{2(\partial_\mu g_{\lambda\nu}) g_{\varrho\sigma} p^\nu p^\varrho p^\sigma - (\partial_\mu g_{\varrho\sigma}) g_{\lambda\nu} p^\nu p^\varrho p^\sigma}{2F^3}.\end{aligned} \tag{10.8}$$

Writing out the left-hand side of equations (10.7) in full, we obtain

$$\frac{\partial F}{\partial \xi^\lambda} - \frac{\partial^2 F}{\partial p^\lambda \partial \xi^\mu} \cdot \frac{d\xi^\mu}{d\tau} - \frac{\partial^2 F}{\partial p^\lambda \partial p^\mu} \cdot \frac{dp^\mu}{d\tau} = 0 \quad (\lambda = 1, 2, \ldots, n). \tag{10.9}$$

Insertion of formulae (10.8) into these equations yields

$$\frac{(\partial_\lambda g_{\mu\nu}) p^\mu p^\nu}{2F} - \frac{(\partial_\mu g_{\lambda\nu}) g_{\varrho\sigma} p^\nu p^\varrho p^\sigma}{F^3} \cdot \frac{d\xi^\mu}{d\tau} + \frac{(\partial_\mu g_{\varrho\sigma}) g_{\lambda\nu} p^\nu p^\varrho p^\sigma}{2F^3} \cdot \frac{d\xi^\mu}{d\tau} -$$
$$- \frac{g_{\lambda\mu}}{F} \cdot \frac{dp^\mu}{d\tau} + \frac{g_{\lambda\nu} g_{\mu\varrho} p^\nu p^\varrho}{F^3} \cdot \frac{dp^\mu}{d\tau} = 0 \quad (\lambda = 1, \ldots, n). \tag{10.10}$$

This system of equations cannot be solved for $dp^\lambda/d\tau$, since the determinant of the coefficients is

$$W = \left| \frac{\partial^2 F}{\partial p^\lambda \partial p^\mu} \right| = 0, \tag{10.11}$$

which follows from the fact that the function F is homogeneous in the variables p^λ. The equations above can be transformed if, following Weierstrass and Bolza, we introduce the function

$$G(\xi; p) \stackrel{\text{def}}{=} -\frac{1}{[\sum_\nu (p^\nu)^2]^2} \begin{vmatrix} \dfrac{\partial^2 F}{\partial p^\lambda \partial p_\mu} & \vdots & p^\lambda \\ \cdots & \cdots & \cdots \\ p^\mu & \vdots & 0 \end{vmatrix}. \tag{10.12}$$

The details of how equations (10.10) are brought to the normal form, as required by the theory of variational calculus, when the p^ν's in them have been replaced by $d\xi^\nu/d\tau$, are given in the book by Bianchi [2]. We confine ourselves merely to the statement that the values of $dp^\mu/d\tau$ calculated from the equations of the geodesics

$$\frac{dp^\mu}{d\tau} + \begin{Bmatrix} \mu \\ \varrho\sigma \end{Bmatrix} p^\varrho p^\sigma = 0 \quad (\mu = 1, 2, \ldots, n), \tag{10.13}$$

do indeed satisfy the Euler–Lagrange equations (10.10).

The parameter τ may be taken to be a (metric) arc because, as we shall see below, the Euler–Lagrange equations are invariant in form under transformations of the parameter for extremals.

Using equations (10.13) and the definition of the Christoffel symbols $\left\{{\mu \atop \varrho\sigma}\right\}$, we can write

$$\frac{dp^\mu}{d\tau} = -\frac{1}{2} g^{\mu\varkappa}\{\partial_\varrho g_{\varkappa\sigma} + \partial_\sigma g_{\varrho\varkappa} - \partial_\varkappa g_{\varrho\sigma}\} p^\varrho p^\sigma .$$

On substituting these values into the left-hand side of equations (10.10) bringing it to a common denominator $2F^3$, and then multiplying both sides by $2F^3$, we obtain

$$L_\lambda = (\partial_\lambda g_{\mu\nu}) g_{\varrho\sigma} p^\varrho p^\sigma p^\mu p^\nu - 2(\partial_\mu g_{\lambda\nu}) g_{\varrho\sigma} p^\nu p^\varrho p^\sigma p^\mu + (\partial_\mu g_{\varrho\sigma}) g_{\lambda\nu} p^\nu p^\varrho p^\sigma p^\mu +$$
$$+ \tfrac{1}{2} g^{\mu\varkappa} [\partial_\varrho g_{\varkappa\sigma} + \partial_\sigma g_{\varrho\varkappa} - \partial_\varkappa g_{\varrho\sigma}] p^\varrho p^\sigma [2 g_{\lambda\mu} g_{\omega\tau} p^\omega p^\tau - 2 g_{\lambda\omega} g_{\mu\tau} p^\omega p^\tau] .$$

By virtue of the relation

$$g^{\mu\varkappa} g_{\lambda\mu} = \delta^\varkappa_\lambda$$

the last term of the sum above when divided by a factor of 2, equals

$$[\partial_\varrho g_{\varkappa\sigma} + \partial_\sigma g_{\varrho\varkappa} - \partial_\varkappa g_{\sigma\varrho}] p^\varrho p^\sigma p^\omega p^\tau [\delta^\varkappa_\lambda g_{\omega\tau} - \delta^\varkappa_\tau g_{\lambda\omega}]$$
$$= [\partial_\varrho g_{\lambda\sigma} + \partial_\sigma g_{\varrho\lambda} - \partial_\lambda g_{\varrho\sigma}] p^\varrho p^\sigma p^\omega p^\tau g_{\omega\tau} - [\partial_\varrho g_{\sigma\tau} + \partial_\sigma g_{\varrho\tau} - \partial_\tau g_{\varrho\sigma}] p^\varrho p^\sigma p^\omega p^\tau g_{\lambda\omega} .$$

We insert this into the expression for L_λ and take the factor $p^\varrho p^\sigma p^\omega p^\tau$ out in front of the braces, which, of course, necessitates an initial change of dummy indices in the first three terms of the expression L_λ, to obtain

$$L_\lambda = p^\varrho p^\sigma p^\omega p^\tau \{(\partial_\lambda g_{\omega\tau}) g_{\varrho\sigma} - 2(\partial_\omega g_{\lambda\tau}) g_{\varrho\sigma} + (\partial_\tau g_{\varrho\sigma}) g_{\lambda\omega} + (\partial_\sigma g_{\lambda\sigma}) g_{\omega\tau} +$$
$$+ (\partial_\sigma g_{\varrho\lambda}) g_{\omega\tau} - (\partial_\lambda g_{\varrho\sigma}) g_{\omega\tau} - (\partial_\varrho g_{\tau\sigma}) g_{\lambda\omega} - (\partial^\sigma g_{\varrho\tau}) g_{\lambda\omega} + (\partial_\tau g_{\varrho\sigma}) g_{\lambda\omega}\} .$$

The first and sixth terms in the braces $\{\ldots\}$ cancel since the product $p^\varrho p^\sigma p^\omega p^\tau$ is left unaltered when the pairs of indices (g, σ) and (ω, τ) are interchanged. For the same reasons, the second term cancels with the fourth and fifth terms. The third term cancels with the seventh, because the factor in front of the braces is not affected when the indices ϱ and τ are interchanged. Finally, the eighth and ninth terms cancel as the factor in front of the braces does not change when the indices σ and τ are switched around. In this way we arrive at the conclusion that $L_\lambda = 0$, which was actually what we had to prove.

Of course, the equations

$$\frac{d^2 \xi^\nu}{d\tau^2} + \left\{{\nu \atop \lambda\mu}\right\} \frac{d\xi^\lambda}{d\tau} \cdot \frac{d\xi^\mu}{d\tau} = 0 \qquad (\nu = 1, 2, \ldots, n)$$

constitute a necessary condition for the arc to possess the property that its length be a minimum. The question of whether this is a sufficient condition, and if so, when, has been investigated for sufficiently close points p, q.

Integral investigations in this area were not started until recently. They constitute a branch of science called the calculus of variations in the large. This is a difficult area, involving knowledge of combinatorial topology and is represented today chiefly by the American and Soviet schools. As far as two-dimensional Riemannian spaces are concerned, most of the results in this field have been obtained thanks to the introduction by Hopf and Rinow of the concept of a "complete differentiable surface".

Let us return for the moment to the left-hand sides of the Euler–Lagrange equations (10.7). In short-hand notation we write

$$F_\lambda \overset{\text{def}}{=} \left[\frac{\partial F}{\partial \xi^\lambda} - \frac{d}{d\tau}\left(\frac{\partial F}{\partial p_\lambda}\right)\right]_{p^\nu = \frac{d\xi^\nu}{d\tau}} \qquad (\lambda = 1, 2, \ldots, n) \tag{10.14}$$

and we assume in general that F represents a function which depends in some arbitrary fashion on the coordinates ξ^λ of a point in space and the coordinates p^λ of a variable contravariant vector. A further assumption is that F is a scalar (absolute invariant), i.e. that it does not change under any transformations of coordinate system. Finally, we assume that the function F is of class C^2 with respect to all its $2n$ variables.

It is our assertion that F_λ represents a field of covariant vectors along the curve (10.3).

To verify this, we must calculate $F_{\lambda'}$ and study the transformation law. When the identity

$$F(\xi^\lambda; p^\lambda) \equiv F(\xi^{\lambda'}; p^{\lambda'}),$$

that is,

$$F(\xi^\lambda; p^{\lambda'} A^\lambda_{\lambda'}) \equiv F(\xi^{\lambda'}; p_{\lambda'})$$

is differentiated throughout, the following relations are obtained

$$\frac{\partial F}{\partial \xi^{\lambda'}} = \frac{\partial F}{\partial \xi^\lambda} A^\lambda_{\lambda'} + \frac{\partial F}{\partial p^\lambda} p^{\mu'} \partial_{\lambda'} A^\lambda_{\mu'},$$

$$\frac{\partial F}{\partial p^{\lambda'}} = \frac{\partial F}{\partial p^\lambda} A^\lambda_{\lambda'}.$$

Thus, we have

$$F_{\lambda'} = \frac{\partial F}{\partial \xi^{\lambda'}} - \frac{d}{d\tau}\left(\frac{\partial F}{\partial p^{\lambda'}}\right) = \frac{\partial F}{\partial \xi^\lambda} A^\lambda_{\lambda'} + \frac{\partial F}{\partial p^\lambda} p^{\mu'} \partial_{\lambda'} A^\lambda_{\mu'} - \frac{d}{d\tau}\left(\frac{\partial F}{\partial p^\lambda} A^\lambda_{\lambda'}\right)$$

$$= \frac{\partial F}{\partial \xi^\lambda} A^\lambda_{\lambda'} + \frac{\partial F}{\partial p^\lambda} p^{\mu'} \partial_{\lambda'} A^\lambda_{\mu'} - \frac{d}{d\tau}\left(\frac{\partial F}{\partial p^\lambda}\right) A^\lambda_{\lambda'} - \frac{\partial F}{\partial p^\lambda} \cdot \frac{d}{d\tau}(A^\lambda_{\lambda'})$$

$$= F_\lambda A^\lambda_{\lambda'} + \frac{\partial F}{\partial p^\lambda} p^{\mu'} \partial_{\lambda'} A^\lambda_{\mu'} - \frac{\partial F}{\partial p^\lambda} (\partial_{\mu'} A^\lambda_{\lambda'}) \frac{d\xi^{\mu'}}{d\tau}$$

$$= F_\lambda A^\lambda_{\lambda'} + \frac{\partial F}{\partial p^\lambda} (\partial_{\lambda'} A^\lambda_{\mu'}) \left[p^{\mu'} - \frac{d\xi^{\mu'}}{d\tau} \right] \qquad (\partial_{\lambda'} A^\lambda_{\mu'} = \partial_{\mu'} A^\lambda_{\lambda'}).$$

Accordingly,

$$[F_{\lambda'}]_{p = \frac{d\xi}{d\tau}} = F_\lambda A^\lambda_{\lambda'},$$

which proves our assertion.

The foregoing theorem implies that the Euler–Lagrange equations are invariant in character. This fact also makes it possible to give an absolute form to the calculus of variations. Important results in this direction have been obtained by the Belgian mathematician, de Donder [19], who introduced tensor symbols into the calculus of variations; this enabled him not only to simplify the reasoning but also to generalize many results to the general n-dimensional case.

104. The Geometry of Embedded Spaces. Suppose that the m-dimensional space Y_m $(1 \leqslant m \leqslant n-1)$ is embedded in the space X_n. If X_n is referred to the coordinate system (λ), the surface Y_m is defined by means of the system of parametric equations

$$\xi^\nu = \xi^\nu(\eta^\alpha) \qquad (\nu = 1, 2, \ldots, n;\ a = 1, 2, \ldots, m), \tag{10.15}$$

where the matrix

$$\left[\frac{\partial \xi^\nu}{\partial \eta^\alpha}\right] \tag{10.16}$$

is assumed to be of rank m in order to exclude singular points. We assume furthermore that, depending on subsequent needs, the functions $\xi^\nu(\eta^a)$ are sufficiently regular. It will also be assumed that on the manifolds Y_m we shall carry out changes $(a) \to (a')$ of curvilinear coordinates, independent of changes of coordinates $(\lambda) \to (\lambda')$ in the surrounding space X_n.

As on p. 150, we set

$$C^\nu_a \overset{\text{def}}{=} \frac{\partial \xi^\nu}{\partial \eta^a} \tag{10.17}$$

and we obtain a split tensor which obeys the transformation law

$$C_{a'}^{\nu'} = C_a^{\nu} A_{\nu}^{\nu'} B_{a'}^{a}. \tag{10.18}$$

Now assume that the space Y_m has been bristled by means of a system of vectors

$$\underset{\alpha}{b}^{\nu} \quad (\alpha = m+1, m+2, \ldots, n) \tag{10.19}$$

which are linearly independent and which together with the vectors

$$\underset{a}{b}^{\nu} \overset{*}{=} C_a^{\nu} \quad (a = 1, 2, \ldots, m) \tag{10.20}$$

constitute a system of n linearly independent vectors.

Assume further that a field of parallel displacement objects $\Gamma_{\mu\lambda}^{\nu}$ is defined in the space X_n in such a way that it becomes a space L_n.

We show that in the space Y_m parallel displacement of quantities can then be defined uniquely (by means of bristling) in the space Y_m, i.e. that Y_m can be turned into a space L_m.

This displacement in L_m will be referred to as an *induced displacement*, and it will also depend on both $\Gamma_{\mu\lambda}^{\nu}$ and on the bristling of the space Y_m. To reach this induced object Γ_{ba}^{c} by a geometric route, let us imagine that in Y_m we are given a curve C and that at some point p on it, we have a contravariant vector $\underset{0}{v}$ lying in Y_m. Let q be a nearby point on the curve C and let v denote the vector obtained by parallel displacement of the vector $\underset{0}{v}$ along C from p to q. The vector v will not in general lie in Y_m; we therefore take its projection $'v$ onto Y_m along the direction determined by the bristling vectors. This vector $'v$ will be called a *vector parallelly displaced* in Y_m.

Integral methods must, of course, be used to determine this induced displacement exactly. The analytic expression can be expressed, as we shall see, very simply in terms of the quantity C_{ν}^{a} (cf. p. 152) and the tensor C_a^{ν}.

Now take a vector field v^a defined in a domain of the manifold Y_m. Its coordinates in the space X_n will be

$$v^{\nu} = v^a C_a^{\nu}.$$

We now extend the field v^a in the "neighbourhood" Y_m in a regular fashion to the whole space X_n [(1)], form the covariant derivative

$$\nabla_{\mu} v^{\nu} = \partial_{\mu} v^{\nu} + \Gamma_{\mu\lambda}^{\nu} v^{\lambda},$$

---

(1) Note that if we take the absolute derivative instead of the covariant derivative the necessity of "extending" the field v^a is avoided.

and take the projectio. of the quantity $\nabla_\mu v^\nu$ onto the space Y_m (cf. also equation (4.71))

$$'\nabla_b v^a = C_b^\mu C_\nu^a \nabla_\mu v^\nu .$$

We obtain

$$'\nabla_b v^a = C_{b\nu}^{\mu a}(\partial_\mu v^\nu + \Gamma_{\mu\lambda}^\nu v^\lambda) = C_\nu^a C_b^\mu \partial_\mu v^\nu + C_{b\nu}^{\mu a} \Gamma_{\mu\lambda}^\nu v^\lambda . \tag{10.21}$$

However,

$$C_b^\mu \partial_\mu v^\nu = \frac{\partial \xi^\mu}{\partial \eta^b} \cdot \frac{\partial v^\nu}{\partial \xi^\mu} = \frac{\partial v^\nu}{\partial \eta^b} = \partial_b v^\nu$$

and we can further write

$$C_\nu^a \partial_b v^\nu = \partial_b [C_\nu^a v^\nu] - v^\nu \partial_b C_\nu^a = \partial_b v^a - v^c C_c^\nu \partial_b C_\nu^a .$$

Inserting this equality into formula (10.21), we obtain

$$'\nabla_b v^a = \partial_b v^a + v^c \{ - C_c^\nu \partial_b C_\nu^a + C_{b\nu c}^{\mu a \lambda} \Gamma_{\mu\lambda}^\nu \} . \tag{10.22}$$

We now make the definition

$$'\Gamma_{bc}^a \stackrel{\text{def}}{=} \Gamma_{\mu\lambda}^\nu C_{\nu bc}^{a\mu\lambda} - C_c^\nu \partial_b C_\nu^a \tag{10.23}$$

and on noting that the second term on the right-hand side is transformed further by the assumption $\partial_b = C_b^\mu \partial_\mu$, we then have

$$'\Gamma_{bc}^a \stackrel{\text{def}}{=} \Gamma_{\mu\lambda}^\nu C_{\nu bc}^{a\mu\lambda} - C_{bc}^{\mu\lambda} \partial_\mu C_\lambda^a . \tag{10.24}$$

It is easily verified that the $'\Gamma$'s so defined do indeed represent an object of parallel displacement for Y_m.

Now consider

$$'\Gamma_{b'c'}^{a'} = \Gamma_{\mu\lambda}^\nu C_{\nu\, b'c'}^{a'\mu\lambda} - C_{b'c'}^{\mu\lambda} \partial_\mu C_\lambda^{a'}$$

and note that

$$C_\nu^{a'} = C_\nu^a B_a^{a'} , \qquad C_{b'}^\mu = C_b^\mu B_{b'}^b , \qquad \text{etc.}^{(1)} .$$

Making use of the formula

$$C_\lambda^a C_c^\lambda = \delta_c^a$$

we can write

$$\begin{aligned} '\Gamma_{b'c'}^{a'} &= \Gamma_{\mu\lambda}^\nu C_{\nu bc}^{a\mu\lambda} B_{a\,b'c'}^{a'b\,c} + C_\lambda^{a'} \partial_{b'} C_{c'}^\lambda \\ &= \Gamma_{\mu\lambda}^\nu C_{\nu bc}^{a\mu\lambda} B_{a\,b'c'}^{a'b\,c} + B_a^{a'} C_\lambda^a B_{b'}^b \partial_b (C_c^\lambda B_{c'}^c) \\ &= \Gamma_{\mu\lambda}^\nu C_{\nu bc}^{a\mu\lambda} B_{a\,b\,c'}^{a'b\,c} + B_{a\,b'}^{a'b} C_{\lambda c}^{a\lambda} \partial_b B_{c'}^c + B_{a\,b'c'}^{a'b\,c} C_\lambda^a \partial_b C_c^\lambda \end{aligned}$$

---

(1) The relation $C_{b'}^\mu = C_b^\mu B_{b'}^b$ follows directly from the theorem concerning the differentiation of composite functions. The relation $C_\nu^{a'} = C_\nu^a B_a^{a'}$, on the other hand, results from equation (4.57) since C_ν^a is a split tensor.

$$= {}'\Gamma^{a}_{bc} B^{a'b\,c}_{a\,b'c'} + B^{a'b}_{a\,b'} C^{a\lambda}_{\lambda c} \partial_b B^{c}_{c'} = {}'\Gamma^{a}_{bc} B^{a'b\,c}_{a\,b'c'} + B^{a'b}_{a\,b'} \partial_b B^{a}_{c'}$$
$$= {}'\Gamma^{a}_{bc} B^{a'b\,c}_{a\,b'c'} - B^{b\,c}_{b'c'}\, \partial_b B^{a'}_{c},$$

which proves our theorem.

Using the parameters $'\Gamma$, we are now able to form the covariant derivative of any quantity in the space Y_m, provided it can be shown (as was shown in the case of fields of contravariant vectors) to be the projection of the covariant derivative of that quantity calculated in the enveloping space L_n.

Since, in the formula

$$'\nabla_b v^a = \partial_b v^a + {}'\Gamma^{a}_{bc} v^c,$$

v^c is defined only on Y_m and only $\partial_b v^a$ appears, the right-hand side of the last formula is independent of the particular way in which the field v^c is extended to v^λ.

The covariant derivative of a field of covariant vectors w_a lying in Y_m is defined by the formula

$$'\nabla_b w_a = \partial_b w_a - {}'\Gamma^{c}_{ba} w_c. \tag{10.25}$$

It is seen from equation (10.24) that if the object Γ is symmetric, so is the induced object $'\Gamma$. This means that under bristling the space Y_m lying in the space A_n becomes the space A_m.

For mixed (split) quantities, e.g. for G^{ν}_{a} or G^{a}_{ν}, which transform by the law

$$G^{\nu'}_{a'} = G^{\nu}_{a} A^{\nu'}_{\nu} B^{a}_{a'},$$

or

$$G^{a'}_{\nu'} = G^{a}_{\nu} A^{\nu'}_{\nu} B^{a}_{a'},$$

the covariant derivative $'\nabla_b G$ is calculated according to the formulae

$$\begin{aligned} '\nabla_b G^{\nu}_{a} &= \partial_b G^{\nu}_{a} + \Gamma^{\nu}_{\mu\lambda} C^{\mu}_{b} G^{\lambda}_{a} - {}'\Gamma^{c}_{ba} G^{\nu}_{c}, \\ '\nabla_b G^{a}_{\nu} &= \partial_b G^{a}_{\nu} + {}'\Gamma^{a}_{bc} G^{c}_{\nu} - \Gamma^{\lambda}_{\mu\nu} C^{\mu}_{b} G^{a}_{\lambda}. \end{aligned} \tag{10.26}$$

These formulae can be arrived at by setting $G^{\nu}_{a} = v^{\nu} u_a$, for instance, and employing the Leibniz rule for differentiating the product and finally using the formulae for $'\nabla_b v$ and $'\nabla_b u_a$, and then generalizing the definition.

It may also be verified *ex post facto* that

$$\begin{aligned} '\nabla_{b'} G^{\nu}_{a'} &= ('\nabla_b G^{a}_{\nu}) B^{ba}_{b'a'} A^{\nu'}_{\nu}, \\ '\nabla_{b'} G^{a'}_{\nu'} &= ('\nabla_b G^{a}_{\nu}) B^{a'b}_{ab'} A^{\nu}_{\nu'}. \end{aligned} \tag{10.27}$$

Indeed,

$$\begin{aligned}
'\nabla_{b'} G^{\nu'}_{a'} &= \partial_{b'} G^{\nu'}_{a'} + \Gamma^{\nu'}_{\mu'\lambda'} C^{\mu'}_{b'} G^{\lambda'}_{a'} - {}'\Gamma^{c'}_{b'a'} G^{\nu'}_{c'} \\
&= B^{b}_{b'} \partial_b [G^{\nu}_{a} A^{\nu'}_{\nu} B^{a}_{a'}] + [\Gamma^{\nu}_{\mu\lambda} A^{\nu'\mu}_{\nu'\mu'}{}^{\lambda}_{\lambda'} - A^{\mu}_{\mu'}{}^{\lambda}_{\lambda'} \partial_\mu A^{\nu'}_{\lambda}] C^{\varrho}_{b} A^{\mu'}_{\varrho} B^{b}_{b'} G^{\sigma}_{a} A^{\lambda'}_{\sigma} B^{a}_{a'} \\
&\quad - G^{\nu}_{d} A^{\nu'}_{\nu} B^{d}_{c'} [{}'\Gamma^{c}_{ba} B^{c'}_{c}{}^{b}_{b'}{}^{a}_{a'} - B^{b}_{b'}{}^{a}_{a'} \partial_b B^{c'}_{a}] \\
&= A^{\nu'}_{\nu} B^{b}_{b'}{}^{a}_{a'} \partial_b G^{\nu}_{a} + G^{\nu}_{a} B^{b}_{b'}{}^{a}_{a'} \partial_b A^{\nu'}_{\nu} + G^{\nu}_{a} A^{\nu'}_{\nu} B^{b}_{b'} \partial_b B^{a}_{a'} \\
&\quad + \Gamma^{\nu}_{\mu\lambda} A^{\nu'}_{\nu}{}^{\mu}_{\mu'}{}^{\lambda}_{\lambda'}{}^{\mu'}_{\varrho}{}^{\lambda'}_{\sigma} B^{b}_{b'}{}^{a}_{a'} C^{\varrho}_{b} G^{\sigma}_{a} - C^{\varrho}_{b} G^{\sigma}_{a} A^{\mu}_{\mu'}{}^{\lambda}_{\lambda'}{}^{\mu'}_{\varrho}{}^{\lambda'}_{\sigma} B^{a}_{a'}{}^{b}_{b'} \partial_\mu A^{\nu'}_{\lambda} \\
&\quad - G^{\nu}_{d} A^{\nu'}_{\nu} B^{d}_{c'}{}^{c'}_{a}{}^{b}_{b'}{}^{a}_{a'}\, {}'\Gamma^{c}_{ba} + G^{\nu}_{d} A^{\nu'}_{\nu} B^{d}_{c'}{}^{b}_{b'}{}^{a}_{a'} \partial_b B^{c'}_{a} \\
&= A^{\nu'}_{\nu}{}^{b}_{b'}{}^{a}_{a'} (\partial_b G^{\nu}_{a} + \Gamma^{\nu}_{\mu\lambda} C^{\mu}_{b} G^{\lambda}_{a} - {}'\Gamma^{c}_{ba} G^{c}_{\nu}) \\
&\quad + [G^{\nu}_{a} B^{b}_{b'}{}^{a}_{a'} \partial_b A^{\nu'}_{\nu} - C^{\mu}_{b} G^{\lambda}_{a} B^{a}_{a'}{}^{b}_{b'} \partial_\mu A^{\nu'}_{\lambda}] \\
&\quad + \{G^{\nu}_{a} A^{\nu'}_{\nu} B^{b}_{b'} \partial_b B^{a}_{a'} + G^{\nu}_{a} A^{\nu'}_{\nu} B^{d}_{c'}{}^{b}_{b'}{}^{a}_{a'} \partial_b B^{c'}_{a}\}.
\end{aligned}$$

But

$$C^{\mu}_{b} \partial_\mu A^{\nu'}_{\lambda} = \partial_b A^{\nu'}_{\lambda},$$

and thus the terms in the brackets cancel. Similarly

$$B^{a}_{a'} \partial_b B^{c'}_{a} = \partial_b (B^{a}_{a'}{}^{c'}_{a}) - B^{c'}_{a} \partial_b B^{a}_{a'} = \partial_b B^{c'}_{a'} - B^{c'}_{a} \partial_b B^{a}_{a'} = - B^{c'}_{a} \partial_b B^{a}_{a'},$$

whence

$$\begin{aligned}
G^{\nu}_{d} A^{\nu'}_{\nu} B^{d}_{c'}{}^{b}_{b'}{}^{a}_{a'} \partial_b B^{c'}_{a} &= - G^{\nu}_{d} A^{\nu'}_{\nu} B^{d}_{c'}{}^{b}_{b'}{}^{c'}_{a} \partial_b B^{a}_{a'} = - G^{\nu}_{d} A^{\nu'}_{\nu} B^{d}_{a} B^{b}_{b'} \partial_b B^{a}_{a'} \\
&= - G^{\nu}_{a} A^{\nu'}_{\nu} B^{b}_{b'} \partial_b B^{a}_{a'},
\end{aligned}$$

so that the terms in braces also cancel and the first formula of (10.27) is therefore proved. The second formula is proved in a similar manner.

The tensor C^{ν}_{a} is the simplest mixed tensor. We introduce the notation

$$H^{\nu}_{ba} \stackrel{\text{def}}{=} {}'\nabla_b C^{\nu}_{a} \tag{10.28}$$

and following Schouten we refer to the tensor H as the (three-index) *curvature tensor* of the embedded space Y_m. Note that in order to define this tensor, the space Y_m must be bristled since, as is implied by the first formula of (10.27), we have

$$H_{ba}{}^{\nu} = \partial_b C^{\nu}_{a} + \Gamma^{\nu}_{\mu\lambda} C^{\mu\lambda}_{ba} - {}'\Gamma^{c}_{ba} C^{\nu}_{c},$$

from which we see that the quantity C^{a}_{λ} (which also appears implicitly in the formulae for ${}'\Gamma^{c}_{ba}$) occurs here as well. Making use of definition (10.24), we obtain the following formulae for the coordinates of the curvature tensor

$$H_{ba}{}^{\nu} = E^{\nu}_{\lambda} \partial_b C^{\lambda}_{a} + \Gamma^{\varrho}_{\mu\lambda} C^{\mu\lambda}_{ba} E^{\nu}_{\varrho}, \tag{10.29}$$

where E^{ν}_{ϱ} is a shorthand symbol for

$$E^{\nu}_{\varrho} = A^{\nu}_{\varrho} - C^{\nu}_{\varrho} = A^{\nu}_{\varrho} - C^{\nu}_{a} C^{a}_{\varrho}. \tag{10.30}$$

It is immediately seen from formulae (10.29) that the tensor H^{ν}_{ba} is symmetric with respect to the inferior suffixes when the object Γ is symmetric and, moreover,

$$\partial_b C^{\nu}_a = \frac{\partial^2 \xi^{\nu}}{\partial \eta^b \partial \eta^a} = \frac{\partial^2 \xi^{\nu}}{\partial \eta^a \partial \eta^b} = \partial_b C^{\nu}_b.$$

Now suppose that the surrounding space is a Riemannian space V_n, whose metric tensor $g_{\mu\lambda}$ defines a positive definite form. Then a preferred object of parallel displacement, given in terms of Christoffel symbols, exists in V_n.

Let $'g_{ab}$ denote the projection of the metric tensor $g_{\lambda\mu}$ onto the space Y_m. The order of the tensor $'g_{ab}$ can be shown to be equal to m. Even more can be shown, viz. that the coefficients $'g_{ab}$ define a positive definite quadratic form.

Indeed, assume that $\sum_a (\lambda^a)^2 > 0$ and calculate $W = 'g_{ab} \lambda^a_{\mathbf{a}} \lambda^b = g_{\nu\mu} C^{\nu\mu}_{ab} \lambda^a \lambda^b = g_{\nu\mu} \varrho^{\nu} \varrho^{\mu}$, where we have set $\varrho^{\nu} = C^{\nu}_a \lambda^a$. If we had $\varrho^{\nu} = 0$ for $\nu = 1, 2, \ldots, n$, then since the matrix C^{ν}_a is of rank m, we would have to have $\lambda^a = 0$ for $a = 1, 2, \ldots, m$, contrary to our assumption. However, since $\sum_{\nu} (\varrho^{\nu})^2 > 0$, the quadratic form $W = g_{\nu\mu} \varrho^{\nu} \varrho^{\mu} > 0$, which proves our assertion. We now show that the metric of the embedded space Y_m induced by the tensor $'g_{ab}$ coincide with the metric established by the tensor $g_{\lambda\mu}$.

Indeed, consider a contravariant vector v^a in the space Y_m. Its length in terms of the tensor $'g_{ab}$ is

$$|v| = ['g_{ab} v^a v^b]^{1/2} = [C^{\lambda}_a C^{\mu}_b g_{\lambda\mu} v^a v^b]^{1/2}. \tag{10.31}$$

Now the vector v^a is equally a vector of the space X_n and as such has coordinates v^{λ} equal to

$$v^{\lambda} = C^{\lambda}_a v^a \tag{10.32}$$

and its length in terms of the tensor $g_{\lambda\mu}$ is

$$|v| = [g_{\lambda\mu} v^{\lambda} v^{\mu}]^{1/2}. \tag{10.33}$$

Inserting the values (10.32) into the right-hand side of equation (10.31) we obtain

$$|v| = [C^{\lambda}_a v^a C^{\mu}_b v^b g_{\lambda\mu}]^{1/2} = [g_{\lambda\mu} v^{\lambda} v^{\mu}]^{1/2},$$

which is in agreement with formula (10.33).

Similarly for the angular metrics: the scalar product of the vectors $\underset{1}{v}$ and $\underset{2}{v}$ in terms of the induced metric is $\underset{1}{v} \cdot \underset{2}{v} = {}'g_{ab}\underset{1}{v}^a \underset{2}{v}^b$, which we rearrange further to get

$$\underset{1}{v} \cdot \underset{2}{v} = C^{\lambda\mu}_{ab} g_{\lambda\mu} \underset{1}{v}^a \underset{2}{v}^b = g_{\lambda\mu} C^{\lambda}_{a} \underset{1}{v}^a C^{b}_{\mu} \underset{2}{v}^b = g_{\lambda\mu} \underset{1}{v}^\lambda \underset{2}{v}^\mu .$$

It is thus equal to the scalar product calculated according to the metric in force in the space V_n, which is what was to be shown.

The following problem now arises. The metric of the embedded space based on the tensor $'g_{ab}$ is independent of how the space Y_m is bristled. The covariant derivative of quantities of the space Y_m depends on the bristling. On the other hand, this covariant derivative can be defined in terms of the tensor $'g_{ab}$, since the rank of its coordinate matrix is maximal and thus equal to m, and the covariant derivative defined by means of the Christoffel symbols $'\left\{ {c \atop ba} \right\}$ will then be made independent of any bristling of the space Y_m. The question is therefore under which bristling will the two derivatives be identical, i.e. when will

$$'\Gamma^c_{ba} = {}'\left\{ {c \atop ba} \right\}. \tag{10.34}$$

It can be shown that if the bristling is orthogonal to Y_m, then relation (10.34) does indeed hold.

It is proved that

$$'\nabla_c {}'g_{ba} = 0 . \tag{10.35}$$

Whence relation (10.34) then follows from the theorem on p. 244. Our space Y_m thus becomes a Riemannian space V_m.

105. A Hyperspace V_{n-1} Embedded in the Space V_n. We now go on to consider the special case when $m = n-1$. Schouten originated [67] the geometry of the subspaces Y_{n-1} for the general case, when the space X_n has a field of symmetric objects of parallel displacement, i.e. when the space is A_n. This geometry is based on the definition of two vectors at every point of the hypersurface Y_{n-1} — a covariant vector t_λ (tangent to Y_{n-1}) and a contravariant one n^λ (the bristling of the hyperspace Y_{n-1}) — and on the introduction of two *normalization conditions*. This theory is interesting although it does not lend itself to a brief presentation so that the interested reader is referred to Schouten's book *Der Ricci-Kalkül* for a concise exposition of it.

Let us now proceed immediately to the case of a Riemannian space based on the tensor $g_{\lambda\mu}$, which gives a positive definite form. In this case, we have available a distinguished bristling by means of a field of vectors orthogonal to Y_{n-1}, e.g. by means of the vector n^λ which is the vector product of the vectors

$$\underset{a}{b}^\lambda \overset{\text{def}}{=} C_a^\lambda \quad (a = 1, 2, \ldots, n-1), \tag{10.36}$$

so that

$$n^\lambda = g^{\lambda\mu} e_{\mu_1 \ldots \mu_{n-1}\mu} \underset{1}{b}^{[\mu_1} \ldots \underset{n-1}{b}^{\mu_{n-1}]}. \tag{10.37}$$

Under such a bristling the space Y_{n-1} becomes the space V_{n-1} with the positive definite tensor

$$'g_{ab} = C_{ab}^{\lambda\mu} g_{\lambda\mu}.$$

If the matrix $[C_a^\lambda]$ is of rank $(n-1)$, then the $(n-1)$-th vector

$$\underset{1}{b}^{[\mu_1} \ldots \underset{n-1}{b}^{\mu_{n-1}]}$$

is not zero and the vector n^λ is linearly independent of the vectors (10.36). We define

$$N^\lambda \overset{\text{def}}{=} \text{unit vector } n^\lambda = \frac{n^\lambda}{|n|} \tag{10.38}$$

and set

$$N_\lambda = g_{\lambda\mu} N^\mu.$$

Next, we put

$$h_{ba} \overset{\text{def}}{=} C_{ba}^{\mu\lambda} \nabla_\mu N_\lambda. \tag{10.39}$$

Here, the vector N depends on the system of curvilinear coordinates (a) in the space Y_{n-1} and, therefore, we cannot state *a priori* that h_{ba} will be a tensor, even though

$$C_{b'a'}^{\mu\ \lambda} = C_{ba}^{\mu\lambda} B_{b'a'}^{b\ a}.$$

Suppose that we take some other coordinate system (a') and calculate

$$\underset{1'}{b}^{[\mu_1} \ldots \underset{(n-1)'}{b}^{\mu_{n-1}]}.$$

Since

$$\underset{a'}{b}^\nu \overset{*}{=} C_{a'}^\nu = C_a^\nu B_{a'}^a \overset{*}{=} B_{a'}^a \underset{a}{b}^\nu,$$

we have

$$\underset{1'}{b}^{[\mu_1} \ldots \underset{(n-1)'}{b}^{\mu_{n-1}]} = \underset{1}{b}^{[\mu_1} B_{1'}^{|1|} \ldots \underset{n-1}{b}^{\mu_{n-1}]} B_{(n-1)'}^{n-1} = \underset{a_1}{b}^{[\mu_1} B_{1'}^{|a_1|} \ldots \underset{a_{n-1}}{b}^{\mu_{n-1}]} B_{(n-1)'}^{a_{n-1}}$$

$$= \underset{1}{b}^{[\mu_1} \ldots \underset{n-1}{b}^{\mu_{n-1}]} \varepsilon_{a_1 \ldots a_{n-1}} B_{1'}^{a_1} \ldots B_{(n-1)'}^{a_{n-1}} = \underset{1}{b}^{[\mu_1} \ldots \underset{n-1}{b}^{\mu_{n-1}]} \det B_{a'}^a.$$

Setting

$$\delta = \det(B_a^{a'})$$

we can write

$$\underset{(a')}{n^\lambda} = \frac{1}{\delta} \underset{(a)}{n^\lambda}.$$

This implies that

$$\underset{(a')}{N^\lambda} = \operatorname{sgn} \delta \cdot \underset{(a)}{N^\lambda},$$

and consequently that

$$\overset{(a')}{N_\lambda} = \operatorname{sgn} \delta \cdot \overset{(a)}{N_\lambda}. \tag{10.40}$$

This leads to a further conclusion: that h_{ab} is a covariant G-tensor of the space V_{n-1}.

It can be shown that the tensor h_{ab} is symmetric in both indices and that it is related to the tensor $H_{ab}{}^{\nu}$, viz.

$$H_{ab}{}^{\nu} = -h_{ab} N^{\nu}. \tag{10.41}$$

The tensor h_{ab} is called the *second fundamental tensor of the hyperspace* V_{n-1}. The rank of the matrix h_{ab} need not be equal to $n-1$, and many be lower, and indeed may even be zero. This last possibility occurs if and only if the hyperspace V_{n-1} is a geodesic space. A space X_m embedded in a space A_n is called a *geodesic space* at the point p if every geodesic of A_n, originating at the point p and having at that point a direction lying in X_m, lies entirely in X_m. If X_m is geodesic at every point p, then it is referred to briefly as a geodesic space.

The tensor h_{ab} plays a major part in many problems of the geometry of the subspace V_{n-1}.

We shall briefly mention only the principal problems, referring the reader to the book by Schouten and Struik [69] and to specialized works for further details.

106. Normal Geodesic Coordinates. Suppose that a hyperspace V_{n-1} is given in Riemannian space V_n by means of the parametric representation

$$\xi^\nu = \xi^\nu(\eta^a) \qquad (\nu = 1, 2, \ldots, n;\ a = 1, 2, \ldots, n-1), \tag{10.42}$$

where we assume that the matrix

$$C_a^\nu = \frac{\partial \xi^\nu}{\partial \eta^a}$$

has a maximal rank of $n-1$.

Let N^λ be a field of unit vectors normal to V_{n-1} (cf. definition (10.38)). Through each point of the hyperspace V_{n-1} we draw a geodesic (of the space V_n) tangent to N. The result is an $(n-1)$-parameter family of geodesics which we call a congruence of geodesics. The vector field N^λ gives an orientation to each geodesic of this congruence. Now take a number ε (positive or negative) and on each geodesic from point p on V_{n-1} take a point q such that the relative length of the geodesic from p to q be ε. The geometric locus of the points q will be a certain hypersurface $\tilde{V}_{n-1}$, provided that the value of ε is not too large. It can be shown that the geodesics of our congruence are orthogonal to the hypersurface $\tilde{V}_{n-1}$.

A rigorous proof of this theorem is not easy. We give the proof for the special case when $V_n = R_n$. Let ξ^λ denote the Cartesian coordinates. The parametric equations of the hypersurface $\tilde{V}_{n-1}$ can then be written as

$$\xi^\nu = \xi^\nu(\eta^a) + N^\nu(\eta^a)\varepsilon. \tag{10.43}$$

In this case we have

$$\tilde{C}_a^\nu = C_a^\nu + \varepsilon \frac{\partial N^\nu}{\partial \eta^a}; \tag{10.44}$$

from which it follows that

$$\begin{aligned} \tilde{n}^\lambda &= g^{\lambda\mu} e_{\mu_1 \ldots \mu_{n-1}\mu} \tilde{C}_1^{[\mu_1} \ldots \tilde{C}_{n-1}^{\mu_{n-1}]} \\ &= g^{\lambda\mu} e_{\mu_1 \ldots \mu_{n-1}\mu} \left(C_1^{[\mu_1} + \varepsilon \frac{\partial N^{[\mu_1}}{\partial \eta^1} \right) \ldots \left(C_{n-1}^{\mu_{n-1}]} + \varepsilon \frac{\partial N^{\mu_{n-1}]}}{\partial \eta^{n-1}} \right). \end{aligned} \tag{10.45}$$

But since the field of vectors N^λ is a field of unit vectors, each of the vectors $\dfrac{\partial N^\lambda}{\partial \eta^a}$ is perpendicular to N^λ, or is a null vector, and we can therefore write

$$\frac{\partial N^\lambda}{\partial \eta^a} = \varrho_a^b C_b^\lambda, \tag{10.46}$$

where the ϱ_a^b are scalars which depend, of course, on η^a. Substitution from these relations into equation (10.45) now yields

$$\tilde{n}^\lambda = g^{\lambda\mu} g e_{\mu_1 \ldots \mu_{n-1}\mu} (C_1^{[\mu_1} + \varepsilon \varrho_1^b C_b^{[\mu_1}) \ldots (C_{n-1}^{\mu_{n-1}]} + \varepsilon \varrho_{n-1}^b C_b^{\mu_{n-1}]}). \tag{10.47}$$

On carrying out the multiplications in the parentheses and omitting the zero terms which occur when the product contains at least two factors $C_a^{\mu_a}$ and $C_b^{\mu_b}$, where $a = b$, on the right-hand side we have

$$C_1^{[\mu_1} \ldots C_{n-1}^{\mu_{n-1}]} \cdot W,$$

where W is a scalar expression of the form

$$W = 1+\varepsilon\varrho_a^a + \ldots + \varepsilon^{n-1}\det(\varrho_b^a). \tag{10.48}$$

Equation (10.47) can thus be written as

$$\tilde{n}^\lambda = g^{\lambda\mu} e_{\mu_1 \ldots \mu_{n-1}\mu} C_1^{[\mu_1} \ldots C_{n-1}^{\mu_{n-1}]} \cdot W = n^\lambda \cdot W. \tag{10.49}$$

Hence the conclusion is that the vector $\tilde{n}^\lambda$, normal to $\tilde{V}_{n-1}$, is proportional to the vector n^λ, which is normal to V_{n-1}, so that n^λ is normal to $\tilde{V}_{n-1}$ and the theorem has therefore been proved.

Let us now confine ourselves to an interval $\langle\alpha, \beta\rangle$ of values of ε such that $W \neq 0$, which interval certainly contains a neighbourhood of zero. The matrix $\tilde{C}_a^\nu$ will then be of rank $n-1$, and the set of geodesics corresponding to the values of the interval $\langle\alpha, \beta\rangle$ will occupy an n-dimensional set Ω. Let (λ') denote a coordinate system in Ω such that

$$\begin{aligned} \xi^{\lambda'} &= \eta^a \cdot \delta_a^{\lambda'} \quad \text{for } \lambda = 1, 2, \ldots, n-1, \\ \xi^{n'} &= \varepsilon, \end{aligned} \tag{10.50}$$

where ε is the relative length of the geodesic originating at a point in V_{n-1} with coordinates η^a. We further assume that it can be arranged, by restricting the interval $\langle\alpha, \beta\rangle$, that the set Ω is such that exactly one geodesic of our congruence passes through each point of the interval. Such a system of coordinates $\xi^{\lambda'}$ will be called a system of *normal geodesic coordinates*, i.e. normal with respect to the hypersurface V_{n-1}. It can easily be shown that the metric tensor of a space V_n in such a coordinate system is of the form

$$\begin{aligned} g_{\lambda'\mu'} &= {}'\tilde{g}_{ab}\delta_{\lambda'}^a \delta_{\mu'}^b \quad \text{for } \lambda, \mu = 1, 2, \ldots, n-1, \\ g_{\lambda'n'} &= g_{n'\lambda'} = 0 \quad \text{for } \lambda = 1, 2, \ldots, n-1, \\ g_{n'n'} &= 1. \end{aligned} \tag{10.51}$$

Recall that

$${}'\tilde{g}_{ab} \overset{\text{def}}{=} \tilde{C}_a^\nu \tilde{C}_b^\mu g_{\nu\mu},$$

whereby $g_{\lambda'\mu'}$ are in general functions of all the $\xi^{\lambda'}$'s, and not only of $\xi^{\lambda'}$ for $\lambda = 1, 2, \ldots, n-1$.

107. The Curvature Vector. The Relative and Forced Curvature Vectors. Given in the space A_n a regular curve V_1

$$\xi^\nu = \xi^\nu(\sigma) \quad (\nu = 1, 2, \ldots, n). \tag{10.52}$$

We have already seen that in the space A_n the equations of geodesics are of the form

$$\frac{d^2\xi^\nu}{d\sigma^2}+\Gamma^\nu_{\mu\lambda}\frac{d\xi^\mu}{d\sigma}\cdot\frac{d\xi^\lambda}{d\sigma}=0 \quad (\nu=1,2,\dots,n). \tag{10.53}$$

We show that for any curve (10.52) the left-hand sides in (10.53) represent the coordinates of a contravariant vector attached at the variable point of the curve (10.52).

Indeed,

$$k^\nu \stackrel{\text{def}}{=} \frac{d^2\xi^\nu}{d\sigma^2}+\Gamma^\nu_{\mu\lambda}\frac{d\xi^\mu}{d\sigma}\cdot\frac{d\xi^\lambda}{d\sigma}, \tag{10.54}$$

being the absolute derivative of the vector field $\frac{d\xi^\nu}{d\sigma}$, must represent a vector. The vector k^ν is said to be the *curvature vector* of the curve (10.52) at the point under consideration, and one can say that geodesic lines are characterized by the property of having a null curvature vector at each point on the curve. A point p for which this vector is a null vector is called a *point of inflexion* (of the first kind).

For a curve in the space V_n the curvature vector is defined in terms of the Christoffel symbols, viz.

$$k^\nu = \frac{d^2\xi^\nu}{d\sigma^2}+\begin{Bmatrix}\nu\\ \mu\lambda\end{Bmatrix}\frac{d\xi^\mu}{d\sigma}\cdot\frac{d\xi^\lambda}{d\sigma}=\frac{Di^\nu}{d\sigma}\quad\left(i^\nu=\frac{d\xi^\nu}{d\sigma}\right),$$

where σ stands for the metric arc.

Now suppose that the curve (10.52) lies on the hypersurface V_{n-1}. Let us construct the projection of the vector k^ν onto V_{n-1} (the vector k^ν will not in general lie in V_{n-1}!), and let us denote that projection by $'k$

$$'k^a \stackrel{\text{def}}{=} k^\nu C^a_\nu;$$

the coordinates of this projection in the space V_n are

$$'k^\nu = C^\nu_a{}'k^a = k^\lambda C^a_\lambda C^\nu_a = k^\lambda C^\nu_\lambda.$$

The vector $'k$ is called the *relative curvature vector*. The difference

$$''k^\nu \stackrel{\text{def}}{=} k^\nu - {}'k^\nu \tag{10.55}$$

is referred to as the *forced curvature vector*.

Since

$$C^a_\nu{}''k^\nu = C^a_\nu k^\nu - C^a_\nu{}'k^\nu = {}'k^a - C^a_\nu{}'k^b C^\nu_b = {}'k^a - \delta^a_b{}'k^b = {}'k^a - {}'k^a = 0,$$

the vector $''k$ has the same direction as the bristling vector, i.e.

$$''k^\nu = \gamma N^\nu. \tag{10.56}$$

The coefficient γ, which is a relative number, is known as the *normal curvature of the curve* (10.52).

Suppose that the curve (10.52) is a geodesic in the space V_{n-1}. This means that

$$\frac{'Di^a}{d\sigma} = 0,$$

i.e. that

$$C^a_\mu \frac{Di^a}{d\sigma} = 0$$

or

$$C^a_\mu k^\mu = 0,$$

i.e. that the relative curvature vector is identically zero; conversely, if this is the case, the curve (10.52) is a geodesic in the space V_{n-1}.

The coefficient γ is expressed in terms of the tensor h_{ab} as follows

$$\gamma = h_{ab} i^a i^b. \tag{10.57}$$

The proof is omitted.

Since the vector $''k^\nu$ lies on a normal to V_{n-1} and since γ depends only on i^a, two curves of the hypersurface, with a common tangent at a point of intersection, have the same forced curvature vector at that point. Moreover, if θ denotes the angle between the vectors k^ν and $''k^\nu$, we have the relation

$$|''k| = |k| \cdot |\cos\theta|, \tag{10.58}$$

which can be regarded as a generalization of the theorem of Meusnier.

The forced curvature vector is at the same time also the curvature vector for geodesics of the space V_{n-1} having a common tangent direction at the point under consideration.

108. Asymptotic Lines. Curves on a hypersurface for which at each point the normal curvature $\gamma = 0$ are called *asymptotic lines* of the space V_{n-1}. The directions i^a for which the relation

$$h_{ab} i^a i^b = 0, \tag{10.59}$$

is satisfied are known as the *asymptotic directions*.

If the coordinate matrix of the tensor h_{ab} is of rank $n-1$ the asymptotic directions form a cone in the local space E_{n-1}. If the form $h_{ab}\lambda^a\lambda^b$ is positive definite, there are no (real) asymptotic directions at the point under consideration.

In addition to ordinary asymptotic curves, defined above, one can introduce concepts of asymptotic curves of various orders (Hlavatý [32], Schouten–Struik [69]).

109. Spherical Points. If the coefficient γ has a constant value for every direction at a particular point p of the hypersurface V_{n-1}, then p is called a *spherical point* (umbilical point).

At such a point we have

$$h_{ab} = \varrho' g_{ab}, \tag{10.60}$$

where ϱ' is a scalar coefficient. If $\varrho' = 0$, then V_{n-1} is a geodesic hypersurface (p. 262) at that point, and p itself is called a *point of flattening*.

It can be shown that if $V_{n-1} \subset R_n$ and if V_{n-1} is geodesic at every point p, then $V_{n-1} = R_{n-1}$.

110. Conjugate Directions. At a point $p \in V_{n-1} \subset V_n$ take two non-zero vectors v^a and $\overset{*}{v}{}^a$ with the property

$$h_{ab} v^a \overset{*}{v}{}^b = 0. \tag{10.61}$$

The directions determined by such vector v and $\overset{*}{v}$ are said to be *conjugate to each other*.

111. Curvature Lines. Let us consider the tensor h_{ab} and assume that it is not proportional to $'g_{ab}$, that is, that the point is not spherical. The directions of non-zero vectors v with the property

$$h_{ab} v^a = \varrho' g_{ab} v^a \tag{10.62}$$

are called the *principal directions of the tensor* h_{ab}. These directions are mutually perpendicular to one another.

The values of ϱ_c for the corresponding principal directions are calculated as the roots of the characteristic equation

$$\det(h_{ab} - \varrho' g_{ab}) = 0. \tag{10.63}$$

If this equation has multiple roots, there are infinitely many principal directions. The values of ϱ_c, called the *principal curvatures of the hypersurface* V_{n-1} at the point under consideration, may also be obtained from the relation

$$h_{ab} = \sum_{c} \varrho_c \overset{c}{e}_a \overset{c}{e}_b, \tag{10.64}$$

where $\underset{a}{e^c}$ and $\overset{a}{e}_c$ constitute, respectively, an orthogonal set of contravariant vectors and the inverse set of covariant vectors.

Lines on the hypersurfaces V_{n-1}, with the property that at each point the vector tangent to the line has the principal direction of the vector v^a which satisfies relation (10.62), are called *curvature lines*.

112. The Equations of Gauss and Mainardi–Codazzi. It is well known from the classical differential geometry of two-dimensional surfaces, embedded in three-dimensional Euclidean space that the coordinates of the metric tensor $'g_{ab}$ of a surface and the coordinates of the tensor h_{ab} are so related that in general for any arbitrary tensors $'g_{ab}$ and h_{ab} given *a priori* there does not exist a surface having these given $'g_{ab}$ and h_{ab} as the corresponding tensors. G. Mainardi (1857) and D. Codazzi (1868) gave the conditions (for $n = 3$ and $V_3 = R_3$) which tensors $'g_{ab}$ and h_{ab} must satisfy in order for a surface corresponding to them to exist. These equations, written in ordinary analytic form, are very complicated, obscure, and non-symmetric. Only with the appearance of tensor calculus on the scene did it become possible for these conditions to be obtained in a simple manner and for them to be written in a concise invariant form. Similarly, the Gauss equation giving the relationship between the curvature tensors of an enveloping space and the embedded space and tensor h_{ab} could not be written concisely without tensor symbols.

In general, by $R_{\omega\mu\lambda\nu}$ we denote the transvection of the curvature tensor onto the metric tensor for the space V_n, i.e.

$$R_{\omega\mu\lambda\nu} \overset{\text{def}}{=} R_{\omega\mu\lambda}{}^{\varrho} g_{\varrho\nu}. \tag{10.65}$$

If $\overline{R}_{abcd}$ is used to denote the curvature tensor of the embedded space V_{n-1}, we have the following generalized Gauss equation

$$\overline{R}_{abcd} = R_{\omega\mu\lambda\nu} C^{\omega\mu\lambda\nu}_{abcd} + 2h_{d[a}h_{b]c} \text{ (1)}. \tag{10.66}$$

(1) Schouten and Struik prefer to use the symbol $h_{[a[c}h_{b]d]}$, where the two square brackets are taken to be independent of each other, with the closing brackets subordinate to the opening brackets "monotonically", i.e. a, $[ab]$, $[cd]$, and accordingly

$$h_{[a[c}h_{b]d]} = \tfrac{1}{2}\{h_{a[c}h_{|b|d]} - h_{b[c}h_{|a|d]}\} = \tfrac{1}{4}\{h_{ac}h_{bd} - h_{ad}h_{bc} - h_{bc}h_{ad} + h_{bd}h_{ac}\}$$
$$= \tfrac{1}{2}\{h_{ac}h_{bd} - h_{ad}h_{bc}\} = h_{a[c}h_{|b|d]},$$

which, for a symmetric tensor, is equal to $-h_{d[a}h_{b]c}$.

The Mainardi–Codazzi equations take the form

$$R_{\omega\mu\lambda\nu} N^{\nu} C_{abc}^{\omega\mu\lambda} = 2' \nabla_{[a} h_{b]c}. \tag{10 67}$$

It is thus seen that the form of equations (10.66) and (10.67) is both invariant and extremely simple.

If in particular $V_n = R_n$, that is, if the enveloping space is Euclidean, then $R_{\omega\mu\lambda}{}^{\nu} = 0$ and, accordingly, $R_{\omega\mu\lambda\nu} = 0$ while equations (10.66) and (10.67) assume the form

$$\bar{R}_{abcd} = 2h_{[d[a} h_{b]c]}, \qquad '\nabla_{[a} h_{b]c} = 0. \tag{10.68}$$

In the special case when $n = 3$ the equations above assume the form familiar to us from the classical theory of the differential geometry of surfaces. There will then be two Mainardi–Codazzi equations for it is sufficient to set $a = 1$, $b = 2$, and either $c = 1$ or $c = 2$; if $a = 2$, $b = 1$, we obtain the same equations, while when $a = b$, we get the identity $0 = 0$. Furthermore these will take on the form

$$\begin{aligned}
\partial_1 h_{21} - \partial_2 h_{11} &= -\left\{ {1 \atop 12} \right\} h_{11} + \left[\left\{ {1 \atop 11} \right\} - \left\{ {2 \atop 12} \right\} \right] h_{12} + \left\{ {2 \atop 11} \right\} h_{22}, \\
\partial_1 h_{22} - \partial_2 h_{12} &= -\left\{ {1 \atop 22} \right\} h_{11} + \left[\left\{ {1 \atop 12} \right\} - \left\{ {2 \atop 22} \right\} \right] h_{12} + \left\{ {2 \atop 12} \right\} h_{22}.
\end{aligned} \tag{10.69}$$

To arrive at the classical form of Gauss equations, note that the relations

$$R_{(\omega\mu)\lambda\nu} = R_{\omega\mu(\lambda\nu)} = 0 \quad \text{and} \quad R_{\omega\mu\lambda\nu} = R_{\lambda\nu\omega\mu} \tag{10.70}$$

hold, just as in the case of the tensor $G_{\omega\mu\lambda\nu}$ on p. 78, i.e. for $n = 2$ the tensor $R_{\omega\mu\lambda\nu}$ has just one essential coordinate, e.g. R_{1212}. We therefore conclude that the tensors $R_{\omega\mu\lambda\nu}$ and $G_{\omega\mu\lambda}{}^{\nu}$ can differ only by a scalar factor

$$R_{\omega\mu\lambda\nu} = \sigma G_{\omega\mu\lambda\nu}.$$

Let us calculate this factor. We know that

$$R_{\omega\mu\lambda\nu} = R_{\omega\mu\lambda}{}^{\varrho} g_{\varrho\nu}.$$

We multiply the relation above by $g^{\nu\omega}$ and, of course, sum over
result is

$$R_{\omega\mu\lambda\nu} g^{\nu\omega} = R_{\omega\mu\lambda}{}^{\varrho} g_{\varrho\nu} g^{\nu\omega} = R_{\omega\mu\lambda}{}^{\varrho} \delta_{\varrho}^{\omega} = R_{\omega\mu\lambda}{}^{\omega} = R_{\mu\lambda}.$$

We thus have

$$\begin{aligned}
R_{\mu\lambda} &= \sigma G_{\omega\mu\lambda\nu} g^{\nu\omega} = \tfrac{1}{2}\sigma g^{\nu\omega}[g_{\omega\mu} g_{\lambda\nu} - g_{\lambda\mu} g_{\omega\nu}] = \tfrac{1}{2}\sigma[\delta_{\mu}^{\nu} g_{\lambda\nu} - 2g_{\lambda\mu}] \\
&= \tfrac{1}{2}\sigma[g_{\lambda\mu} - 2g_{\lambda\mu}] = -\tfrac{1}{2}\sigma g_{\lambda\mu} = -\tfrac{1}{2}\sigma g_{\mu\lambda}.
\end{aligned} \tag{10.71}$$

If, as in Section 81, we set

$$R \overset{\text{def}}{=} R_{\mu\lambda} g^{\mu\lambda}, \tag{10.72}$$

we obtain

$$R = -\tfrac{1}{2}\sigma g_{\mu\lambda} g^{\mu\lambda} = -\tfrac{1}{2}\sigma n = -\tfrac{1}{2}\sigma\cdot 2 = -\sigma. \tag{10.73}$$

Accordingly, we have the relation

$$R_{\omega\mu\lambda\nu} = -RG_{\omega\mu\lambda\nu} = -Rg_{[\omega|\mu|}g_{\lambda]\nu}. \tag{10.74}$$

Whence, it follows in particular that

$$R_{1212} = -RG_{1212} - \tfrac{1}{2}R\mathfrak{g} \qquad (\mathfrak{g} = g_{11}g_{22} - g_{12}^2). \tag{10.75}$$

The quantity

$$\varkappa \overset{\text{def}}{=} \frac{R}{n(n-1)} \tag{10.76}$$

has been referred to as the scalar curvature of the space V_n (at some particular point) and for $n = 2$, also as the Gaussian curvature. We plainly have $\varkappa = R/2$ for $n = 2$.

Thus, for $n = 2$, noting that it suffices to consider only the combination of indices $a = 1$, $b = 2$, $c = 1$, $d = 2$ as other combinations yield either the same equation or the identity $0 = 0$, the Gaussian equation (10.68) becomes

$$\overline{R}_{1212} = h_{21}{}^2 - h_{11}h_{22} = -(h_{11}h_{22} - h_{12}{}^2) \tag{10.77}$$

or, making use of (10.75),

$$\varkappa({}'g_{11}{}'g_{22} - {}'g_{12}{}^2) = h_{11}h_{22} - h_{12}{}^2. \tag{10.78}$$

This last equation is just the classical form for Gaussian curvature, expressed in terms of the quotient of determinants of the second and first quadratic forms of the surface.

113. The Differential Parameters of Beltrami. In the theory of the differential geometry of surfaces a major role is played by the Beltrami parameters which bear the name of their creator. These parameters can be very simply defined by means of tensor calculus and, moreover, it follows immediately from their definition that they are invariants.

Given a scalar field σ in the space V_n, it is known that $\partial_\lambda \sigma$ is then a field of covariant vectors. We now form the expression

$$\varrho = g^{\lambda\mu}(\partial_\lambda \sigma)\partial_\mu \sigma. \tag{10.79}$$

which is an invariant. If we set

$$v^{\lambda} \overset{\text{def}}{=} g^{\lambda\mu}\partial_{\mu}\sigma, \tag{10.80}$$

then

$$|v|^2 = g_{\nu\mu}v^{\nu}v^{\mu} = g_{\nu\mu}g^{\nu\tau}\partial_{\tau}\sigma \cdot g^{\mu\omega}\partial_{\omega}\sigma = \delta^{\tau}_{\mu}\partial_{\tau}\sigma \cdot g^{\mu\omega}\partial_{\omega}\sigma = g^{\mu\omega}(\partial_{\mu}\sigma)\partial_{\omega}\sigma = \varrho,$$

i.e. the scalar ϱ represents the squared length of the vector (10.80), or in other words the squared length of the gradient of the field σ. Beltrami denoted the invariant (10.79) as

$$\varDelta_1\sigma \overset{\text{def}}{=} g^{\lambda\mu}(\partial_{\lambda}\sigma)\partial_{\mu}\sigma; \tag{10.81}$$

this scalar is called the *first Beltrami differential parameter* (of the field σ).

If in addition to the scalar field σ we have a second scalar field τ, then the invariant

$$\varDelta(\sigma, \tau) \overset{\text{def}}{=} g^{\lambda\mu}(\partial_{\lambda}\sigma)\partial_{\mu}\tau \tag{10.82}$$

is called the *mixed parameter of Beltrami* for the fields σ and τ. The geometric interpretation of $\varDelta(\sigma, \tau)$ is also very simple: it is just the scalar product of the gradients of the fields σ and τ.

Note that

$$\varDelta_1(\sigma) = \varDelta(\sigma, \sigma). \tag{10.83}$$

In order to define the second Beltrami differential parameter (for the scalar field σ), we take the tensor $\nabla_{\lambda}\nabla_{\mu}\sigma$ into account and contract it with the tensor $g^{\lambda\mu}$

$$\varDelta_2\sigma \overset{\text{def}}{=} g^{\lambda\mu}\nabla_{\lambda}\nabla_{\mu}\sigma. \tag{10.84}$$

The quantity $\varDelta_2\sigma$ is the *second Beltrami parameter*. It is plainly a scalar, and is therefore an invariant.

If $V_n = R_n$ and if system (i) is a rectangular Cartesian system, then

$$g^{ij} \overset{*}{=} \delta^{ij}, \qquad \begin{Bmatrix} k \\ i\,j \end{Bmatrix} \overset{*}{=} 0.$$

Consequently,

$$\nabla_i\nabla_j\sigma = \partial_{ij}\sigma,$$

whence

$$\varDelta_2\sigma = \delta^{ij}\partial_{ij}\sigma = \sum_{i=1}^{n} \partial_i^2\sigma \tag{10.85}$$

and the parameter $\varDelta_2$ reduces to the ordinary *Laplacian* of the field σ. The Laplacian thus is an invariant although form (10.85) is invariant only under the group G_m of coordinate transformations. On passing to other coordinates (e.g. polar), the Laplacian assumes a more complicated form.

As an application of the parameter Δ_1, consider in the space R_3 a surface V_2 given by the equations

$$\xi^\nu = \xi^\nu(\eta^1, \eta^2) \qquad (\nu = 1, 2, 3)$$

and assume that the net

$$\eta^a = \text{const} \qquad (a = 1, 2)$$

is orthogonal, so that

$$'g_{12} = 0, \tag{10.86}$$

and also assume that the curves

$$\eta^2 = \text{const} \tag{10.87}$$

constitute a family of geodesics. The equations of the geodesics are

$$\frac{d^2\eta^a}{d\sigma^2} + {}'\begin{Bmatrix} a \\ bc \end{Bmatrix} \frac{d\eta^b}{d\sigma} \cdot \frac{d\eta^c}{d\sigma} = 0 \qquad (a = 1, 2),$$

which for $a = 2$ becomes by virtue of equation (10.87),

$$'\begin{Bmatrix} 2 \\ bc \end{Bmatrix} \frac{d\eta^b}{d\sigma} \cdot \frac{d\eta^c}{d\sigma} = 0,$$

that is,

$$'\begin{Bmatrix} 2 \\ 11 \end{Bmatrix} \left(\frac{d\eta^1}{d\sigma} \right)^2 = 0,$$

whence

$$'\begin{Bmatrix} 2 \\ 11 \end{Bmatrix} = 0. \tag{10.88}$$

Now

$$\begin{Bmatrix} 2 \\ 11 \end{Bmatrix} = \tfrac{1}{2}\,'g^{2j}\{\partial_1 'g_{j1} + \partial_1 'g_{1j} - \partial_j 'g_{11}\}.$$

However, since

$$'g^{21} = -\frac{'g_{12}}{|'\mathfrak{g}|} = 0,$$

therefore

$$\begin{Bmatrix} 2 \\ 11 \end{Bmatrix} = \tfrac{1}{2}\,'g^{22}\{2\partial_1 'g_{12} - \partial_2 'g_{11}\} = -\tfrac{1}{2}\,'g^{22}\partial_2 'g_{11}.$$

Equation (10.88) accordingly becomes

$$'g^{22}\partial_2 'g_{11} = 0.$$

Hence, since $'g^{22} > 0$, we conclude that

$$'g_{11} = f(\eta^1).$$

If we set

$$\eta^{1'} = \int \sqrt{f(\eta^1)}\, d\eta^1,$$

then for an element of arc we get the expression

$$d\sigma^2 = (d\eta^{1'})^2 + 'g_{2'2'}(d\eta^{2'})^2. \tag{10.89}$$

The problem of finding geodesic lines on surfaces (which, as we know, is difficult as it entails solving a system of non-linear equations of the second order) can be simplified significantly when we find on the surface a one-parameter family of geodesic lines. This latter problem leads to the form (10.89) of the squared element of arc and it can easily be shown that this in turn means solving the equation

$$\Delta_1[\theta(\eta^1, \eta^2)] = 1. \tag{10.90}$$

EXAMPLE 1. As an example we calculate the first and second tensors for surfaces of revolution in the space R_3 and the formulae for the curvature tensor and the Ricci tensor.

In the space R_3 we take a rectangular Cartesian coordinate system and we choose the axis ξ^3 to be the axis of rotation of the surface of revolution. The surface of revolution is defined when we give the equation of a meridian, e.g. in the plane of the axes (ξ^1, ξ^3). We write this equation in the parametric form

$$\xi^1 = \varphi(\eta^1), \qquad \xi^3 = \psi(\eta^1)$$

regarding the parameter η^1 as the arc. In this case, the relation

$$\varphi'^2 + \psi'^2 = 1 \tag{10.91}$$

is satisfied identically. The meridian is assumed furthermore to be a curve of class C^2; in this event, in addition to the identity (10.91) the relation

$$\varphi'\varphi'' + \psi'\psi'' = 0 \tag{10.92}$$

will also be satisfied. The parametric equations of the surface of revolution under consideration will be

$$\xi^1 = \varphi(\eta^1)\cos\eta^2, \qquad \xi^2 = \varphi(\eta^1)\sin\eta^2, \qquad \xi^3 = \psi(\eta^1)$$

and the formulae found earlier can be used if we set

$$g_{\lambda\mu} \overset{*}{=} \delta_{\lambda\mu}.$$

However, if differential geometry formulae written in vector form are employed, we can calculate the coordinates g_{ab} of the first tensor from the formulae

$$g_{ab} = r_a r_b$$

as the scalar products of the vectors

$$r_a = \frac{\partial r}{\partial \eta^a} \qquad (a = 1, 2),$$

where r is the radius vector of the surface, i.e. a vector which is attached at the origin and whose terminal point varies over the surface.

In our case we have, where for convenience we write θ instead of η^2

$$r_1(\varphi' \cos\theta, \varphi' \sin\theta, \psi'), \qquad r_2(-\varphi \sin\theta, \varphi \cos\theta, 0).$$

From this we have

$$g_{11} = \varphi'^2 + \psi'^2 = 1,$$

$$g_{12} = 0,$$

$$g_{22} = \varphi^2,$$

$$\mathfrak{g} = g_{11} g_{22} - g_{12}^2 = \varphi^2.$$

Next, let us calculate the contravariant coordinates of the metric tensor

$$g^{1i} = \frac{g_{22}}{\mathfrak{g}} = \frac{\varphi^2}{\varphi^2} = 1,$$

$$g^{12} = -\frac{g_{12}}{\mathfrak{g}} = 0,$$

$$g^{22} = \frac{g_{11}}{\mathfrak{g}} = \frac{1}{\varphi^2}.$$

We can now calculate the Christoffel symbols. If we note that

$$\partial_a g_{11} = \partial_a g_{12} = 0 \qquad (a = 1, 2),$$

$$\partial_1 g_{22} = 2\varphi\varphi', \qquad \partial_2 g_{22} = 0$$

we obtain successively

$$\begin{Bmatrix} 1 \\ 11 \end{Bmatrix} = \tfrac{1}{2} g^{1l} \{\partial_1 g_{1l} + \partial_1 g_{1l} - \partial_l g_{11}\}$$

$$= \tfrac{1}{2} g^{11} \{\partial_1 g_{11} + \partial_1 g_{11} - \partial_1 g_{11}\} = 0,$$

$$\begin{Bmatrix} 1 \\ 12 \end{Bmatrix} = \tfrac{1}{2} g^{11} \{\partial_1 g_{12} + \partial_2 g_{11} - \partial_1 g_{12}\} = 0,$$

$$\begin{Bmatrix} 1 \\ 22 \end{Bmatrix} = \tfrac{1}{2}g^{11}\{\partial_2 g_{12}+\partial_2 g_{12}-\partial_1 g_{22}\} = -\varphi\varphi',$$

$$\begin{Bmatrix} 2 \\ 11 \end{Bmatrix} = \tfrac{1}{2}g^{22}\{\partial_1 g_{12}+\partial_1 g_{12}-\partial_2 g_{11}\} = 0,$$

$$\begin{Bmatrix} 2 \\ 12 \end{Bmatrix} = \tfrac{1}{2}g^{22}\{\partial_1 g_{22}+\partial_2 g_{12}-\partial_2 g_{12}\} = \frac{\varphi'}{\varphi},$$

$$\begin{Bmatrix} 2 \\ 22 \end{Bmatrix} = \tfrac{1}{2}g^{22}\{\partial_2 g_{22}+\partial_2 g_{22}-\partial_2 g_{22}\} = 0.$$

The next step is to calculate the coordinates of the tensor h_{ab}. It is known from classical differential geometry that they are equal to the scalar products

$$h_{ab} = r_{ab}N,$$

where N is a unit vector normal to the surface, while

$$r_{ab} = \frac{\partial^2 r}{\partial\eta^a\partial\eta^b} \qquad (a, b = 1, 2).$$

In our case, we have

$$r_{11}(\varphi''\cos\theta,\ \varphi''\sin\theta,\ \psi''),$$

$$r_{12}(-\varphi'\sin\theta,\ \varphi'\cos\theta,\ 0),$$

$$r_{22}(-\varphi\cos\theta,\ -\varphi\sin\theta,\ 0).$$

If r_3 denotes the vector product

$$r_3 = [r_1, r_2],$$

then it will have the coordinates

$$r_3(-\varphi\psi'\cos\theta,\ -\varphi\psi'\sin\theta,\ \varphi\varphi').$$

From this, we have

$$|r_3|^2 = \varphi^2\psi'^2+\varphi^2\varphi'^2 = \varphi^2,$$

that is,

$$|r_3| = \varphi$$

(we may assume without loss of generality that $\varphi \geqslant 0$). Thus, since

$$N = \frac{r_3}{|r_3|},$$

therefore

$$N(-\psi'\cos\theta,\ -\psi'\sin\theta,\ \varphi').$$

Accordingly, we have

$$h_{11} = -\psi'\varphi''+\varphi'\psi'',$$
$$h_{12} = 0,$$
$$h_{22} = \varphi\psi'.$$

We can now calculate the Gaussian curvature K and the mean curvature H of the surface by the formulae

$$K = \frac{h_{11}h_{22}-h_{12}^2}{g_{11}g_{22}-g_{12}^2}, \qquad H = h_{ab}g^{ab}.$$

In our case, since $g_{12} = h_{12} = 0$,

$$K = \frac{h_{11}h_{22}}{g_{11}g_{22}} = \frac{\varphi\psi'(\varphi'\psi''-\psi'\varphi'')}{\varphi^2} = \frac{\psi'(\varphi'\psi''-\psi'\varphi'')}{\varphi},$$

$$H = g^{11}h_{11}+g^{22}h_{22} = h_{11}+\frac{h_{22}}{\varphi^2} = \varphi'\psi''-\psi'\varphi''+\frac{\psi'}{\varphi}$$
$$= \frac{\psi'+\varphi(\varphi'\psi''-\psi'\varphi'')}{\varphi}.$$

Note that since only the derivative of ψ and not the function itself appears in the formulae above, both curvatures K and H can be expressed in terms of the function φ alone; indeed, by virtue of relations (10.91) and (10.92) we have

$$\psi' = \varepsilon[1-\varphi'^2]^{1/2}, \qquad \psi'' = \frac{-\varepsilon\varphi'\varphi''}{[1-\varphi'^2]^{1/2}}. \tag{10.93}$$

Inserting these expressions into the formulae for K and H, we get

$$K = -\frac{\varphi''}{\varphi}, \qquad H = \varepsilon\frac{1-\varphi'^2-\varphi\varphi''}{\varphi[1-\varphi'^2]^{1/2}}.$$

We now calculate the coordinates of the curvature tensor

$$R_{abc}{}^{d} = 2\partial_{[a}\left\{{}_{b]c}^{\;d}\right\}+2\left\{{}_{[a|e|}^{\;\;d}\right\}\left\{{}_{b]c}^{\;e}\right\}. \tag{10.94}$$

In view of the fact that $R_{aac}{}^{d} = 0$ and $R_{bac}{}^{d} = -R_{abc}{}^{d}$, the curvature tensor can have at most four different coordinates, viz.

$$R_{121}{}^{1}, \quad R_{121}{}^{2}, \quad R_{122}{}^{1}, \quad R_{122}{}^{2}.$$

When the values previously found for the Christoffel symbols are inserted into formula (10.94), we get

$$R_{121}{}^1 = 0, \quad R_{121}{}^2 = \frac{\varphi''}{\varphi}, \quad R_{122}{}^1 = -\varphi\varphi'', \quad R_{122}{}^2 = 0.$$

For the Ricci tensor R_{ab} we thus find that

$$R_{ab} = R_{cab}{}^c = R_{1ab}{}^1 + R_{2ab}{}^2,$$

$$R_{11} = R_{111}{}^1 + R_{211}{}^2 = -R_{121}{}^2 = -\frac{\varphi''}{\varphi},$$

$$R_{12} = R_{112}{}^1 + R_{212}{}^2 = -R_{122}{}^2 = 0,$$

$$R_{21} = R_{121}{}^1 + R_{221}{}^2 = R_{121}{}^1 = 0,$$

$$R_{22} = R_{122}{}^1 + R_{222}{}^2 = R_{122}{}^1 = -\varphi\varphi''.$$

Accordingly, we have further

$$R = R_{ab}g^{ab} = R_{11}g^{11} + R_{22}g^{22} = -\frac{\varphi''}{\varphi} - \frac{\varphi''}{\varphi} = -\frac{2\varphi''}{\varphi}.$$

Hence by formula (10.76) the Gaussian curvature $\varkappa$ (= K) can be calculated for the second time when it is found to be

$$\varkappa = \frac{R}{n(n-1)} = -\frac{\varphi''}{\varphi}.$$

We now look for the spherical points on the surface of revolution. Owing to the equality $h_{12} = g_{12} = 0$, the condition

$$h_{ab} = \varrho g_{ab}$$

assumes the simple form

$$\begin{vmatrix} h_{11} & g_{11} \\ h_{22} & g_{22} \end{vmatrix} = 0,$$

that is,

$$\varphi^2(\varphi'\psi'' - \psi'\varphi'') - \varphi\psi' = 0$$

or

$$\varphi[\varphi(\varphi'\psi'' - \psi'\varphi'') - \psi'] = 0.$$

This equation yields the alternatives: either $\varphi = 0$ or $\varphi(\varphi'\psi'' - \psi'\varphi'') - \psi' = 0$.

The points on the meridian for which $\varphi = 0$ are points which also lie on the axis of rotation. Then $\mathfrak{g} = \varphi^2 = 0$, and these are consequently singular points with respect to the parametric representation under consideration. Such a point will be regular if $\psi' = 0$ and is then, of course, a spherical point. To discover at what points on the meridian the equation

$$\varphi[\varphi'\psi'' - \psi'\varphi''] - \psi' = 0,$$

is satisfied, we make use of formulae (10.93). The equation above then goes over into the equivalent equation

$$\varphi\varphi'' + 1 - \varphi'^2 = 0.$$

Let us find the centre of curvature of the meridian

$$\xi^1 = \varphi(s), \qquad \xi^3 = \psi(s) \qquad (s = \eta').$$

If the coordinates of this centre are denoted by (ζ^1, ζ^3), then by the familiar formulae of classical differential geometry

$$\zeta^1 = \varphi - \psi' \frac{\varphi'^2 + \psi'^2}{\varphi'\psi'' - \psi'\varphi''}, \qquad \zeta^3 = \psi + \varphi' \frac{\varphi'^2 + \psi'^2}{\varphi'\psi'' - \psi'\varphi''}.$$

There upon

$$\zeta^1 = \varphi - \frac{\psi'}{\varphi'\psi'' - \psi'\varphi''} = \frac{\varphi[\varphi'\psi'' - \psi'\varphi''] - \psi'}{\varphi'\psi'' - \psi'\varphi''},$$

which means that the spherical points are those meridional points whose centre of curvature lies on the axis of rotation.

Since we already have calculated the Christoffel symbols, we can write down the equations of geodesic lines on the surfaces of revolution.

These equations are

$$\frac{d^2\eta^a}{d\sigma^2} + \begin{Bmatrix} a \\ bc \end{Bmatrix} \frac{d\eta^b}{d\sigma} \cdot \frac{d\eta^c}{d\sigma} = 0 \qquad (a = 1, 2),$$

i.e. writing s instead of η^1 and θ instead of η^2 and bearing in mind that σ is the arc along a geodesic line,

$$\frac{d^2s}{d\sigma^2} + \begin{Bmatrix} 1 \\ 11 \end{Bmatrix} \left(\frac{ds}{d\sigma}\right)^2 + 2\begin{Bmatrix} 1 \\ 12 \end{Bmatrix} \frac{ds}{d\sigma} \cdot \frac{d\theta}{d\sigma} + \begin{Bmatrix} 1 \\ 22 \end{Bmatrix} \left(\frac{d\theta}{d\sigma}\right)^2 = 0,$$

$$\frac{d^2\theta}{d\sigma^2} + \begin{Bmatrix} 2 \\ 11 \end{Bmatrix} \left(\frac{ds}{d\sigma}\right)^2 + 2\begin{Bmatrix} 2 \\ 12 \end{Bmatrix} \frac{ds}{d\sigma} \cdot \frac{d\theta}{d\sigma} + \begin{Bmatrix} 2 \\ 22 \end{Bmatrix} \left(\frac{d\theta}{d\sigma}\right)^2 = 0;$$

If use is made of the formulae for the Christoffel symbols calculated above, these equations become

$$\frac{d^2s}{d\sigma^2} - \varphi\varphi' \left(\frac{d\theta}{d\sigma}\right)^2 = 0, \qquad \frac{d^2\theta}{d\sigma^2} + \frac{2\varphi'}{\varphi} \cdot \frac{ds}{d\sigma} \cdot \frac{d\theta}{d\sigma} = 0. \qquad (10.95)$$

This system of equations can be replaced by a single differential equation by eliminating the parameter σ.

We note that the curves $\theta = \text{const}$, $s = \sigma$ (meridians of the surface of revolution under consideration) satisfy the system of equations (10.95), so that they are geodesic lines. Accordingly, in order to find further geodesics we can assume

$$s = s(\theta),$$

and we then have

$$\frac{ds}{d\theta} = \frac{\dfrac{ds}{d\sigma}}{\dfrac{d\theta}{d\sigma}}, \qquad \frac{d^2 s}{d\theta^2} = \frac{\dfrac{d\theta}{d\sigma}\cdot\dfrac{d^2 s}{d\sigma^2} - \dfrac{ds}{d\sigma}\cdot\dfrac{d^2\theta}{d\sigma^2}}{\left(\dfrac{d\theta}{d\sigma}\right)^3}.$$

Insertion of the appropriate values from equation (10.95) into the right-hand side of the second formula above yields the easily integrable equation

$$\frac{d^2 s}{d\theta^2} = \frac{\varphi\varphi'\left(\dfrac{d\theta}{d\sigma}\right)^3 + \dfrac{2\varphi'}{\varphi}\left(\dfrac{ds}{d\sigma}\right)^2\dfrac{d\theta}{d\sigma}}{\left(\dfrac{d\theta}{d\sigma}\right)^3} = \varphi\varphi' + \frac{2\varphi'}{\varphi}\left(\frac{ds}{d\theta}\right)^2.$$

EXAMPLE 2. Suppose that a sphere S_{n-1} of radius R and centre at the origin is given in the Euclidean space R_n. Its equation is

$$\sum_{\lambda} (\xi^\lambda)^2 = R^2. \tag{10.96}$$

One of the parametric representations of this sphere S_{n-1} is the system

$$\begin{aligned} \xi^1 &= R\cos\eta^1, \\ \xi^\nu &= R\prod_{j=1}^{\nu-1}\sin\eta^j\cos\eta^\nu \qquad \text{for } \nu = 2, 3, \ldots, n-1, \\ \xi^n &= R\prod_{j=1}^{n-1}\sin\eta^j. \end{aligned} \tag{10.97}$$

The ranges of the parameters η^a $(a = 1, 2, \ldots, n-1)$ are

$$\begin{aligned} &0 \leqslant \eta^1 \leqslant \pi, \\ &0 \leqslant \eta^a < 2\pi \qquad \text{for } a = 2, 3, \ldots, n-1; \end{aligned} \tag{10.98}$$

To find those points for which this representation is non-singular, we have to construct the matrix

$$C_a^\nu = \frac{\partial \xi^\nu}{\partial \eta^a} \qquad (\nu = 1, \dots, n, \; a = 1, \dots, n-1) \tag{10.99}$$

and discover for what sets of values η^a this matrix is of rank $n-1$. The simplest way to do this one which avoid calculating n determinants of degree $n-1$ extracted from this matrix, is to form the scalar products

$$g_{ab} \stackrel{\text{def}}{=} r_a \cdot r_b = \sum_\nu C_a^\nu C_b^\nu \tag{10.100}$$

and then to calculate the Gram determinant

$$G \stackrel{\text{def}}{=} \det g_{ab} . \tag{10.101}$$

The places where G equals zero give the singular points of representation (10.97).

The calculations, details of which we omit, yield

$$g_{ab} = g_{ba} = 0, \qquad \text{provided that } a \neq b. \tag{10.102}$$

However,

$$g_{11} = R^2, \qquad g_{22} = R^2 \sin^2 \eta^1, \qquad g_{33} = R^2 \sin^2 \eta^1 \sin^2 \eta^2$$

and, generally,

$$g_{aa} = R^2 \sum_{j=1}^{a-1} \sin^2 \eta^j \qquad \text{for } a > 1. \tag{10.103}$$

Relations (10.102) tell us that the net of parametric lines

$$\eta^j = \tau, \qquad \eta^a = \text{const} \qquad \text{for } a \neq j$$

is an orthogonal net. Accordingly, we have

$$G = R^{2(n-1)} \sin^2 \eta^{n-2} \sin^2 \eta^{n-3} \dots \sin^{2(n-1)} \eta^1 . \tag{10.104}$$

It is thus seen that G is zero when $\sin \eta^1 = 0$ or $\sin \eta^2 = 0, \dots,$ or $\sin \eta^{n-2} = 0$. By virtue of equations (10.97), this set of alternatives is equivalent to the relations

$$\xi^2 = \dots = \xi^n = 0 \quad \text{or} \quad \xi^3 = \dots = \xi^n = 0 \quad \dots, \quad \text{or} \quad \xi^{n-1} = \xi^n = 0.$$

Since each of the planes above is contained in the next, we are left with the set

$$\xi^{n-1} = \xi^n = 0,$$

which defines an $(n-2)$-dimensional plane. The intersection of this plane with the sphere (10.96) gives

$$\sum_{\lambda=1}^{n-2} (\xi^\lambda)^2 = R^2, \tag{10.105}$$

i.e., an $(n-2)$-dimensional sphere S_{n-2}. The metric tensor g_{ab} of the sphere S_{n-1} degenerates at the points of a subsphere S_{n-2}. For $n = 3$, S_1 consists of two points.

114. The Frenet Equations. These equations, also known as the *Frenet–Serret equations,* were initially derived for curves in R_3, but were subsequently generalized in various directions: e.g. by Jordan for curves in R_n, by Kühne and Blaschke for curves in V_n, by Hlavatý for curves in L_n.

Given a sufficiently regular curve C in the space V_n, where the metric tensor defines a positive definite quadratic form [1], we represent this curve parametrically by

$$\xi^\nu = \xi^\nu(\sigma) \qquad (\nu = 1, 2, \ldots, n), \tag{10.106}$$

where σ is the arc (metric). We now set

$$\underset{1}{t}^\nu \overset{\text{def}}{=} \frac{d\xi^\nu}{d\sigma}. \tag{10.107}$$

The vector $\underset{1}{t}$ is, as we know, a unit vector tangent to C. We further set

$$\underset{k}{\bar{t}}^\nu \overset{\text{def}}{=} \frac{D\,\underset{k-1}{\bar{t}}^\nu}{d\sigma} = D_\sigma \underset{k-1}{\bar{t}}^\nu \qquad (\underset{1}{\bar{t}} = \underset{1}{t};\ k = 2, 3, \ldots). \tag{10.108}$$

Suppose that the vectors

$$\underset{1}{\bar{t}}^\nu, \ldots, \underset{m}{\bar{t}}^\nu \tag{10.109}$$

are linearly independent but that the vector

$$\underset{m+1}{\bar{t}}^\nu \tag{10.110}$$

is linearly dependent on the vectors (10.109). We naturally exclude from our discussion points at which the latter property occurs in an isolated manner, i.e. where it does not hold for neighbouring points (such points are points of inflexion of various orders). The number m must of course

[1] Struik [69] carries out this reasoning in a more general setting, viz. he does not assume that the form $g_{\lambda\mu}x^\lambda x^\mu$ is necessarily definite.

satisfy the inequality

$$1 \leqslant m \leqslant n. \tag{10.111}$$

We let $\underset{2}{t}^{\nu}$ denote the unit vector of the vector $D_\sigma \underset{1}{\bar{t}}$, that is

$$\underset{2}{t}^{\nu} = \frac{D_\sigma \underset{1}{\bar{t}}^{\nu}}{|D_\sigma \underset{1}{\bar{t}}^{\nu}|}, \tag{10.112}$$

so that we can then write

$$D_\sigma \underset{1}{t}^{\nu} = \underset{1}{\varkappa}\, \underset{2}{t}^{\nu}. \tag{10.113}$$

Thus, we have $\underset{1}{\varkappa} > 0$. We next form the vector $D_\sigma \underset{2}{t}^{\nu}$. This vector is then decomposed into a component along $\underset{1}{t}$ and a remainder r^{ν}: it will have no component along $\underset{2}{t}$ since $\underset{2}{t}$ is a unit field. Hence we have

$$D_\sigma \underset{2}{t}^{\nu} = \alpha \underset{1}{t}^{\nu} + r^{\nu}. \tag{10.114}$$

It can then be shown that $a = -\underset{1}{\varkappa}$ since covariant differentiation of the scalar product $\underset{1}{t} \cdot \underset{2}{t}$, which is equal to zero, gives the result

$$\underset{1}{t} \cdot D_\sigma \underset{2}{t} + \underset{2}{t} \cdot D_\sigma \underset{1}{t} = 0.$$

Since r^{ν} is perpendicular to $\underset{1}{t}^{\nu}$, it follows from (10.114) that

$$\underset{1}{t} \cdot D_\sigma \underset{2}{t} = \alpha.$$

On the other hand, from equation (10.113) we have

$$\underset{2}{t} \cdot D_\sigma \underset{1}{t} = \underset{1}{\varkappa},$$

whence $\alpha + \underset{1}{\varkappa} = 0$, or $\alpha = -\underset{1}{\varkappa}$.

Let $\underset{2}{\varkappa}$ denote the length of the vector r^{ν} and $\underset{3}{t}^{\nu}$ its unit vector. If we had $\underset{2}{\varkappa} = 0$, the vector $D_\sigma \underset{2}{t}$ would be linearly dependent on $\underset{1}{t}$ and $\underset{2}{t}$, and it could then be readily shown to be linearly dependent on $\underset{1}{\bar{t}}$ and $\underset{2}{\bar{t}}$, which cannot be the case unless $m = 2$. Continuing in a similar manner, we further arrive at the formula

$$D_\sigma \underset{3}{t}^\nu = -\underset{2}{\varkappa}\underset{2}{t}^\nu + \underset{3}{\varkappa}\underset{4}{t}^\nu,$$

where $\underset{4}{t}$ is a unit vector perpendicular to the trivector $\underset{1}{t}^{[\lambda}\underset{2}{t}^{\mu}\underset{3}{t}^{\nu]}$, whereas $\underset{3}{\varkappa}$ is a positive scalar coefficient. Reasoning inductively we find

$$\begin{aligned} D_\sigma \underset{1}{t}^\nu &= \underset{1}{\varkappa}\underset{2}{t}^\nu, \\ D_\sigma \underset{j}{t}^\nu &= -\underset{j-1}{\varkappa}\;\underset{j-1}{t}^\nu + \underset{j}{\varkappa}\;\underset{j+1}{t}^\nu \qquad (j = 2, \ldots, m-1), \\ D_\sigma \underset{m}{t}^\nu &= -\underset{m-1}{\varkappa}\;\underset{m-1}{t}^\nu . \end{aligned} \tag{10.115}$$

If $m = n$, then this is the *complete system of Frenet equations*. Suppose, on the other hand, that $m < n$ and set $p = n-m$. Then, choosing at any point $\underset{0}{p}$ of the curve C a system of p unit vectors

$$\underset{m+2}{t}^\nu, \underset{m+2}{t}^\nu, \ldots, \underset{n}{t}^\nu \tag{10.116}$$

such that the system

$$(\underset{1}{t}, \ldots, \underset{n}{t}) \tag{10.117}$$

is an orthogonal system (for $p > 1$ such a choice can be made in an infinite number of ways), we carry out the parallel displacement of the vectors (10.116) to all other points of the curves. We know that lengths and angles are preserved under such displacement, so that the entire system remains orthonormal.

Accordingly, we obtain the additional equations

$$D_\sigma \underset{j}{t}^\nu = 0 \qquad (j = m+1, \ldots, n), \tag{10.118}$$

and setting

$$\underset{0}{\varkappa} = \underset{m}{\varkappa} = \underset{m+1}{\varkappa} = \ldots = \underset{n-1}{\varkappa} = \underset{n}{\varkappa} = 0 \tag{10.119}$$

we can in every case write the system (10.115) in the more general and uniform form

$$D_\sigma \underset{j}{t}^\nu = -\underset{j-1}{\varkappa}\;\underset{j-1}{t}^\nu + \underset{j}{\varkappa}\;\underset{j+1}{t}^\nu \qquad (j = 1, 2, \ldots, n); \tag{10.120}$$

these are the generalized Frenet equations.

Note that if $m = n$, then

$$\underset{1}{\varkappa} > 0, \quad \ldots, \quad \underset{n-2}{\varkappa} > 0, \tag{10.121}$$

while the last vector $\underset{n}{t}$ can be directed so that the n-vector

$$\underset{1}{t}^{[\lambda_1} \dots \underset{n}{t}^{\lambda_n]} \tag{10.122}$$

has a positive fundamental coordinate, the scalar coefficient $\underset{n-1}{\varkappa}$ being an algebraic quantity as it is.

The coefficients

$$\underset{j}{\varkappa}(\sigma) \qquad (j = 1, 2, \dots, n-1) \tag{10.123}$$

are known as the first, second, ..., $(n-1)$-th curvature of the curve C. They are scalar functions of the arc σ and are therefore invariants associated with the variable point of the curve C. For a geodesic curve we have $D_\sigma \underset{1}{t} \equiv \equiv 0$, so that all the curvatures are zero.

Just as for curves in a three-dimensional Euclidean space R_3 we have a theorem which states that curves for which the first and second curvature are given functions of the arc, are defined up to motions of the curve, so in the general case for curves in a space V_n we have the following general theorem

THEOREM [1]. *If in a space V_n with a positive definite metric tensor we are given $n-1$ functions of the parameter σ*

$$f_1(\sigma), \dots, f_{n-1}(\sigma) \qquad (0 \leqslant \sigma \leqslant L),$$

where the function $f_{n-j}(\sigma)$ is of class C^j $(j = 1, 2, \dots, n-1)$ and, moreover,

$$f_j(\sigma) > 0 \quad \textit{for } j = 1, 2, \dots, n-2.$$

and if in addition at some point p_0 of the space, a system of n orthonormal vectors

$$\underset{j}{u}^\nu \qquad (j = 1, 2, \dots, n), \tag{10.124}$$

is also given, then there exists exactly one curve C: $\xi^\nu = \xi^\nu(\sigma)$, with the following properties:

1. *the point p_0 corresponding to the value $\sigma = 0$,*

2. *σ represents the length of the arc of the curve C measured from the point p_0,*

(1) Hlavatý [32].

3. *the relations*

$$\underset{j}{\varkappa} = f_j(\sigma) \qquad (j = 1, 2, \ldots, n-1) \tag{10.125}$$

hold,

4. *the system of vectors* $\underset{j}{t^\nu}$ *of the Frenet n-hedral for* $\sigma = 0$ *coincides with the system* (10.124) *given a priori.*

For the proof of this theorem, the reader is referred to more extensive works (e.g. Schouten–Struik [69]).

Equations (10.125) are called the *natural equations* of the curve. In general, if the space V_n does not admit any group of isometric transformations, it is vacuous to speak of congruent (or rigid) displacements of curves or of any other figures consisting of a continuum of points.

It has been shown by V. Hlavatý that the Frenet formulae can be generalized to spaces L_n as well. In this case it is necessary, firstly, to use the affine arc σ in place of the metric arc, which is no longer available. Secondly, we must abandon orthonormal systems attached at a variable point on the curve because this also entails the use of a metric.

We represent the curve C in L_n by the equations $\xi^\nu = \xi^\nu(\sigma)$, where σ now denotes an affine parameter (defined up to linear transformations, p. 208) and, in keeping with the notation employed earlier in the case of Riemannian space, we introduce the notation

$$\underset{1}{t^\nu} = \frac{d\xi^\nu}{d\sigma} = D_\sigma \xi^\nu \tag{10.126}$$

and then set

$$\underset{k+1}{t^\nu} \overset{\text{def}}{=} D_\sigma \underset{k}{t} \qquad (k = 1, 2, \ldots). \tag{10.127}$$

Suppose that the vectors

$$\underset{1}{t}, \underset{2}{t}, \ldots, \underset{m}{t} \tag{10.128}$$

are again linearly independent, while the vector $\underset{m+1}{t}$ is linearly dependent on the vectors (10.128). In this event we have

$$D_\sigma \underset{m}{t} + \sum_{j=1}^{m} \underset{j}{\varkappa} \underset{j}{t} = 0, \tag{10.129}$$

where the $\underset{j}{\varkappa}$'s are scalar coefficients. Making use of the fact that σ is an affine arc, it can be shown that $\underset{m}{\varkappa} = 0$, i.e. that equation (10.129) reduces to the (somewhat simpler) equation

$$D_\sigma \underset{m}{t}^\nu + \sum_{j-1}^{m-1} \underset{j}{\varkappa}\underset{j}{t}^\nu = 0 \qquad (\nu = 1, 2, \ldots, n). \tag{10.130}$$

The coefficients $\underset{j}{\varkappa}$ $(j = 1, 2, \ldots, m-1)$ are called the *affine curvatures* of the curve C. They are functions of the parameter σ, or in other words are associated with the variable point p on the curve C. They are not, however, absolute invariants since when the arc σ undergoes an affine transformation, the curvatures $\underset{j}{\varkappa}$ are multiplied by constant factors, which factors are not even the same for all the j's. It is plain that the number m can, at most, be equal to n.

There is, however, a fundamental difference between the metric curvatures of a curve lying in V_n and the affine curvatures of a curve lying in L_n. If in the metric case $\underset{j}{\varkappa} \equiv 0$, then also

$$\underset{j+1}{\varkappa} \equiv \underset{j+1}{\varkappa} \equiv \ldots \equiv \underset{n-1}{\varkappa} \equiv 0.$$

On the other hand, for affine curvatures we may even have

$$\underset{1}{\varkappa} \equiv \underset{2}{\varkappa} \equiv \ldots \equiv \underset{n-1}{\varkappa} \equiv 0,$$

although $m = n$. Hence no conclusion as to the order of skewness of the curve, if this is what we call m, can be drawn from the fact that its curvatures are zero. The curve with the highest possible order of skewness, i.e. equal to n, may have all its affine curvatures identically equal to zero.

It is therefore necessary to give in addition to the values of $\underset{1}{\varkappa}(\sigma)$, ... also the skewness order m for the natural equations of a curve in L_n.

It follows from the discussion above that affine curvatures are not a generalization of metric curvatures in the sense that the Hlavatý equations

$$\begin{aligned} D_\sigma \underset{1}{t} &= \underset{k+1}{t} \qquad (k = 1, 2, \ldots, m-1), \\ D_\sigma \underset{m}{t} &+ \sum_{j=1}^{m-1} \underset{j}{\varkappa}\underset{j}{t} = 0 \end{aligned} \tag{10.131}$$

become the Frenet equations (10.120) when σ represents a metric arc.

In addition to the above Frenet equations for curves in L_n, it is possible to obtain another system of equations if we require all the curvatures to be differential invariants of the lowest possible order.

Results along these lines have been obtained by W. Blaschke, A. Winternitz and V. Hlavatý. More recently, L. Tamàssy [151] also obtained such a generalization of the Frenet equations for the case of a space having an object Γ of linear connection, in which these equations reduce to the Frenet equations in the special case of the space being Riemannian or Euclidean.

CHAPTER XI

OTHER APPLICATIONS

115. The Lie Derivative. In 1931 Ślebodziński [149] introduced a very important concept which has turned out to have a host of applications in geometry and in physics. Later called the Lie derivative, it was subsequently generalized to a field of any arbitrary geometric objects. This concept, however, calls for a preliminary discussion of another concept, which was introduced later, in 1933, by Schouten and van Kampen [146] in connection with the theory of the deformation of fields of geometric objects.

The concept of a field dragged along.

We are given in the space X_n a field of geometric objects

$$\Omega(\xi) \tag{11.1}$$

together with the invertible point transformation

$$q = f(p), \tag{11.2}$$

within the domain D of ξ, which associates points of the domain D with other points of the domain D. If we set up an allowable coordinate system $\underset{i}{B}$, we can write transformation (11.2) as

$$\eta^i = \overset{i}{f}(\xi^1, \ldots, \xi^n). \tag{11.3}$$

The functions $\overset{\lambda}{f}$ in another coordinate system $\underset{\lambda}{B}$ will, of course, differ from the functions $\overset{i}{f}$. It can easily be shown that if transition from the system $\underset{\lambda}{B}$ to $\underset{i}{B}$ is given by the transformation

$$\xi^i = \varphi^{i\lambda}(\xi^\mu) \quad ^{(1)}, \tag{11.4}$$

(1) Here, λ is not a running index; the letter of the relevant alphabet merely indicates the coordinate system.

then the functions $\overset{\lambda}{f}, \overset{i}{f}, \varphi^{i\lambda}$ are related by

$$\overset{i}{f} = \varphi^{i\lambda} \circ \overset{\lambda}{f} \circ \varphi^{\lambda i}, \tag{11.5}$$

where $\circ$ denotes the composition of functions and $\varphi^{\lambda i}$ are functions which define the inverse of transformation (11.3), i.e.

$$\xi^{\lambda} = \varphi^{\lambda i}(\xi^{j}). \tag{11.6}$$

We now define a new field of objects

$$\omega(\xi) \tag{11.7}$$

of the same type as $\Omega(\xi)$, or with the same transformation law, which depend on the given field $\Omega(\xi)$ and on the point transformation (11.2).

The definition of $\omega(\xi)$ is as follows. We start with some coordinate system $\underset{i}{B}$. Let $\underset{\lambda}{B}$ denote the (uniquely defined) coordinate system which we arrive at by means of the coordinate transformation

$$\xi^{\lambda} = \varphi^{\lambda}(\xi^{j}), \tag{11.8}$$

where

$$\varphi^{\lambda}(\xi^{j}) \overset{\text{def}}{=} \delta^{\lambda}_{i} \overset{i}{f}(\xi^{j}). \tag{11.9}$$

Expressing this in words, the transformation of the coordinates ξ^{i} into ξ^{λ} is identical (in $\underset{i}{B}$) with the point transformation (11.2). We thus see that one and the same transformation appears here in a double role, once as a point transformation (transforming points into points) and again as a coordinate transformation (not affecting points, but changing their coordinates).

The new object ω is to have the same coordinates at the point p in the system $\underset{\lambda}{B}$ as the object Ω has at the point $q = f(p)$ in $\underset{i}{B}$.

This definition does, of course, need justification as to the invariant definition of ω, i.e. that it depends only on the original object Ω and on the point transformation (11.2). Such a proof can be given [120].

The newly defined field $\omega(\xi)$ is called a *field generated by dragging the field* $\Omega(\xi)$ by means of transformation (11.2).

The concept of a field dragged along is closely bound up with the concept of Lie derivative.

Suppose that we are given in our space a field of contravariant vectors

$$v(\xi) \tag{11.10}$$

of a suitable class of regularity (at least of class C^1). Such a field defines an "infinitesimal point transformation"

$$\xi^i \to \xi^i + v^i(\xi)\,dt, \tag{11.11}$$

which, of course, is not invariant in character. However, if we pass to the limit as $dt \to 0$ we do obtain invariant results.

Now suppose that we are given a field of geometric objects $\Omega(\xi)$ of sufficient regularity together with a field of contravariant vectors (11.10).

We fix dt and the coordinate system $\underset{i}{B}$, and consider the point transformation

$$\eta^i = \xi^i + v^i(\xi)\,dt. \tag{11.12}$$

We denote by $\omega(\xi)$ the dragged along field obtained from $\Omega(\xi)$ by means of transformation (11.12).

We form the difference

$$\omega(\xi) - \Omega(\xi).$$

REMARK. If $\Omega^a(\xi)$ denote the coordinates of the object Ω, there is no general theorem to the effect that $\omega^a - \Omega^a$ always define a geometric object, even though ω and Ω have the same transformation law. In fact, no such theorem can be true, notwithstanding the fact that it does hold for many known objects.

Next, we calculate the limit

$$\lim_{dt\to 0} \frac{\omega(\xi) - \Omega(\xi)}{dt}. \tag{11.13}$$

The limiting object so obtained is called the *Lie derivative of the object Ω with respect to the field v* and we denote it by the symbol

$$\underset{v}{Ł}\Omega(\xi). \tag{11.14}$$

This name was introduced by van Dantzig [98], although it should really be called the Ślebodziński derivative since it was he who introduced this concept, albeit for certain special fields of objects $\Omega(\xi)$, when in 1931 he defined so-called affine motions in spaces with affine connection [150].

It must be shown of course, that, as defined above, $\underset{v}{Ł}\Omega$ is invariant in character, regardless of the original coordinate system $\underset{i}{B}$. Note, moreover, that the dragged along field $\omega(\xi)$ also depends on the fixing of the increment dt.

The question arises as to whether the object $\underset{v}{\pounds}\Omega$ is always a geometric object. The answer to this question is in the negative. It can be shown [89] that a necessary and sufficient condition for the Lie derivative of a geometric object Ω to be itself a geometric object is that Ω be a linear object.

A general formula can be derived for the Lie derivative $\underset{v}{\pounds}\Omega$ when $\Omega(\xi)$ is a field of geometric objects of a special class r. This formula is quite complicated (formula (2.9) on p. 20 of the monograph by Yano [89]). It shows that the derivative depends on $\Omega(\xi)$ and on the first-order partial derivatives of the field Ω and, furthermore, on the auxiliary field v as well as on its partial derivatives of order up to and including r. It has therefore to be assumed for the definition of the Lie derivative that the given field Ω is of regularity class one, while the class of regularity of the vector field v must be at least as great as the class r of the object Ω.

In brief, it may be said that the Lie derivative of a field Ω is the differential concomitant of a pair of fields (Ω, v). When the object Ω is of the first class, this comitant is a differential comitant of the first order. This fact persuaded me to attempt in 1958 to define the Lie derivative by isolating uniquely from all concomitants the differential comitants of a particular order by imposing certain additional conditions. A satisfactory solution to this problem was not obtained until recently. A. Szybiak defines the Lie derivative $\underset{v}{\pounds}\Omega$ by means of the two axioms

$$\begin{aligned} &\text{I} \quad \underset{v}{\pounds}\Omega^a(\xi) = v^i\partial_i\Omega^a + \psi^a(\Omega; \partial v, \ldots, \partial^r v), \\ &\text{II} \quad \overline{\underset{v}{\pounds}\Omega^a} = \underset{v}{\pounds}\Omega^b \cdot D^a_b, \end{aligned} \tag{11.15}$$

where II simply gives the transformation law for the Lie derivative, and the coefficients D are related to the transformation law of the given object Ω in the following manner. If

$$\bar{\Omega}^a = \Phi^a(\Omega; A) \tag{11.16}$$

is the transformation law for Ω, where the single symbol A denotes the totality of parameters of an object of the r-th class, then

$$D^a_b \overset{\text{def}}{=} \frac{\partial \Phi^a}{\partial \Omega^b}. \tag{11.17}$$

With regard to postulate I, it must be further assumed that the unknown functions ψ^a are homogeneous in all the variables appearing after the semicolon.

Note that although it has as many coordinates as the object itself, the Lie derivative of the object Ω^a, need not be of the same type, although in many cases it is.

Below we list the formulae for the Lie derivatives of the principal objects.

If σ is a scalar field, then

$$\underset{v}{£}\sigma = v^i \partial_i \sigma, \tag{11.18}$$

so that the Lie derivative is again a scalar field.

If $\mathfrak{g}$ is a field of densities (of weight p), then

$$\underset{v}{£}\mathfrak{g} = v^i \partial_i \mathfrak{g} + p\mathfrak{g}\partial_i v^i \tag{11.19}$$

and it can be verified that the Lie derivative is again a density of weight p.

If w^i is a field of contravariant vectors, then in this case we have

$$\underset{v}{£}w^i = v^k \partial_k w^i - w^k \partial_k v^i \tag{11.20}$$

and it can be shown that the Lie derivative is again a field of contravariant vectors.

In particular, we see that the Lie derivative of the field v^i itself vanishes identically.

If u_i is a field of covariant vectors, then we have

$$\underset{v}{£}u_i = v^k \partial_k u_i + u_k \partial_i v^k \tag{11.21}$$

and again the Lie derivative is itself a field of covariant vectors.

For a tensor field g_{ij}, we obtain the following formula

$$\underset{v}{£}g_{ij} = v^k \partial_k g_{ij} + g_{ik} \partial_j v^k + g_{kj} \partial_i v^k. \tag{11.22}$$

and the Lie derivative $\underset{v}{£}g_{ij}$ is again a tensor field of rank two. The formula above can be put in the simpler form

$$\underset{v}{£}g_{ij} = 2\nabla_{(i}v_{j)}, \tag{11.23}$$

where $v_j = v^k g_{ij}$ and where the covariant derivative ∇ is taken with the Christoffel symbols as the parameters of connection.

For tensors, and more generally for tensor densities, the Lie derivative always has the same transformation law.

If Γ^k_{ij} represent a field of objects of parallel displacement (linear connection), then the Lie derivative is given by the formula

$$\underset{v}{\pounds}\Gamma_{ij}^{k} = \nabla_i \nabla_j v^k + R_{lij}{}^{k} v^l, \tag{11.24}$$

where $R_{ijl}{}^{k}$ is the curvature tensor associated with the connection Γ, and the symbol ∇ denotes the covariant derivative. We see here that the Lie derivative is a tensor (of rank three) and is thus of a type different from that of the object Γ itself. The reason for this lies in the fact that the difference of two objects of linear connection is a tensor.

As an illustrative example we should mention that the Lie derivative of the object of anholonomicity vanishes identically

$$\underset{v}{\pounds}\Omega_{ij}{}^{k} = 0, \tag{11.25}$$

and that the Lie derivative of a field of Pentsov objects is not a field of geometric objects since the transformation law for these objects is not linear.

The concept of Lie derivative has found extensive applications, as mentioned earlier. One particular application concerns the investigaton of the number of essential parameters in groups of motions in Riemannian space and affine motion groups in spaces with an affine connection [89]. The concept of Lie derivative has also penetrated various physical theories. The first application to the thermo-hydrodynamics of ideal fluids was given by van Dantzig. Now further applications for the concept have developed in the theory of elasticity.

With the aid of the concept of the Lie derivative it is possible to classify spaces either with linear connection or with Riemannian metric.

As in example of this, below we list several special cases which can be characterized by using the concept of Lie derivative and hence, with the aid of the vector field v^i which appears in the definition of the Lie derivative. The symbol $\Gamma_{ij}{}^{k}$ below denotes either an independent linear connection or a Christoffel symbol of the second kind when we have to deal with a Riemannian space:

$\pounds\Gamma_{ij}{}^{k} = 0$	Affine collineation
$\pounds\Gamma_{ij}{}^{k} - A_i^k u_j - A_j^k u_i = 0$	Projective motion
$\nabla_l \pounds\Gamma_{ij}{}^{k} = 0$	Special curvature collineation
$\pounds R_{ijk}{}^{l} = 0$	Curvature collineation
$\pounds g_{ij} = 0$	Motion
$\pounds g_{ij} = 0,\ g_{ij} v^i v^j = \text{const}$	Translation

$\pounds g_{ij} = \sigma g_{ij}$	Conformal motion
$\pounds g_{ij} = \sigma_{ij},\ \sigma = \text{const}$	Homothetic motion
$\pounds g_{ij} = \sigma g_{ij},\ \nabla_{ij}\sigma = 0$	Special conformal motion
$\pounds \Gamma_{ij}{}^{k} - A_i^k \partial_j \sigma - A_j^k \partial_i \sigma + g_{ij} g^{kl} \partial_l \sigma = 0$	Special conformal collineation
$\pounds \Gamma_{ij}{}^{k} - A_i^k \partial_j \sigma - A_j^k \partial_i \sigma = 0,\ \nabla_{ij}\sigma = 0$	Special projective collineation

116. The Concept of Differentiable Manifold. Our entire discussion up to now has been of a local character. We have considered either geometric objects attached at a particular point of the n-dimensional space X_n, or fields of objects defined in bounded domains, most frequently in the neighbourhood of a particular point.

Even classical differential geometry in due course embraced certain global or integral problems as we call them. These include the famous theorems, dating back to the days of Gauss, which are known as the *curvatura integra* and concern the value of a surface integral with Gaussian curvature over a closed two-dimensional surface embedded in a three-dimensional Euclidean space. This theorem was generalized to closed surfaces which are not topological images of a two-dimensional sphere and later also to one-sided closed surfaces. Topological invariants associated with the closed surface (necessarily, regular so that a Gaussian curvature does exist) being studied appear in the formulae obtained in this generalization. The continued development of topology fostered the development of geometric investigations of a global character. A more detailed exposition of only the foundations for these investigations would exceed the scope of this book. Accordingly, we shall content ourselves with giving the definition of an n-dimensional manifold, and of a differential manifold, and with citing several theorems which belong to global differential geometry. We have chosen as examples those theorems in which tensor calculus appears in a very essential role.

Suppose that we are given a topological space E with the property that every neighbourhood of every point is a topological image of the interior of an n-dimensional sphere. Such a space will be called an *n-dimensional manifold*. A manifold is said to be *compact* if a finite covering can be chosen from each family of neighbourhoods which constituting a covering of the whole space.

Now suppose that we have a family of neighbourhoods which constitute a covering and that in each neighbourhood we take a local coordinate system which, we recall, means that we have a unique continuous mapping of the points of that neighbourhood into sequence of real numbers $\{\xi^\lambda\}$ with the property

$$\sum_{\lambda=1}^{n} (\xi^\lambda)^2 < \delta^2, \qquad \text{where } \delta > 0.$$

Then, if some point p belongs to the part common to two neighbourhoods U_i and U_j and its coordinates are ξ^λ in U_i and $\xi^{\lambda'}$ in U_j, there exists a continuous and invertible transformation with the property that

$$\xi^{\lambda'} = \varphi^{\lambda'}(\xi^\lambda) \qquad (\lambda = 1, \ldots, n, \lambda' = 1', \ldots, n') \tag{11.26}$$

defined in $U_i \cap U_j$, which gives a transition from the coordinate system $\{\xi^\lambda\}$ in U_i to the coordinate system $\{\xi^{\lambda'}\}$ in U_j. The set of all such transformations is called the *atlas* of the given manifold.

A manifold is said to be *of class* C^r $(r \geqslant 1)$ if there exists an atlas such that all transformations (11.26) are of the class of regularity C^r.

A manifold (with $r \geqslant 1$) is called a *Riemannian manifold* if it is at least of class C^1 and if there exists for it an atlas such that a field of metric tensors $g_{\lambda\mu}(\xi)$ is defined on every map of the the atlas; in the parts common to $U_i \cap U_j$ the components of these fields are to be related by

$$g_{\lambda'\mu'} = g_{\lambda\mu} A^\lambda_{\lambda'} A^\mu_{\mu'}. \tag{11.27}$$

The possibility of constructing a Riemannian metric in a given differentiable manifold has been investigated by Whitney [46].

A manifold is said to be *orientable* if there exists an atlas such that at all "joints" of the individual maps we have

$$J = \det A^{\lambda'}_\lambda > 0. \tag{11.28}$$

The following theorem has been given by a mathematician of Polish origin, Bochner [92].

THEOREM. *Let an n-dimensional Riemannian manifold V_n which is compact, orientable, and of class C^3 be given. Suppose that a vector field v^λ of class C^1 is defined in the whole manifold. Let*

$$\nabla_\mu v^\mu$$

denote the divergence of this field, where covariant differentiation ∇_μ is defined by means of the Christoffel object $\left\{ {\nu \atop \lambda\mu} \right\}$. We then have

$$\int_{V_n} \nabla_\mu v^\mu \sqrt{g}\, d\xi^1 \ldots d\xi^n = 0. \tag{11.29}$$

REMARK. The integral above should, of course, be interpreted in an appropriate way since no "general" coordinate system ξ^λ exists for the entire manifold V_n.

Two important corollaries follow from this theorem.

Suppose that in V_n we have some scalar field σ of class C^2 and let $\Delta_2\sigma$ denote the second differential Beltrami parameter of the field σ:

$$\Delta_2\sigma = g^{\lambda\mu}\nabla_\lambda\nabla_\mu\sigma. \tag{11.30}$$

We then have

COROLLARY 1. *For a compact orientable Riemannian manifold we always have (for every σ)*

$$\int_{V_n}\Delta_2\sigma\sqrt{\mathfrak{g}}\,d\xi^1 \ldots d\xi^n = 0. \tag{11.31}$$

COROLLARY 2. *If we have $\Delta_2\sigma \geqslant 0$ throughout the compact orientable manifold V_n, then $\sigma =$* const.

117. Harmonic Tensors. In Section 98 we introduced the concept of the rotation of a field of covariant p-vectors $w_{\lambda_1\ldots\lambda_p}$:

$$\operatorname{Rot} w_{\lambda_1\ldots\lambda_p} \stackrel{\text{def}}{=} (p+1)\,\partial_{[\mu}w_{\lambda_1\ldots\lambda_p]}, \tag{11.32}$$

and the concept of the divergence of a field of p-vectors — densities $\mathfrak{w}^{\lambda_1\ldots\lambda_p}$ (of weight $+1$)

$$\operatorname{Div}\mathfrak{w}^{\lambda_1\ldots\lambda_p} \stackrel{\text{def}}{=} \partial_\mu\mathfrak{w}^{\mu\lambda_2\ldots\lambda_p}. \tag{11.33}$$

Note that Rot w is a tensor, albeit with increased valence, and that Div $\mathfrak{w}$ is a tensor density of the same weight but with reduced valence. We also recall that these quantities exist in the amorphous space X_n which has neither an object of connection $\Gamma^\nu_{\lambda\mu}$ nor metric tensor $g_{\lambda\mu}$.

If the space X_n is furnished with a metric which renders it a space V_n, then the rotation and divergence operators can be written as follows

$$\begin{aligned}\operatorname{Rot} w_{\lambda_1\ldots\lambda_p} &= (p+1)\nabla_{[\mu}w_{\lambda_1\ldots\lambda_p]},\\ \operatorname{Div}\mathfrak{w}^{\lambda_1\ldots\lambda_p} &= \nabla_\mu\mathfrak{w}^{\mu\lambda_2\ldots\lambda_p}.\end{aligned} \tag{11.34}$$

In this case the divergence operator can be defined not only for a field of vector densities $\mathfrak{w}$ but also for a field of contravariant p-vectors $v^{\lambda_1\ldots\lambda_p}$, viz.

$$\operatorname{Div} v^{\lambda_1\ldots\lambda_p} = \nabla_\mu v^{\mu[\lambda_1\ldots\lambda_p]}. \tag{11.35}$$

Moreover, since we can raise indices, for a given field of p-vectors $w_{\lambda_1\ldots\lambda_p}$

we can define the tensor

$$w^{\lambda_1 \dots \lambda_p} \stackrel{\text{def}}{=} w_{\mu_1 \dots \mu_p} g^{\lambda_1 \mu_1} \dots g^{\lambda_p \mu_p}, \tag{11.36}$$

which will be a contravariant p-vector.

DEFINITION. A field of p-vectors $w_{\lambda_1 \dots \lambda_p}$ is said to be *harmonic* if its rotation and its divergence both vanish identically, i.e.

$$\begin{aligned} \nabla_{[\mu} w_{\lambda_1 \dots \lambda_p]} &= 0, \\ \nabla_\mu w^{\mu \lambda_2 \dots \lambda_p} &= 0. \end{aligned} \tag{11.37}$$

We quote two theorems concerning fields of harmonic tensors.

THEOREM (de Rham and Kodaira [62]). *A necessary and sufficient condition for a field w to be harmonic is that*

$$\mathrm{Div}\,\mathrm{Rot}\, w + \mathrm{Rot}\,\mathrm{Div}\, w = 0.$$

THEOREM (Yano [89]). *A harmonic tensor which is the rotation of another tensor is identically zero.*

CONCLUSION

The choice of material presented in this book has on the one hand been dictated by the desire to expound the fundamental concepts of tensor calculus and, on the other hand, has stemmed from certain subjective appraisals of what parts have the most frequent, or the most important, applications (chiefly in the field of geometry). The theoretical foundations of tensor calculus themselves belong par excellence to geometry inasmuch as the concepts of this calculus are to be invariant, i.e. independent of the choice of any particular allowable coordinate system. This fact, which seems to us to be of fundamental importance, has been strongly emphasized throughout the entire exposition. It is our belief that after studying this book the reader will not confuse the concept of a tensor of rank two with that of a matrix or the concept of a vector with that of a sequence of numbers.

Although the book is entitled briefly *Tensor Calculus*, it in fact constitutes an introduction to the calculus, a study of which should enable one to proceed to more profound studies in this field, e.g. reading the work by Schouten and Struik [69] which bears the modest title *An Introduction to Newer Methods of Differential Geometry*, or reading other, more extensive works on the subject.

BIBLIOGRAPHY

A. Textbooks and Monographs

[1] A. Aczél, and S. Gołąb, *Funktionalgleichungen der Theorie der geometrischen Objekte*, Warsaw 1960.

[2] L. Bianchi, *Lezioni di geometria differenziale*, Vol. I, Pisa 1922, Vol. II, Part 1, Bologna 1927, Part 2, Bologna 1930.

[3] W. Blaschke and H. Reichardt, *Einführung in die Differentialgeometrie*, Berlin 1960.

[4] G. Bouligand, *Leçons de géométrie vectorielle*, Paris 1924.

[5] L. Brandt, *Vector and Tensor Analysis*, New York 1947.

[6] L. Brillouin, *Les tenseurs en mécanique et en elasticité*, Paris 1938.

[7] A. Buhl, *Formules stokiennes*, Paris 1926.

[8] E. Cartan, *Les espaces de Finsler*, Paris 1934.

[9] — *Leçons sur la théorie des spineurs*, Vols. I and II, Paris 1938.

[10] — *Les spineurs de l'espace à $n > 3$ dimensions. Les spineurs en géométrie riemannienne*, Paris 1938.

[11] — *Leçons sur la géométrie de Riemann*, Paris 1951.

[12] V. Cisotti, *Cenni sui fondamenti del calcolo tensoriale con applicazioni alla teoria dell'elasticità*, Milan 1932.

[13] A. McConnell, *Applications of the Absolute Differential Calculus*, London 1947.

[14] E.M. Corson, *Introduction to Tensors, Spinors and Relativistic Wave Equations*, London 1953.

[15] H.V. Craig, *Vector and Tensor Analysis*, New York 1943.

[16] R. Crowell and R. Williamson, *Calculus of Vector Functions*, Englewood Cliffs, N.Y. 1962.

[17] P. Delens, *La métrique angulaire des espaces de Finsler et la géométrie differentielle projective*, Paris 1934.

[18] M. Denis-Papin and C.A. Kaufmann, *Cours de calcul tensoriel appliqué* (*Géométrie differentielle absolue*), Paris 1953.

[19] Th. de Donder, *Théorie invariantive du calcul des variations*, Paris 1935.

[20] Ya. S. Dubnov, *The Fundamentals of Tensor Calculus*, Moscow–Leningrad 1950 (Russian).

[21] H. Dürrie, *Vektoren*, Munich 1941.

[22] A. Duschek and A. Hochrainer, *Grundzüge der Tensorrechnung in analytischer Darstellung*, I Teil: *Tensoralgebra*, Vienna 1948, II Teil: *Tensoranalysis*, Vienna 1950, III Teil: *Anwendungen in Physik und Technik*, Vienna 1955.

[23] A. Duschek and W. Mayer, *Lehrbuch der Differentialgeometrie* B. II, Leipzig–Berlin 1950.

[24] A. Einstein, *The Meaning of Relativity*, Princeton 1950.

[25] L. P. Einsenhart, *Non-Riemannian Geometry*, New York 1927.

[26] — *Riemannian Geometry*, Princeton 1949.

[27] R. Gans, *Vector Analysis*, London 1947.

[28] A. Goetz, T. Huskowski, R. Krasnodębski, H. Pidek-Łopuszańska and M. Rochowski, *Exterior Differential Forms and Some Applications*, Warsaw 1965 (Polish).

[29] G. B. Gurevich, *The Theory of Algebraic Invariants*, Moscow–Leningrad 1948 (Russian).

[30] F. Hausdorff, *Mengenlehre*, Berlin–Leipzig 1927.

[31] V. Hlavatý, *Differentialgeometrie der Kurven und Flächen und Tensorrechnung*, Groningen 1939.

[32] — *Les courbes de la variété générale à n dimensions*, Paris 1934.

[33] A. Hoborski, *Differential Geometry*, Part 2: *Surface Theory and an Outline of Tensor Theory*, Cracow 1928 (Polish).

[34] — *The Theory of Curves*, Parts 1 and 2, Cracow 1933 (Polish).

[35] — *The Theory of Dyadics* (*A Supplement to Vector Theory*), Cracow 1935 (Polish).

[36] W. Hurewicz and H. Wallmann, *Dimension Theory*, Princeton 1941.

[37] G. Juvet, *Introduction au calcul tensoriel et au calcul différentiel absolu*, Paris 1922.

[38] V. F. Kagan, *The Fundamentals of Surface Theory*, Part 1, Moscow–Leningrad 1947, Part 2, Moscow–Leningrad 1948 (Russian).

[39] E. Karaśkiewicz, *An Outline of the Theory of Vectors and Tensors*, Warsaw 1964 (Polish).

[40] N. E. Kochin, *Vector Calculus and Elementary Tensor Calculus*, Moscow 1951 (Russian).

[41] K. Kuratowski, *Topologie*, Vols. I and II, Warsaw 1948–1961.

[42] M. Kucharzewski and M. Kuczma, *Basic Concepts of the Theory of Geometric Objects*, Warsaw 1964.

[43] R. Lagrange, *Calcul différential absolu*, Paris 1926.

[44] G. Lamé, *Leçons sur les coordonnées curvilignes et leurs diverses applications*, Paris 1859.

[45] H. Lass, *Vector and Tensor Analysis*, New York 1950.

[46] D. Laugwitz, *Differentialgeometrie*, Teubner V. 1960.

[47] T. Levi-Civita, *Der absolute Differentialkalkül*, Berlin 1928.

[48] A. Lichnerowicz, *Éléments du calcul tensoriel*, Paris 1950.

[49] — *Théorie globale des connexions et des groupes d'holonomie*, Rome 1955.

[50] — *Algèbre et analyse linéares*, Paris 1956.

[51] A. Lotze, *Vektor und Affinor-Anylysis*, Munich 1950.

[52] A. Michal, *Matrix and Tensor Calculus with Applications to Mechanics, Elasticity and Aeronautics*, New York 1948.
[53] A. Mostowski and M. Stark, *Higher Algebra*, Part 1, Warsaw 1953 (Polish).
[54] F. Nevanlinna and N. Nevanlinna, *Absolute Analysis*, Berlin 1959.
[55] A. Nijenhuis, *Theory of the Geometric Object*, Amsterdam 1952.
[56] O. Nikodym, *The Theory of Tensors with Applications to Geometry and Mathematical Physics*, Vol. I, Warsaw 1938 (Polish).
[57] F. Ollendorf, *Die Welt der Vektoren. Einführung in die Theorie und Anwendung der Vektoren, Tensoren und Operatoren*, Vienna 1950.
[58] A. Z. Petrov, *Einstein Spaces*, Moscow 1961 (Russian).
[59] H. B. Phillips, *Vector Analysis*, New York 1933.
[60] P. K. Rashevskiĭ, *Riemannian Geometry and Tensor Anylysis*, Moscow 1953 (Russian).
[61] H. Reichardt, *Vorlesungen über Vektor und Tensorrechung*, Berlin 1957.
[62] G. de Rham and K. Kodaira, *Harmonic Integrals*, Princeton 1950.
[63] H. Rund, *The Differential Geometry of Finsler Spaces*, Berlin 1959.
[64] A. Rubinowicz, *Vectors and Tensors*, Warsaw–Wrocław 1950 (Polish).
[65] V. Rysavy, *Vectors and Tensors*, Prague 1949 (Czech).
[66] J. A. Schouten, *Die direkte Analysis zur neuen Relativitätstheorie*, Amsterdam 1918.
[67] — *Ricci-Calculus*, Berlin 1954.
[68] — *Tensor Analysis for Physicists*, Oxford 1951.
[69] — and D. J. Struik, *Einführung in die neueren Methoden der Differentialgeometrie*, Vol. I, Groningen 1935, Vol. II, Groningen 1938.
[70] H. Schmidt, *Einführung in die Vektor- und Tensorrechnung*, Berlin 1953.
[71] P. A. Shirokov, *Tensor Analysis*, Part 1: *Tensor Algebra*, Leningrad–Moscow 1934 (Russian).
[72] W. Ślebodziński, *Formes extérieures et leurs applications*, Vol. I, Warsaw 1954, Vol. II, Warsaw 1963.
[73] I. S. Sokolnikoff, *Tensor Analysis. Theory and Applications to Geometry and Mechanics of Continua*, New York 1964.
[74] B. Spain, *Tensor Calculus*, Edinburgh 1953.
[75] D. J. Struik, *Theory of Linear Connections*, Berlin 1934.
[76] J. L. Synge and A. Schild, *Tensor Calculus*, Toronto 1956.
[77] J. H. Taylor, *Vector Analysis with an Introduction to Tensor Analysis*, New York 1946.
[78] T. Y. Thomas, *The Elementary Theory of Tensors with Applications to Geometry and Mechanics*, New York 1931.
[79] — *The Differential Invariants of Generalized Spaces*, Cambridge 1934.
[80] — *Concepts from Tensor Analysis and Differential Geometry*, New York 1961.
[81] L. Tonelli, *Fondamenti di calcolo delle variazioni*, Vol. II, Bologna 1923.
[82] T. Trajdos-Wróbel, *Introduction to Vector Analysis*, Warsaw 1959 (Polish).
[83] O. Veblen, *Invariants of Quadratic Differential Forms*, Cambridge 1933.
[84] — and J. H. C. Whitehead, *The Foundations of Differential Geometry*, Cambridge 1932.

[85] W. Voigt, *Die fundamentalen physikalischen Eigenschaften der Kristalle*, Leipzig 1898.

[86] G. Vranceanu, *Leçons de géométrie différentielle*, Vols. I and II, Bucharest 1957.

[87] W. Waliszewski, *Categories, Groupoids, Pseudogroups and Analytical Structures*, Warsaw 1965.

[88] R. Weitzenböck, *Invariantentheorie*, Groningen 1923.

[89] K. Yano, *The Theory of Lie Derivatives and its Applications*, Groningen 1955.

[90] A. Zajtz, *Komitanten der Tensoren zweiter Ordnung*, Scien. Comm. Jagellonian University, Math. Ser. 8 (1964).

B. Specialized Works

[91] A. Bielecki and S. Gołąb, *Sur un problème de la métrique angulaire dans les espaces de Finsler*, Ann. Polon. Soc. Math. 18 (1945), p. 134–144.

[92] S. Bochner, *Remarks on the Theorem of Green*, Duke Math. J. 3 (1927), p. 334–338.

[93] E. Bompiani, *La géométrie des espaces courbes et le tenseur d'énergie d'Einstein*, C. R. Paris 174 (1922), p. 737–739.

[94] — *Significato del tensore di torsione in una connesione affine*, Boll. Un. Mat. Ital. 6 (1951), p. 273–276.

[95] H. Brandt, *Über eine Verallgemeinerung des Gruppenbegriffs*, Math. Ann. 96 (1926), p. 360–366.

[96] L. E. J. Brouwer, *Über den natürlichen Dimensionsbegriff*, J. Math. 142 (1913), p. 146–157.

[97] E. Cartan, *Sur une généralisation de la notion de courbure de Riemann et les espaces à torsion*, C. R. Paris 174 (1922), p. 593–595.

[98] D. van Dantzig, *Zur allgemeinen projektiven Differentialgeometrie* I, II, Kon. Ak. v. Wetensch. Amsterdam 35 (1932), p. 524–534 and p. 535–542.

[99] L. Dubikajtis, *Sur la notion de pseudogroupe de transformations.*

[100] Ya. S. Dubnov, *Integration covariante dans les espaces de Riemann à deux et à trois dimensions*, Trudy Sem. Vektor. Tenzor. Anal. 2–3 (1935).

[101] — *Über Tensoren mit nichtskalaren Komponenten*, ibidem 1 (1933), p. 196–222.

[102] B. Eckmann and A. Frölicher, *Sur l'intégrabilité de structures presque complexes*, C. R. Paris 232 (1951), p. 2284–2286.

[103] Ch. Ehresmann, *Gattungen von lokalen Strukturen*, Jber. Deutsch. Math.-Verein. 60 (1957).

[104] E. Fermi, *Sopra i fenomeni che avengono in vicinanza di una linea oraria*, Rend. d. Lincei 31 (1922), p. 21–23 and 51–52.

[105] S. Gołąb, *Ein Beitrag zur konformen Abbildung von zwei Riemannschen Räumen aufeinander*, Ann. Soc. Polon. Math. 13 (1930), p. 13–19.

[106] — *Sopra la connesioni lineari generali. Estensione d'un teorema di Bompiani nel caso più generale*, Ann. Mat. Pura Appl. 8 (1930–31), p. 283–291.

[107] — *Über das Anholonomitätsobjekt von Schouten und van Dantzig*, Proc. Math. Congr. Oslo 1936.

[108] — *Über die Möglichkeit einer absoluten Auszeichnung der Gruppe von Koordinantensystemen in verschiedenen Räumen*, Abh. Inter. Math. Congr. Zurich 1932.

[109] — *Über die Klassifikation der geometrischen Objekte*, Math. Z. 44 (1938), p. 104–114.

[110] — *Über den Begriff der "Pseudogruppe von Transformationen"*, Math. Ann. 116 (1939), p. 768–780.

[111] — *Sur la théorie des objets géométriques*, Ann. Soc. Pol. Math. 20 (1947), p. 10–27.

[112] — *Généralisation des équations de Bonnet–Kowalewski dans l'espace à un nombre arbitraire de dimensions*, ibidem 22 (1949), p. 97–156.

[113] — *On the Angular Metrics in General Spaces*. I. *The Chasles Principle*, Wiadom. Mat. 38 (1933), p. 1–12 (Polish).

[114] — *La notion de similitude parmi les objets géométriques*, Bull. Int. Acad. Polon. Sci. (1950), p. 1–7.

[115] — *La relation d'équivalence et les objets géométriques*, Fund. Math. 50 (1962), p. 381–386.

[116] — *Sur quelques points concernant la notion du comitant*, Ann. Soc. Pol. Math. 17 (1938), p. 177–192.

[117] — *Sur une condition suffisante (et nécessaire) pour qu'un espace riemannien à trois dimensions soit euclidien*, Sci. Comm., Jagell. Univ. Math. Ser. 9 (1963), p. 27–29.

[118] — and M. Kucharzewski, *Zur Theorie der geometrischen Objekte*, Ann. Polon. Math. 2 (1955), p. 215–218.

[119] — *Über die Invarianz gewisser Eigenschaften von Affinoren bei Transformationen der entsprechenden Untergruppen der allgemeinen affinen Gruppe*, Tensor 8 (1958), p. 1–7.

[120] — A. Jakubowicz and P. Kucharczyk, *Sur la notion du "champ traîné"*, Mathematica 4 (1962), p. 247–252.

[121] — A. Jakubowicz, M. Kucharzewski and M. Kuczma, *Sur l'objet géométrique représentant une direction munie d'un sens*, Ann. Polon. Math. 15 (1964), p. 233–236.

[122] F. Graiff, *Sull'integrazione tensoriale negli spazi di Riemann a curvatura costante*, Ist. Lombardo Accad. Sci. Lett. Rend. A. 84 (1951), p. 155–163.

[123] — *Sulla possibilità di costruire parallelogrammi chiusi in alcune varietà a torsione*, Boll. Un. Mat. Ital. 7 (1952), p. 132–135.

[124] J. Haantjes and G. Laman, *On the definition of geometric objects*, I, II, Indag. Math. 15 (1953), p. 208–215 and 216–222.

[125] V. Hlavatý, *The elementary basic principles of the unified theory of relativity*, Proc. Nat. Inst. Sci. 38 (1952), p. 243–247.

[126] — *The holonomy group* III. *Metrisable spaces*, J. Math. Mech. 9 (1960), p. 89–122.

[127] S. Hokari, *Über die Übertragungen, die der erweiterten Transformationsgruppe angehören*, J. Fac. Sci. Hokkaido Univ. Ser. I. 3 (1935), p. 15–28 and 4 (1935), p. 41–50.

[128] A. Jakubowicz, *On some geometric objects equivalent to affinors*, Sci. Comm. Szczecin Techn. Univ. (1960) (Polish).

[129] — *Über die Metrisierbarkeit der affin-zusammenhängenden Räume*, Tensor 14 (1963), p. 132–137.

[130] — *On the metrizability of n-dimensional space with affine connection*, Sci. Comm. Szczecin Techn. Univ. (1967) (Polish).

[131] A. Kawaguchi, *The foundation of the theory of displacement*, II, Proc. Imp. Acad. 10 (1934), p. 45–58.

[132] F. Klein, *Vergleichende Betrachtungen über neuere geometrische Forschungen*, Math. Ann. 43 (1893), p. 65–100.

[133] R. König and E. Pöschl, *Axiomatischer Aufbau der Operationen in Tensorraum*, Berich. Sachs. Akad. 86 (1934), p. 129–154.

[134] T. Levi-Civita, *Nozione di parallelismo in una varietà qualunque e conseguente spezificazione geometrica della curvatura riemanniana*, Rend. Palermo 42 (1917), p. 173–205.

[135] A. Lichnerowicz, *Généralisations de la géométrie kählerienne globale*, Coll. d. geom. diff. Louvain (1951), p. 99–122.

[136] A. Lopschitz, *Integrazione tensoriale in una varietà riemanniana a due dimensioni*, Trudy Sem. Vektor. Tenzor. Anal. 2–3 (1935), p. 200–211.

[137] J. Makai, *Über geometrische Objekte, die aus adjungierten n-Beinen gebildet sind*, Acta Math. Acad. Sci. Hungar. 13 (1962), p. 219–222.

[138] A. D. Michal, *Functionals of R-dimensional manifolds admitting continuous groups of transformations*, Trans. Amer. Math. Soc. 29 (1927), p. 612–646.

[139] A. Nijenhuis, *X_{n-1}-forming sets of eigenvectors*, Kon. Akad. v. Wetensch. Amsterdam 54 (1951), p. 200–212.

[140] — *A note on geometric objects of second order*, Tensor 16 (1963), p. 4–7.

[141] Yu. Pentsov, *Classification of one-component differential objects of class v*, Dokl. Akad. Nauk SSSR 54 (1946), p. 567–570 (Russian).

[142] H. Pidek, *Sur les objets géométriques de la classe zéro qui admettent une algèbre*, Ann. Soc. Polon. Math. 24, II (1951), p. 111–128.

[143] J. A. Schouten, *Über die Einordnung der Affingeometrie in die Theorie der höheren Übertragungen*, Math. Z. 17 (1923), p. 161–182 and 183–188.

[144] — *Über die verschiedenen Arten der Übertragung die einer Differentialgeometrie zu Grunde gelegt werden können*, Math. Z. 13 (1922), p. 56–81 and 15, 168.

[145] — *Über unitäre Geometrie*, Kon. Akad. v. Wetensch. Amsterdam 32 (1929), p. 457–465.

[146] — and E. R. van Kampen, *Zur Einbettungs- und Krümmungstheorie nichtholonomer Gebilde*, Math. Ann. 103 (1930), p. 752–783.

[147] — and D. van Dantzig, *Über unitäre Geometrie*, Math. Ann. 103 (1930), p. 319–346.

[148] — and D. van Dantzig, *Generelle Feldtheorie I, Zum Unifizierungsproblem der Physik*, Kon. Akad. v. Wetensch. Amstersdam 35 (1932), p. 642–656.

[149] — and J. Haantjes, *On the theory of the geometric object*, Proc. London Math. Soc. 42 (1937), p. 356–376.

[150] W. Ślebodziński, *Sur les équations canoniques de Hamilton*, Bull. Acad. Roy. de Belg. 17 (1931), p. 864–870.

[151] L. Tamàssy, *Frenetsche Formeln für Kurven in affinzusammenhängenden Räumen*, Publ. Math. Debrecen 8 (1961), p. 147–159.

[152] O. Veblen, *Normal coordinates for the geometry of paths*, Proc. Nat. Acad. Sci. U.S.A. 8 (1922), p. 192–197.

[153] H. Weyl, *Reine Infinitesimalgeometrie*, Math. Z. 2 (1918), p. 348–411.

[154] — *Zur Infinitesimalgeometrie: Einordnung der projektiven und der konformen Auffassung*, Nachr. Akad. Wiss. Göttingen Math.-Phys. Kl. II, (1921), p. 99–112.

[155] J. Weyssenhoff, *Duale Grössen, Grossrotation, Grossdivergenz und die Stokes–Gauss'schen Sätze in allgemeinen Räumen*, Ann. Soc. Polon. Math. 16 (1938), p. 127–144.

[156] J. H. C. Whitehead, *Convex regions in the geometry of paths*, Quart. J. Math. Oxford Ser. 3 (1932), p. 33–42.

[157] W. Wrona, *Conditions nécessaires et suffisantes qui déterminent les espaces einsteiniens, conformément-euclidiens et de courbure constante*, Ann. Soc. Polon. Math. 20 (1947), p. 28–80.

[158] A. Wundheiler, *Rheonome Geometrie, Absolute Mechanik*, Prace Mat.-Fiz. 40 (1932), p. 97–142.

[159] — *Objekte, Invarianten und Klassifikation der Geometrien*, Trudy Sem. Vektor. Tensor. Anal. 4 (1937), p. 366–375.

LIST OF PRINCIPAL FORMULAE AND SYMBOLS

$\varrho(p, q)$ metric of Fréchet space 4–5

X_n n-dimensional analytic space 6

$\mathfrak{G}$ pseudogroup of transformations 8

J Jacobian of coordinate transformation 12

G_r pseudogroup of transformations $y = \varphi(x)$ of class C^r 13

G_a affine group 14

E_n space based on group G_a 14

G_c centered affine group 15

G_u unimodular group 15

G_{cu} centered unimodular group 15

G_{su} special unimodular group 15

G_m orthogonal (metric) group 15

δ_{ij} Kronecker symbols 16

R_n space based on group G_m 18

G_{mc} orthogonal centered group 18

G_0 rotation group 18

E_p linear subspaces 19

$\varphi[\lambda \to \lambda']$ symbol for transformation of coordinate system 23

$A_\lambda^{\lambda'}$ partial derivatives of coordinate transformation 23

$J = \det A_\lambda^{\lambda'}$ Jacobian of transformation $\varphi[\lambda \to \lambda']$ 24

$A_{\mu_1 \dots \mu_q}^{\lambda_1 \dots \lambda_q} = A_{\mu_1}^{\lambda_1} A_{\mu_2}^{\lambda_2} \dots A_{\mu_q}^{\lambda_q}$ 26

$\partial_\lambda = \partial / \partial \xi^\lambda$ 27

$\partial_\lambda A_{\mu'}^\lambda = -\partial_{\mu'} \ln |J|$ 28

$\underset{0}{A}{}_\lambda^{\lambda'} = [A_\lambda^{\lambda'}]_{\xi^\lambda = \underset{0}{\xi}{}^\lambda}$ 34

$v^{\lambda'} = \underset{0}{A}{}^{\lambda'}_{\lambda} v^{\lambda}$ transformation law for contravariant vectors 34

$\omega' = F(\omega, T)$ transformation law for geometric objects 34

$\omega_{\alpha'} = F_{\alpha'}[\omega_{\alpha}; T(\lambda \to \lambda')]$ transformation law for geometric objects 38

$u_{\lambda'} = \underset{0}{A}{}^{\lambda}_{\lambda'} u_{\lambda}$ transformation law for covariant vectors 41

$\sigma = u_{\lambda} v^{\lambda}$ transvection of contravariant vector onto covariant vector 48

$w^{\lambda_1' \ldots \lambda_p'}{}_{\mu_1' \ldots \mu_p'} = A^{\lambda_1' \ldots \lambda_p' \mu_1 \ldots \mu_q}_{\lambda_1 \ldots \lambda_p \mu_1' \ldots \mu_q'} w^{\lambda_1 \ldots \lambda_p}{}_{\mu_1 \ldots \mu_q}$ transformation law for tensor of valence (p, q) 60

A^{λ}_{μ} unit tensor 61

$\cos\varphi = \dfrac{\sum \underset{1}{v}{}^{\lambda} \underset{2}{v}{}^{\lambda}}{\sqrt{\sum (\underset{1}{v}{}^{\lambda})^2} \sqrt{\sum (\underset{2}{v}{}^{\lambda})^2}}$ formula for cosine of angle between vectors in Euclidean space 72

$(\ )$ symbol for mixing of indices 74–75

$[\ ]$ symbol for skewing of indices 75

$w^{\lambda_1 \ldots \lambda_p}$ p-vector 76

$|\ |$ symbol for elimination of index in mixing or skewing operation 78

$\varepsilon^{\lambda_1 \ldots \lambda_n}, \varepsilon_{\lambda_1 \ldots \lambda_n}$ Ricci symbols 82–83

$\omega' = f(\omega, J)$ transformation law for objects of type J 83

$\mathfrak{w}' = J \cdot \mathfrak{w}$ transformation law for density 83

$\mathfrak{g}' = \varepsilon |J|^r \cdot \mathfrak{g}$ transformation law for G- and W-densities of weight $(-r)$ 84

$\int\limits_{D(\lambda)} \mathfrak{g}^{-1} d\xi^1 \ldots d\xi^n$ measure of oriented domain 89

$\tilde{\mathfrak{g}}' = \operatorname{sgn} J \cdot \tilde{\mathfrak{g}}$ transformation law for W-scalars 90

$\mathfrak{T}^{\lambda_1' \ldots \lambda_p'}{}_{\mu_1' \ldots \mu_p'} = \operatorname{sgn} J \cdot |J|^r A^{\lambda_1' \ldots \lambda_p' \mu_1 \ldots \mu_q}_{\lambda_1 \ldots \lambda_p \mu_1' \ldots \mu_q'} \mathfrak{T}^{\lambda_1 \ldots \lambda_p}{}_{\mu_1 \ldots \mu_q}$ transformation law for tensor G-density 93

$b^{\lambda'} = b^{\lambda} \Delta^{\lambda'}_{\lambda}$ transformation law for definors 99

$\mathfrak{g}' = J^{-2} \cdot \mathfrak{g}$ transformation law for $\mathfrak{g} = \det g_{\lambda\mu}$ 102

$g^{\lambda\nu} = \mathfrak{g}^{-1} \cdot \text{minor } g_{\nu\lambda}$ 103

$g_{\lambda\mu} = \mathfrak{g} \cdot \text{minor } g^{\mu\lambda}$ 103

$|v| = |g_{\lambda\mu} v^{\lambda} v^{\mu}|^{1/2}$ length of vector v 106

$\sigma[\underset{1}{v}, \underset{2}{v}] = g_{\lambda\mu} \underset{1}{v}{}^{\lambda} \underset{2}{v}{}^{\mu}$ scalar product of vectors $\underset{1}{v}, \underset{2}{v}$ 109–110

$\cos\varphi = \dfrac{\sigma[\underset{1}{v}, \underset{2}{v}]}{|\underset{1}{v}| \cdot |\underset{2}{v}|}$ formula for cosine of angle between vectors in Riemannian space 110

$b^{\lambda\mu} = \underset{1}{v}^{[\lambda}\underset{2}{v}^{\mu]}$ simple bivector 112

$v^* = v|v|^{-1}$ unit vector of vector 113

V_n Riemannian space 115

$e^{\lambda_1 \ldots \lambda_n}, e_{\lambda_1 \ldots \lambda_n}$ Ricci n-vectors 116

$[\underset{1}{v}, \ldots, \underset{n-1}{v}]$ vector product 118

$\tilde{T}^{\lambda'_1 \ldots \lambda'_p}{}_{\mu'_1 \ldots \mu'_q} = \operatorname{sgn} J \cdot A^{\lambda'_1 \ldots \lambda'_p \mu_1 \ldots \mu_q}_{\lambda_1 \ldots \lambda_p \mu'_1 \ldots \mu'_q} \tilde{T}^{\lambda_1 \ldots \lambda_p}{}_{\mu_1 \ldots \mu_q}$ transformation law for G-tensors 128

$\underset{K}{u}^{\nu}$ base of anholonomic system 133, 136

$\partial_K = A^\lambda_K \partial_\lambda$ 139

$\Omega_{IJ}{}^K = A^{\lambda\mu}_{IJ} \partial_{[\lambda} A^K_{\mu]}$ object of anholonomicity 140

$\Omega_{I'J'}{}^{K'} = A^{IJK'}_{I'J'K} \Omega_{IJ}{}^K - A^K_{[I'} \partial_{J']} A^{K'}_K$ transformation law for $\Omega_{IJ}{}^K$ 142

$\Omega_{I'} = A^I_{I'} \Omega_I - A^K_{[K'} \partial_{I']} A^{K'}_K$ transformation law for Ω_I 144

$\omega' = \dfrac{A^{2'}_1 + A^{2'}_2 \omega}{A^{1'}_1 + A^{1'}_2 \omega}$ transformation law for Pentsov object 146

$\omega^{\lambda'} = \dfrac{A^{\lambda'}_1 + A^{\lambda'}_\mu \omega^\mu}{A^{1'}_1 + A^{1'}_\mu \omega^\mu}$ transformation law for generalized Pentsov object 147

$\mathfrak{T}_{a_1 \ldots a_p} = \mathfrak{T}_{\lambda_1 \ldots \lambda_p} C^{\lambda_1 \ldots \lambda_p}_{a_1 \ldots a_p}$ projection of covariant quantity onto subspace 152

$\dfrac{D\mathfrak{v}}{d\tau} = \dfrac{d\mathfrak{v}}{d\tau} + r\mathfrak{v}\omega$ absolute derivative of field of vector density 168

$\dfrac{Dv^\lambda}{d\tau} = \dfrac{dv^\lambda}{d\tau} + \Gamma^\lambda_{\mu\nu} v^\nu \dfrac{d\xi^\mu}{d\tau}$ absolute derivative of field of contravariant vectors 171

$\Gamma^{\lambda'}_{\mu'\nu'} = A^{\lambda'\mu\nu}_{\lambda\mu'\nu'} \Gamma^\lambda_{\mu\nu} - A^{\mu\,\nu}_{\mu'\nu'} \partial_\mu A^{\lambda'}_\nu$ transformation law for object $\Gamma^\lambda_{\mu\nu}$ 172

$\Gamma_\mu = \Gamma^\lambda_{\mu\lambda}, \quad \Lambda_\nu = \Gamma^\lambda_{\lambda\nu}$ 174

$\Gamma_{\mu'} = A^\mu_{\mu'} \Gamma_\mu - \partial_{\mu'} \ln|J|$ transformation law for object Γ_μ 175

$S_{MN}{}^K = \Gamma^K_{[MN]} + \Omega_{MN}{}^K$ torsion tensor 177, 221

$\overset{s}{\Gamma}{}^L_{MN} = \Gamma^L_{(MN)} + \Omega_{MN}{}^K$ symmetric part of object Γ 177

$S_{\mu\nu}{}^\lambda = S_{[\mu} A^\lambda_{\nu]}$ semisymmetric object Γ 177

L_n space equipped with object $\Gamma^\lambda_{\mu\nu}$ 178

A_n space equipped with object $\overset{s}{\Gamma}{}^\lambda_{\mu\nu}$ 178

$\nabla_\mu v^\lambda = \partial_\mu v^\lambda + \Gamma^\lambda_{\mu\nu} v^\nu$ covariant derivative of field v 191

$\nabla_\mu u_\lambda = \partial_\mu u_\lambda - \Gamma^\nu_{\mu\lambda} u_\nu$ covariant derivative of field u_λ 193

$\nabla_\nu T^{\lambda_1\ldots\lambda_p}{}_{\mu_1\ldots\mu_q} = \partial_\nu T^{\lambda_1\ldots\lambda_p}{}_{\mu_1\ldots\mu_q} + \Gamma^{\lambda_1\ldots\lambda_p}_{\nu\varrho_1\ldots\varrho_p} T^{\varrho_1\ldots\varrho_p}{}_{\mu_1\ldots\mu_q} - \Gamma^{\sigma_1\ldots\sigma_q}_{\nu\mu_1\ldots\mu_q} T^{\lambda_1\ldots\lambda_p}{}_{\sigma_1\ldots\sigma_q}$ covariant derivative of tensor field T 194

$\Gamma^{\alpha_1\ldots\alpha_m}_{\nu\beta_1\ldots\beta_m} = \sum_{i=1}^{m} \Gamma^{\alpha_i}_{\nu\beta_i} \delta^{\alpha_1\ldots\alpha_{i-1}\alpha_{i+1}\ldots\alpha_m}_{\beta_1\ldots\beta_{i-1}\beta_{i+1}\ldots\beta_m}$ 194

$\nabla_\mu \mathfrak{w} = \partial_\mu \mathfrak{w} + r\Gamma_\mu \mathfrak{w}$ covariant derivative of field of densities of weight $(-r)$ 196

$\nabla_\mu \mathfrak{T}^{\lambda_1\ldots\lambda_p}{}_{\mu_1\ldots\mu_q} = \partial_\mu \mathfrak{T}^{\lambda_1\ldots}{}_{\mu_1\ldots} + \Gamma^{\lambda_1\ldots}_{\mu\varrho_1\ldots} \mathfrak{T}^{\varrho_1\ldots}{}_{\mu_1\ldots} - \Gamma^{\sigma_1\ldots}_{\mu\mu_1\ldots} \mathfrak{T}^{\lambda_1\ldots}{}_{\sigma_1\ldots} + r\Gamma_\mu \mathfrak{T}^{\lambda_1\ldots}{}_{\mu_1\ldots}$ covariant derivative of tensor density 198

$\nabla_\mu A^\lambda_\nu = 0$ 198

$\nabla_\mu \varepsilon_{\lambda_1\ldots\lambda_n} = \nabla_\mu \varepsilon^{\lambda_1\ldots\lambda_n} = 0$ 199

$\frac{d^2\xi^\nu}{d\sigma^2} + \Gamma^\nu_{\lambda\mu} \frac{d\xi^\lambda}{d\sigma} \cdot \frac{d\xi^\mu}{d\sigma} = 0$ equation of geodesics when σ is natural parameter 201

$(\Gamma^\lambda_{\mu\nu})_p \overset{*}{=} 0$ locally geodesic coordinates 208

$R_{\varrho\mu\nu}{}^\lambda = 2\partial_{[\varrho}\Gamma^\lambda_{\mu]\nu} + 2\Gamma^\lambda_{[\varrho|\omega|}\Gamma^\omega_{\mu]\nu}$ curvature tensor of space L_n 215–216

$R_{JIM}{}^K = 2\partial_{[J}\Gamma^K_{I]M} + 2\Gamma^K_{[J|L|}\Gamma^L_{I]M} + 2\Gamma^K_{LM}\Omega_{IJ}{}^L$ 218

$\nabla_{[J}\nabla_{I]}v^K = \frac{1}{2}R_{JIM}{}^K v^M - S_{JI}{}^M \nabla_M v^K$, $\nabla_{[J}\nabla_{I]}w_K = -\frac{1}{2}R_{JIK}{}^M w_M - S_{JI}{}^M \nabla_M w_K$ 218

$V_{IJ} = R_{IJK}{}^K = 2\partial_{]I}\Gamma_{J]} + 2\Gamma_M \Omega_{IJ}{}^M$ 219

$\nabla_{[\mu}\nabla_{\nu]}\sigma = -S_{\mu\nu}{}^\lambda \partial_\lambda \sigma$ 219

$R_{\lambda\mu} = R_{\nu\lambda\mu}{}^\nu$ Ricci tensor 224

$V_{\lambda\mu} - W_{\lambda\mu} = 2\nabla_\nu S_{\lambda\mu}{}^\nu + 4\nabla_{[\lambda} S_{\mu]\nu}{}^\nu + 4S_{\lambda\mu}{}^\varrho S_{\varrho\nu}{}^\nu$ 224

$R_{[\omega\mu\nu]}{}^\lambda = 2\nabla_{[\omega} S_{\mu\nu]}{}^\lambda - 4S_{[\omega\mu}{}^\varrho S_{\nu]\varrho}{}^\lambda$ 229

$R_{\omega\mu(\nu\lambda)} = 0$ 230

$R_{\omega\mu\nu\lambda} = R_{\nu\lambda\omega\mu}$ 230

$R = R_{\lambda\mu} g^{\lambda\mu}$ 231

$\varkappa = \frac{R}{n(n-1)}$ scalar curvature of space V_n 231

$\nabla_{[\mu} R_{\nu\omega]\lambda}{}^\varrho = 2S_{[\mu\nu}{}^\sigma R_{\omega]\sigma\lambda}{}^\varrho$ Bianchi identity 233

$\nabla_{[\mu} R_{\nu\omega]\lambda}{}^\varrho = 0$ Bianchi identity for symmetric connection 233

$\nabla^\alpha (R_{\alpha\mu} - \frac{1}{2} R g_{\alpha\mu}) = 0$ Einstein's field equation 235

$\left\{ {\varrho \atop \mu\nu} \right\} = \frac{1}{2} g^{\varrho\lambda} (\partial_\mu g_{\nu\lambda} + \partial_\nu g_{\lambda\mu} - \partial_\lambda g_{\mu\nu})$ Christoffel symbols of the second kind 239

$\nabla_\mu g_{\lambda\nu} = 0$, if $\Gamma^\lambda_{\mu\nu} = \left\{ {\lambda \atop \mu\nu} \right\}$ 240

$R_{\omega\mu\lambda}{}^{\nu} = 0$ affine-Euclidean spaces 250

$\Lambda^{\nu}_{\lambda\mu} = \underset{K}{u^{\nu}} \partial_\lambda \overset{K}{u_\mu}$ object of teleparallelism 252

$\tilde{\gamma}^{K}_{LM} = \underset{[L\ \ M]}{u^{\alpha} u^{\beta}} \partial_\beta \overset{K}{u_\alpha}$ generalized Ricci coefficients of rotation 256

$\overset{*}{\gamma}_{KLM} = \underset{L}{u^{\alpha}} \underset{M}{u^{\beta}} \nabla_\beta \underset{K}{u_\alpha}$ Ricci coefficients of rotation 257

W_n Weyl space ($\nabla_\mu g_{\lambda\nu} = -Q_\mu g_{\lambda\nu}$) 258

$R_{\lambda\mu} = \tau g_{\lambda\mu}$ Einstein space 261

$R_{\omega\mu\lambda\nu} = \sigma g_{[\omega|\lambda|} g_{\mu]\nu}$ space of constant curvature 263

$\nabla_\varrho R_{\omega\mu\lambda}{}^{\nu} = k_\varrho R_{\omega\mu\lambda}$ space of recurrent curvature 263

$\nabla_\varrho R_{\omega\mu\lambda}{}^{\nu} = 0$ symmetric space 263

$\nabla_\nu R_{\lambda\mu} = k_\nu R_{\lambda\mu}$ Ricci recurrent space 263

$W_{\omega\mu\lambda}{}^{\nu} = R_{\omega\mu\lambda}{}^{\nu} - 2W_{[\omega\mu]} A^{\nu}_{\lambda} + 2A^{\nu}_{[\omega} W_{\mu]\lambda}$ projective Weyl curvature tensor, where

$W_{\lambda\mu} = \dfrac{1}{1-n^2} [(n+1) R_{\lambda\mu} + V_{\lambda\mu}]$ 266

$\Pi^{\nu}_{\lambda\mu} = \Gamma^{\nu}_{\lambda\mu} - \dfrac{2}{n+1} A^{\nu}_{(\lambda} \Gamma_{\mu)}$ Thomas object of projective connection 268

$C_{\omega\mu\lambda\nu} = R_{\omega\mu\lambda\nu} - \dfrac{1}{n-2} [g_{\omega\lambda} L_{\mu\nu} + g_{\mu\nu} L_{\omega\lambda} - g_{\mu\lambda} L_{\omega\nu} - g_{\omega\nu} L_{\mu\lambda}]$ tensor of conformal curvature, where $L_{\lambda\mu} = -R_{\lambda\mu} + \dfrac{R g_{\lambda\mu}}{2(n-1)}$ 272, 273

$T^{\nu}_{\lambda\mu} = \left\{ {\nu \atop \lambda\mu} \right\} - \dfrac{1}{n} \left[\left\{ {\omega \atop \lambda\omega} \right\} A^{\nu}_{\mu} + \left\{ {\omega \atop \mu\omega} \right\} A^{\lambda}_{\nu} - g^{\nu\omega} g_{\lambda\mu} \left\{ {\tau \atop \omega\tau} \right\} \right]$ Thomas object of conformal connection 274

$N^{c}_{ab} = 2F^{d}_{[a} \{ \partial_{|d|} F^{c}_{b]} - \partial_{b]} F^{c}_{d} \}$ Nijenhuis tensor 278

$\mathrm{Rot}_\mu w_{\lambda_1 \ldots \lambda_p} = \partial_{[\mu} w_{\lambda_1 \ldots \lambda_p]}$ rotation of tensor field 281

$\mathrm{Div}\, \mathfrak{v} = \partial_\mu \mathfrak{v}^{\mu\lambda_2 \ldots \lambda_p}$ divergence of field of vector densities 283

$\int_{X^*_m} \partial_{\lambda_1} w_{\lambda_1 \ldots \lambda_m} d\tau^{\lambda_1 \ldots \lambda_m} = \int_{X^*_{m-1}} w_{\lambda_1 \ldots \lambda_{m-1}} d\tau^{\lambda_1 \ldots \lambda_{m-1}}$ Gauss–Stokes formula 293

${}'\Gamma^{a}_{bc} = \Gamma^{\nu}_{\mu\lambda} C^{a\mu\lambda}_{\nu bc} - C^{\mu\lambda}_{bc} \partial_\mu C^{a}_{\lambda}$ object of induced linear connection in subspace $Y_m \subset X_n$ 306

$H_{ba}{}^{\nu} = \partial_b C^{\nu}_{a} + \Gamma^{\nu}_{\mu\lambda} C^{\mu\lambda}_{ba} - {}'\Gamma^{c}_{ba} C^{\nu}_{c}$ curvature tensor of embedded space 308

$h_{ba} = C^{\mu\lambda}_{ba} \nabla_\mu N_\lambda$ second fundamental tensor of space 311–312

$k^\nu = \frac{d^2\xi^\nu}{d\sigma^2} + \Gamma^\nu_{\mu\lambda}\frac{d\xi^\lambda}{d\sigma}\cdot\frac{d\xi^\mu}{d\sigma}$ curvature vector of curve 315

$h_{ab}i^a i^b = 0$ equations of asymptotic lines 316

$\bar{R}_{abcd} = R_{\omega\mu\lambda\nu} C^{\omega\mu\lambda\nu}_{abcd} + 2h_{d[a} h_{b]c}$ generalized Gaussian equations 318

$R_{\omega\mu\lambda\nu} N^\nu C^{\omega\mu\lambda}_{abc} = 2'\nabla_{[a} h_{b]c}$ generalized Mainardi–Codazzi equations 319

$\Delta_1\sigma = g^{\lambda\mu}(\partial_\lambda\sigma)\partial_\mu\sigma$ Beltrami's first differential parameter 321

$\Delta_2\sigma = g^{\lambda\mu}\nabla_\lambda\nabla_\mu\sigma$ Beltrami's second differential parameter 321

$D_\sigma \underset{j}{t} = -\underset{j-1}{\varkappa}\,\underset{j-1}{t} + \underset{j}{\varkappa}\,\underset{j+1}{t}$ $(\varkappa_0 = \varkappa_n = 0)$ Frenet equations in V_n 333

$D_\sigma \underset{m}{t} + \sum_{j-1}^{m-1} \underset{j}{\varkappa}\,\underset{j}{t} = 0$ Frenet equations for curves in L_n 335

$\underset{v}{Ł}\sigma = v^i\partial_i\sigma$ Lie derivative of scalar field 342

$\underset{v}{Ł}g = v^i\partial_i\mathfrak{g} + p\mathfrak{g}\partial_i v^i$ Lie derivative of density field 342

$\underset{v}{Ł}w^i = v^k\partial_k w^i - w^k\partial_k v^i$ Lie derivative of field of vectors w^i 342

$\underset{v}{Ł}u_i = v^k\partial_k u_i + u_k\partial_i v^k$ Lie derivative of field of vectors u_i 342

$\underset{v}{Ł}g_{ij} = v^k\partial_k g_{ij} + g_{ik}\partial_j v^k + g_{kj}\partial_i v^k$ Lie derivative of metric tensor 342

$\underset{v}{Ł}\Gamma^k_{ij} = \nabla_i\nabla_j v^k + R_{lij}{}^k v^l$ Lie derivative of object of connection 343

$\underset{v}{Ł}_v\Omega^k_{ij} = 0$ 343

$\int_{V_n} \nabla_\mu v^\mu \sqrt{\mathfrak{g}}\, d\xi^1 \dots d\xi^n = 0$ Bochner theorem 345

$\nabla_{[\mu} w_{\lambda_1\dots\lambda_p]} = 0, \quad \nabla_\mu w^{\mu\lambda_2\dots\lambda_p} = 0$ harmonic tensor 347

AUTHOR INDEX

SUBJECT INDEX